An Introduction to Political and Social Data Analysis (With R)

For Olivia and Clayton

An Introduction to Political and Social Data Analysis (With R)

Thomas M. Holbrook

University of Wisconsin-Milwaukee

Sage

Sage

FOR INFORMATION:

2455 Teller Road
Thousand Oaks, California 91320
E-mail: order@sagepub.com

1 Oliver's Yard
55 City Road
London, EC1Y 1SP
United Kingdom

Unit No. 323-333, Third Floor, F-Block
International Trade Tower
Nehru Place, New Delhi – 110 019
India

18 Cross Street #10-10/11/12
China Square Central
Singapore 048423

Printed in Germany
Libri Plureos GmbH
Friedensallee 273
22763 Hamburg

Library of Congress Cataloging-in-Publication Data

ISBN: 9781071929421

Acquisitions Editor: Leah Fargotstein

Editorial Assistant: Jennifer Milewski

Production Editor: Aparajita Srivastava

Copy Editor: Karin Rathert

Typesetter: diacriTech

Cover Designer: Gail Buschman

Marketing Manager: Victoria Velasquez

This book is printed on acid-free paper.

BRIEF CONTENTS

DETAILED CONTENTS

PREFACE

ORIGIN STORY

I have probably used a half a dozen or so different textbooks over the years in my *Political Data Analysis* course, always thinking that I had finally found the right one, only to inevitably decide that I needed to find something else. I suppose I was a bit like Goldilocks: Most books had many things to recommend them, but none were quite right. Some books had a social science orientation but not much to say about political science. Some books were overly technical, while others were not technical enough. Some books were essentially programming manuals with little to no research design or statistics instruction, while others were very good on research design or statistics but offered very little on the computing side. In the back of my mind, I figured I would eventually have to put up or shut up: I was going to have to put together my own textbook.

This book started as a collection of lecture notes that I put online for my undergraduates in the wake of the COVID-19 outbreak in March of 2020. I had always posted a rough set of online notes for my classes, but the pandemic pivot meant that the notes had to be much more detailed and thorough than before. This, coupled with my inability to find a textbook that was right for my class, led me to somewhat hastily cobble together my topic notes into something that started to look like a "book" during the summer of 2021. I used this new set of notes for the fall 2021 hybrid section of my undergraduate course, and it worked out pretty well. I then spent the better part of spring 2022 (a sabbatical semester) expanding and revising (and revising, and revising, and revising) the content, as well as trying to master Bookdown, so I could stitch together the multiple R Markdown files to look more like a book than just a collection of lecture notes. I won't claim to have mastered either Bookdown or Markdown, but I did learn a lot as I used them to put together the first few drafts of this book.

WHAT THIS BOOK IS (AND ISN'T) ABOUT

This book has two purposes: provide students with a comprehensive, hands-on, accessible overview of how to do and interpret quantitative political and social data analysis, while also providing a gentle introduction to using the R programming environment to address those issues. I cover this in greater detail in Chapter 1, but it is worth reinforcing here that this is not a statistics textbook. Statistics are used and discussed, but the focus is on practical aspects of doing data analysis. At the same time, while this book gives students a good initiation to using R in social science research, it is not a comprehensive R manual. Also, although issues of research design and the scientific process are discussed in Chapter 1 and best practices are encouraged throughout, this is not the primary focus of this book. Finally, while the techniques and methods covered here are used across a number of disciplines in the social sciences, many of the examples and illustrations are likely to fit better with courses in political science, sociology, and kindred disciplines. Instructors of courses in disciplines that rely heavily on experimental methods may need to use supplemental materials.

This book relies on base R and a handful of additional packages. By choice, this book does not incorporate or emphasize `tidyverse`. I think there is something to be gained by learning to work with the `tidyverse` tools, but I also think doing so is much easier once you have at least a basic level of familiarity with R. In general, my approach is that you should sit up before you crawl, crawl before you walk, and walk before you run. While `tidyverse` is not emphasized here, astute readers might notice that a couple of `tidyverse` packages are used in a few places: The `Import Dataset` tab in RStudio uses `Readxl` to import datasets and, as a result, creates tibbles, and `dplyr` is used to sample data in a couple of different places.

At the end of every semester in my Political Data Analysis course, I tell students that I think it would be appropriate for them to list *Data Analysis and R*, with a proficiency level of Basic, as skills on their resume, if they receive a grade of B+ or higher. This summarizes what students can get from this textbook, a good, solid introduction to political and social data analysis using R.

HOW TO USE THIS BOOK

The structure of the book is motivated by an approach to education that is especially appropriate for teaching data analysis and related topics: *learn by doing.* In this spirit, virtually every graph and statistical table shown in this book is accompanied by the R code that produced the output. For a few tables and graphs, the code is a bit too complicated for new R users, so it is not included but is available as part of the online resources for this book. Generally, these are the numbered and titled figures and tables.

Students can (and should) "follow along" by running the R code on their own as they work through the chapters. Ideally, they will not encounter problems and will obtain the same results when running the R code. However, anyone who has used R before and especially anyone who can remember how things worked when they were first learning R knows that "ideally" is not how things always work out. This is one of the great things about this approach to teaching data analysis—learning from errors.

Each of the chapters includes a few exercises that can be used for class assignments. First, there are *Concepts and Calculations* exercises. These require students to apply key concepts to problems and examples, calculate and interpret statistics, or interpret R output. Most of these problems lean much more to the "Concepts" than to "Calculations" side, and the calculations tend to be pretty simple. Second, there are *R Problems* that require students to use R commands shown earlier in the chapter to analyze data and interpret the results. In some cases, there is a cumulative aspect to these problems, meaning that students need to use R code they learned in previous chapters. Students who follow along and run the R code as they read the chapters will have a relatively easy time with the R problems. My own preference is to assign both types of problems, but the specific goals and circumstances of individual courses may dictate using more of one type than the other.

KEYS TO STUDENT SUCCESS

For instructors, there are a couple of things that can help put students on the path to success. First, the preparatory chapters are very important to student success, and it is worth taking time to make sure students get through this material successfully. For students who have never had a course on data analysis or used R before (most of my students), this material can be challenging and intimidating. Early on, the process of downloading and installing R and RStudio and then running the first bits of R code can be especially frustrating and intimidating. In fact, this is where some students are lost for the semester. In my experience, a little extra attention and a helping hand at this point can make a big difference to student success with the rest of the book. One thing to consider to avoid many of the problems students typically encounter in the first couple of weeks of the semester is to use the cloud version of RStudio (Posit.cloud). This is discussed at greater length in Chapter 2.

One of the interesting developments I have observed over the past decade or so is that students are increasingly disconnected from the internal structure of their laptops and personal computers. The concepts of directories, drives, folders, and so forth, seem to have lost meaning to many students. A classic example of this comes up almost every semester when I give instructions for loading one of the first data files I use in class, anes20.rda. Students take my sample text, load("<filepath>/anes20.rda"), and try to run it verbatim, as if <filepath> is the name of an actual place on their device. I bring this up not to poke fun at students but to highlight a very real issue that may cause problems early on for some students. Patience and a helping hand are encouraged. Again, these sorts of problems can be avoided if students are using Posit.cloud, especially if instructors are able to pre-load all required packages and datasets. Alternatively, if students download RStudio directly to their own device, it is important to have them open a new folder, perhaps named for the course they are in, and use it as their working directory, where they can store their datasets, script (command) files, and other course-related materials.

An important feature of RStudio is that it is relatively easy to open and work in either a Quarto or R Markdown window, if you want your students to use one of these platforms for doing homework. A key benefit of doing this is that students can run R commands within their document to see if they work and then add text that appears after the resulting R output (statistical results and graphs) rather than copying the output from the RStudio window and pasting it into a different document, which typically requires some additional formatting. A brief guide to using Quarto is provided in the online resources for this book.

One of the most important things instructors can do is encourage students to follow along and run the R code as they work through the chapters. Running the code in this low stakes context will help prepare them for the end-of-chapter exercises and any other assignments that instructors might require. It is also important to encourage students to type in the commands rather than copying and pasting them, as I think this reinforces the connection between the students and what R is doing. Each chapter begins with a "Get Ready" section that tells students which datasets and packages are used in the chapter. Many of the error messages students receive can be traced back to not loading the required datasets or not attaching the required libraries.

For students, the keys to success are largely the same as for most other subject areas. However, for students using this textbook, it is particularly important that they don't fall behind. The material builds cumulatively, and it will become increasingly difficult for students to do well if they fall behind. This applies to both aspects of the book, learning about data analysis and learning how to use R. As alluded to previously, it is important for students to hunker down and get their work done in the first part of the book, where they will learn some data analysis basics and get their first exposure to using R. Also, for students, ask for help! Your instructors want you to succeed, and one of the keys to success is letting the experts (your instructors) help you!

CHAPTER CONTENTS

The 18 chapters listed in the following table constitute four broadly defined parts of the book. The *Preparatory* chapters (1,2, and 4) help orient students to key concepts and processes involved in political and social data analysis and also provide a foundation for using R in the other chapters. The *Descriptive* chapters (3, 5, 6, and 7) cover most of the important descriptive statistics, with an emphasis on matching the appropriate statistics and graphing techniques to the type of data being used. The middle chunk of the book (Chapters 8–13) focuses on different aspects of *Statistical Inference and Hypothesis Testing*, again emphasizing the match between data type and appropriate statistical tests. This section also emphasizes the important role of *effect size* as a complement to measures of statistical significance. The final section (Chapters 14–18) focuses on *Correlation and Regression*.

Chapter Topics	
1. Introduction to Research and Data	**10.** Hypothesis Testing With Two Groups
2. Using R to Do Data Analysis	**11.** Hypothesis Testing With Multiple Groups (ANOVA)
3. Frequencies and Basic Graphs	**12.** Hypothesis Testing With Non-numeric Variables (Crosstabs)
4. Data Preparation	**13.** Measures of Association
5. Measures of Central Tendency	**14.** Correlation and Scatterplots
6. Measures of Dispersion	**15.** Simple Regression
7. Probability	**16.** Multiple Regression
8. Sampling and Inference	**17.** Advanced Regression Topics
9. Hypothesis Testing	**18.** Regression Assumptions

One of the goals in putting this book together was to provide broad enough coverage of topics that it could fit a number of different course needs. For courses focused on "the basics," instructors may not be interested in the chapters on regression analysis and might focus on the preparatory and descriptive chapters, perhaps adding the chapters on crosstabs, measures of association, and correlation and scatterplots. For courses that emphasize regression analysis, instructors might want to skip the chapters on crosstabs and measures of association or some other chapters and make sure to cover the four regression chapters. Some courses do not emphasize statistical inference as much as others, so those instructors might assign fewer of the statistical inference and hypothesis testing chapters.

SUPPLEMENTAL RESOURCES

Instructor Materials

This text includes an array of instructor teaching materials designed to save you time and to help you keep students engaged. To learn more, visit sagepub.com or contact your Sage representative at sagepub.com/findmyrep.

Datasets and Codebooks

The primary datasets used in the text and for end-of-chapter problems are listed in the table that follows:

Dataset	Description
anes20	A collection of individual-level attitudinal, behavioral, and demographic variables taken from the *2020 American National Election Study*; 221 variables with over 8000 respondents.
countries2	A collection of country-level measures of demographic, health, economic, and political outcomes; 49 variables from 196 countries.
county20large	A collection of U.S. county-level measures of demographic, health, economic, and political outcomes; 96 variables from over 3100 counties.
states20	A collection of U.S. state-level measures of demographic, health, economic, and political outcomes; 87 variables taken from 50 states.

These and other datasets can be found for download among the aforementioned instructor materials (you might have to right-click on the link to open the directory in a new tab or window), and the codebooks can be found in Appendix A of this book. Only a handful of variables from each of these datasets are used in the text, leaving many others for homework, research projects, or paper assignments.

Tutorial for Quarto Documents

A brief tutorial on accessing and using Quarto in the RStudio environment is provided in Appendix B. Instructors are encouraged to work through this tutorial and experiment with Quarto a bit on their own before teaching students how to use it.

Advanced R Code

The R code for some of the more complex tables and graph is available for instructors in Appendix C. In some cases, depending on the class and the preparedness of the students, it may beneficial to share this code with students.

ACKNOWLEDGMENTS

I need to acknowledge a number of different people who have, over the years, shaped my own learning about data analysis and, hence, the way I have come to teach about it. First in line is Michael Lewis-Beck, whose undergraduate course on data analysis at the University of Iowa in the early 1980s put me on this path. Among other things, that course provided a model for teaching potentially complicated material in a relatively uncomplicated, unintimidating manner. Thanks, Mike. This book has also been shaped by the hundreds and hundreds of students who have taken some form of my Political Data Analysis course over the years. Mostly, these were students at the University of Wisconsin-Milwaukee (UWM), a university with a diverse (in many ways) student body, making it a perfect place to learn how to teach a potentially intimidating topic to students from a wide variety of backgrounds. Almost without fail, every time I taught the data analysis course, at least one student approached me at the end of the semester to tell me how surprised they were that they could do this type of work and that they actually kind of enjoyed it. I took that as motivation to make the class work for all students, which I hope is reflected in this book.

Another important influence is the books I used when I first started teaching this material, back in the late 1980s. Sam Kachigan's *Statistical Analysis* (1986) was a lifesaver back then. It is one thing to know how to use statistics and do data analysis but quite another to try to teach those things to others. This is when you find yourself asking, "Do I really know how this works?" (Just think about the first time you had to explain the concept of a sampling distribution to students.) Kachigan's book helped me understand, in a relatively straightforward way, a bit more about what was going on "under the hood" for many of the topics I was teaching for the first time. Alongside this, of course, is Mike Lewis-Beck's "little green book" on regression analysis, *Applied Regression: An Introduction* (1980), which also bailed me out countless times when I got to the regression section of my course. The general, accessible approach of these books shaped how I teach data analysis and how I wrote this book.

There are a lot of people to thank for their more direct contributions to this book. Very early on, when COVID-19 pushed us all to online teaching and I was first starting to organize and rewrite my lecture notes as "chapters" to post for my class, the first proofreaders were my children Clayton and Olivia. Clayton proofread the text, and Olivia verified the R code. They weren't too tough on me, but they did catch quite a few things and helped me get off to a good start. Thanks, kids! My research assistants also helped out quite a bit, both by gathering data and proofreading. Shin Young Park was the first to read the whole manuscript and was a thorough and careful proofreader. More than that, Shin Young had a keen eye for things that seemed to work and things that didn't, and she gave important advice on style and substance. She had an important impact on this book. Chan Song Kim read a later version of the manuscript and was still able to identify a lot of things that escaped what I thought was a close reading on my part. A number of people saw the online version of early drafts and offered encouragement (and, sometimes, corrections) along the way: Michael Hansen, Amanda Heideman, William Hicks, Brad Lockerbie, Kenneth Mulligan, and Peter Wielhouwer. Thanks to all of you.

Naturally, I got a bit of help here and there figuring out how to do things in R. This was especially true when I first started using it in the classroom, which was also about the time I first tried to learn R myself. Dave Armstrong, Nick Davis, and Patrick Kraft were especially helpful at this time. And, of course, the internet-wide R community came to the rescue more than a few times!

Of course, this book would not have happened without the capable group at Sage Publications. A special thanks to Leah Fargotstein, an acquisitions editor at Sage, who has been supportive of this manuscript from the get-go. Leah understood what I was trying to do with this book and saw a place for it at Sage. Also, Jennifer Milewski and many others at Sage shepherded the manuscript through the production process with aplomb, keeping me and the book on track. Finally, Emmalyn McWagner's contributions were also critical when it came time to wrangle the figures into publishable pixelation while maintaining their original appearance.

Many reviewers helped to shape the contents of this book, and I appreciate their valuable contributions:

Daniel Ashwood, *Miami University*
Paul R. Babbitt, *Southern Arkansas University*
William Christiansen, *Mount St. Mary's University*
Renato Corbetta, *University of Alabama at Birmingham*
Robert J. Franzese Jr., *The University of Michigan, Ann Arbor*
Jeffrey M. Glas, *University of Georgia*
Wendy L. Hansen, *University of New Mexico*
Scott Liebertz, *University of South Alabama*
Brad Lockerbie, *East Carolina University*
Aaron Sparks, *Elon University*
Salih Yasun, *Virginia Military Institute*

Finally, a big thanks to my spouse, Kathy Dolan, who has supported this project in all ways possible. Without Kathy's support, this, and all the other things I do in my life, would not be possible. xoxo

ABOUT THE AUTHOR

Thomas Holbrook is Emeritus Professor at the University of Wisconsin-Milwaukee, where he was a Distinguished Professor and the Wilder Crane Professor of Government in the political science department. He is a former editor of *American Politics Research* and the author of *Do Campaigns Matter* (Sage, 1996), *Altered States* (Oxford, 2016), and dozens of articles on various aspects of voting behavior and elections in the United States, most recently focusing on local politics. Professor Holbrook has taught undergraduate courses on data analysis and survey research for the past three decades and has integrated R into his data analysis courses for the past several years.

1 INTRODUCTION TO RESEARCH AND DATA

Welcome to the wonderful world of data analysis! Here, depending on how your instructor organizes your course, you will learn the "nuts and bolts" of doing data analysis, including topics ranging from when and how to use both relatively simple and somewhat more complicated statistics, to the important role of different types of graphics, to how to write relatively simple computer commands that connect you to the data you need for your analysis. This chapter introduces several terms and concepts that provide a framework for easing into these and other topics related to political and social data analysis.

POLITICAL AND SOCIAL DATA ANALYSIS

While I am writing this book as a political scientist and much of the data used in it address *political* outcomes, not all topics, theories, ideas, or data sources explored here are strictly political in nature. In fact, this book serves equally as well as an introduction to *social* data analysis. As you might imagine, many political outcomes are influenced by social forces, just as many social outcomes are the byproduct of political structures and processes. Are there meaningful differences between political and social data? The short answer is that, in many cases, there is not a real difference. For instance, in a study of state-level U.S. presidential election outcomes in the 2020 election, it might make perfect sense to look at how certain state population characteristics, such as religiosity, correlate with the state election outcomes. The dependent variable (let's assume it's something like Joe Biden's percentage of the total vote in 2020) is very clearly a political outcome, but the independent variable (measured something like "percentage of the population who attend religious services regularly") would usually be considered a "social" influence, one that might be of great interest to sociologists and kindred social scientists. In this context, the focus of the research is squarely political, but not all data used are strictly political in nature.

Typically, researchers who focus on analyzing political data (probably political scientists) are interested in some aspect of *the structures and processes that influence or are the byproduct of political systems*. This definition is sufficiently broad to capture most of what we usually think of as political data. Researchers whose focus is on social data (perhaps sociologists) tend to be interested in *human interactions and behaviors and the outcomes they produce*. And, of course, there are many other types of data in addition to political and social data, including but not limited to economic, geographic, biometric, genetic, and atmospheric data. All of these types of data have found their way into studies of both political and social outcomes!

You can look at the tables of contents in leading political science and sociology journals to gain an appreciation for the similarities and differences in the types of topics addressed using social and political data. For instance, Table 1.1 provides a sample of the topics covered by articles in the April 2021 issue of *American Journal of Political Science*, a top political science journal, and the June 2021 issue of *Social Forces*, a leading sociology journal.

TABLE 1.1 ■ A Selection of Topics Covered by Articles in Top Political Science and Sociology Journals

American Journal of Political Science (April 2021)	*Social Forces* (June 2021)
Democratization and representation in Indonesia's civil service	The impact of rhetorical strategies on contraceptive use
Ideology and views on political representation	Racial identity and racial dating preferences
Gender diversification in Latin American courts	Sex bias in job interviews
Preventive wars	The mobilization of online political networks
Financial markets and political preferences	Suicide risk in Iceland
State formation in Latin America	Political engagement of undocumented immigrants
Economic news and support for incumbent candidates	Work schedules and material hardship

If you want to gain a further understanding of the types of questions addressed by different social science disciplines, go to the web page of almost any college or university and look at faculty profiles in social science departments (e.g., anthropology, economics, geography, political science, psychology, sociology, to name just a few). Dig a bit deeper, and you should be able to find a list of faculty publications. Make sure to look at a few profiles from different departments to gain a real appreciation for the breadth of topics covered and types of data used by quantitative researchers in those departments.

DATA ANALYSIS OR STATISTICS?

This textbook provides an introduction to data analysis and should not be confused with a statistics textbook. Technically, it is not possible to separate data analysis from statistics, as statistics is a field of study that emerged specifically for the purpose of providing techniques for analyzing data. In fact, with the exception of this chapter, most of the material in this book addresses the use of statistical methods to analyze political and social outcomes; but that is different from a focus on statistics for the sake of learning statistics. Most straight-up texts on statistics highlight the sometimes abstract, mathematical foundations of statistical techniques, with less emphasis on concrete applications. This textbook, along with most undergraduate books on data analysis, focuses much more on *concrete applications of statistical methods that facilitate the analysis of political and social outcomes.*

Some students may be breathing a sigh of relief, thinking, "Oh good, there's no math or formulas!" Not quite. The truth of the matter is that while a lot of the math underlying the statistical techniques can be intimidating and off-putting to people with a certain level of math anxiety (including the author of this book!), some of the key formulas for calculating statistics are very intuitive, and taking time to focus on them (even at a surface level) helps students gain a solid understanding of how to interpret statistical findings. So, while there is not much math in the pages that follow, there are a few statistical formulas, sometimes with funny-looking Greek letters. The main purpose of presenting formulas, though, is to facilitate understanding and to make statistics more meaningful to the reader. If you can add, subtract, divide, multiply, and follow instructions, the "math" in this book should not present a problem.

USES OF DATA ANALYSIS

Contemporary society is swimming in data, and examples of data analysis abound. In fact, if you look at the world around you, you will find plenty of applications of data analysis, encompassing everything from sports analytics (including some of the silly statistics reported during broadcasts of sporting events), to discussions of the prevalence of COVID-19 and the effectiveness of vaccines, to monthly reports of inflation or quarterly updates of economic growth, to weather forecasts, to analyzing the impact of public policies, to models used to select jurors, and many more interesting applications.

Whatever the substantive focus, data analysis can take a number of different forms. Sometimes the goal of data analysis is simply to describe something, to show how measures of some outcomes are distributed. This is referred to as **descriptive** analysis and can be a very interesting and useful part in understanding the state of the world. One example of this that you may be most familiar with is media coverage of public opinion polls, especially those related to elections. By reporting the percentage of voters who intend to vote for one candidate or the other, media outlets are *describing* what political preferences look like at the time of the poll.

Descriptive analysis can take many different forms, including reporting simple results, as in the current polling example, or reporting more advanced statistics that help clarify how the outcomes are distributed, or perhaps using any number of graphing tools that *visualize* the outcomes. Descriptive analysis is the focus of the first several chapters of this book and makes important contributions to many of the other chapters.

Other times, the goal of research is not just to describe outcomes but also to explain why or how those outcomes occurred. This is referred to as **explanatory** analysis. For instance, when reporting the results of an election poll, you might sometimes see media outlets focus on how characteristics such as racial/ethnic identity, sex, ideology, or issue positions are related to which candidates people support. In this sort of very simple form, there is not a lot of difference between descriptive and explanatory analysis, as the goal is to *describe* what sorts of people support the candidates. A more formal type of explanatory analysis that involves developing and testing specific hypotheses about how outcomes on some variables are influenced by outcomes on others is discussed in the next section and given much fuller treatment beginning in Chapter 8.

Whether researchers are interested in descriptive or explanatory analysis, one important aspect of all forms of data analysis is related to finding valid and reliable ways to develop (usually) quantitative measures of the outcomes of interest. This concern with **measurement** is central to all forms of empirical political and social data analysis. Issues related to measurement are covered in the next section and are also interwoven into some of the later chapters.

THE RESEARCH PROCESS

Though the emphasis in this book is squarely on data analysis, that is just one important part of the broader research enterprise. In fact, *data analysis* on its own is not very meaningful, *especially if explanatory analysis is the focus* of the research. Instead, for data analysis to produce meaningful and relevant results, it must take place in the context of a set of expectations and be based on a host of decisions related to those expectations.

Social science research can be thought of as a process, where there is a beginning and an end, although it is also possible to imagine the process as cyclical and ongoing. What is presented in the following section is an idealized version of the research process. There are a couple of important things to understand about this description. First, this process is laid out in four very broadly defined categories for ease of understanding. The text elaborates on a lot of important details that should not be skipped. Second, in the real world of the research, there can be a bit of jumping around from one part of the process to another, not always in the order shown. Finally, this is just one of many different ways of describing the research process. In fact, if you consult ten different books on research methods in the social sciences, you will probably find ten somewhat different depictions of the research process. Having said this, though, at least in points of emphasis, all ten depictions should have a lot in common. The major parts of this process are presented in Figure 1.1.

Interests and Expectations

The foundation of the research process is research interests or research ideas. College students are often asked to write research papers, and one of the first parts of the writing assignment is to identify their research interests or, if they are a bit farther along, their research topic. Students frequently start with something very broad, perhaps, "I want to study elections," or something along the lines of "LGBTQ rights." This is good to know, but it is still too overly general to be very helpful. What is it about elections or LGBTQ rights that you want to know? The key to really kicking things off is to narrow the focus to a more useful research question. Maybe a student interested in elections has observed that some presidents are reelected more easily than others and settles on a more manageable goal, explaining the determinants of incumbent success in presidential elections. Maybe the student interested in LGBTQ rights has observed that some states offer several legal protections for LGBTQ residents, while other states do not. In this case, the student might limit their research interest to explaining variation in LGBTQ rights across the fifty states. The key here is to move from a broad subject area to a narrower research question that gives some direction to the rest of the process.

FIGURE 1.1 ■ An Idealized Description of the Research Process

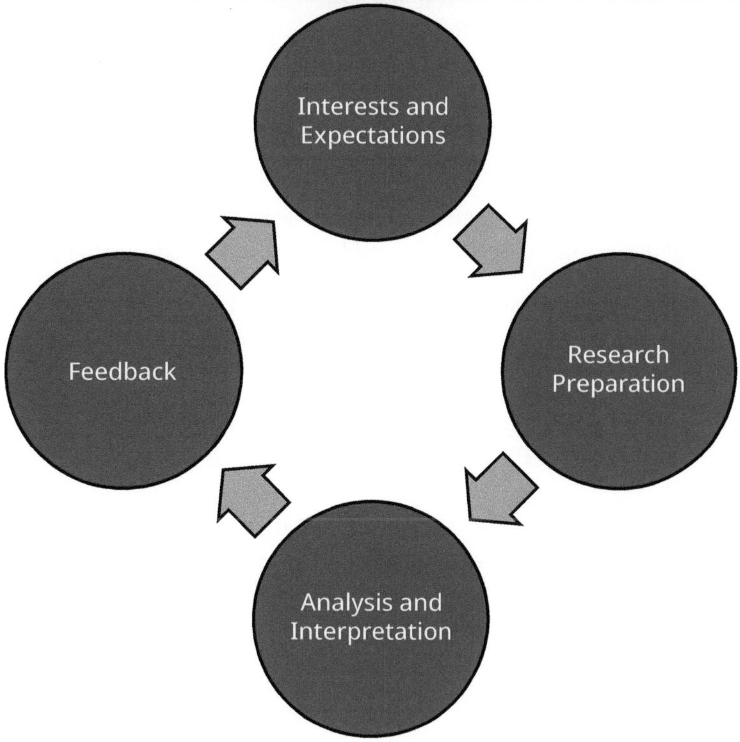

Theory

Still, even if a researcher has narrowed their research interest to a more manageable topic, they need to do a bit more thinking before they can really get started; they need a set of expectations to guide their research. In the case of studying sources of incumbent success, for instance, it is still not clear where to begin. Students need to think about their expectations. What are some ideas about the things that might be related to incumbent success? Do these ideas make sense? Are they reasonable expectations? What **theory** is guiding your research?

"Theory" is one of those terms whose meaning we all understand at some level and perhaps even use in everyday conversations (e.g., "My theory is that . . ."), but it has a fairly specific meaning in the research process. In this context, *theory* refers to *a set of logically connected propositions (or ideas/statements/assumptions) that we take to be true and that, together, can be used to explain a given outcome or set of outcomes.* Think of a theory as a rationale or plausible framework for testing ideas about things that might explain the outcomes of interest.

As an example, a *theory of retrospective voting* can be used to explain support and opposition to incumbent presidents. The retrospective model was developed in part in reaction to findings from political science research that showed that U.S. voters did not know or care very much about ideological issues and, hence, could not be considered "issue voters." Political scientist Morris Fiorina's work on retrospective voting countered that voters don't have to know a lot about issues or candidate positions on issues to be issue voters.[1] Instead, he argued that the standard view of issue voting is too narrow and a theory based *retrospective* issues does a better job of describing the American voter. Some of the key tenets of the retrospective model are the following:

- Elections are referendums on the performance of the incumbent president and their party.
- Voters don't need to understand or care about the nuances of foreign and domestic policies of the incumbent president to hold their administration accountable.
- Voters only need to be aware of the results of those policies—that is, have a sense of whether things have gone well on the international (war, trade, crises, etc.) and domestic (economy, crimes, scandals, etc.) fronts.
- When times are good, voters are inclined to support the incumbent party; when times are bad, they are less likely to support the incumbent party.

This is an explicitly reward–punishment model. It is referred to as *retrospective* voting because the emphasis is on looking back on how things have turned out under the incumbent administration rather than comparing details of policy platforms to decide if the incumbent or challenging party has the best plans for the future.

Hypotheses

The next step in this part of the research process is developing hypotheses that logically flow from the theory. A **hypothesis** is *speculation about the state of the world.* Research hypotheses are based on theories and usually assert that variations in one variable are associated with, result in, or cause variation in another variable. Typically, hypotheses specify an independent variable and a dependent variable. **Independent (explanatory) variables**, often represented as X, are best thought of as the variables that influence or shape outcomes in other variables. They are referred to as *independent* because we are not assuming that their outcomes depend on the values of other variables. **Dependent (response) variables**, often represented as Y, measure the thing we want to explain. These are the variables that we think are affected by the independent variables. One shortcut to recalling this is to remember that the outcome of the *dependent* variable *depends* upon the outcome of the independent variable.

Based on the theory of retrospective voting, for instance, it is reasonable to hypothesize that economic prosperity is positively related to the level of popular support for the incumbent president and their party. Support for the president should be higher when the economy is doing well than when it is not doing well. In social science research, hypotheses sometimes are set off and highlighted separately from the text, just so it is clear what they are:

> H_1: Economic prosperity is positively related to the level of popular support for the incumbent president and their party. Support for the president should be higher when the economy is doing well than when it is not doing well.

In this hypothesis, the independent and dependent variables are represented by two important concepts, *economic prosperity* and *support for the incumbent president*, respectively. **Concepts** are *broad, abstract ideas* that define theoretically relevant phenomena and help us understand the meaning of the theory a bit more clearly. But while concepts such as these help us understand the expectations embedded in the hypothesis, they are sufficiently broad and abstract that we are not quite ready to analyze the data.

Research Preparation

In this stage of the research process, a number of important decisions need to be made regarding the measurement of key concepts, the types of data that will be used, and how the data will be obtained. The preceding hypothesis asserts that two concepts—economic prosperity and support for the incumbent president—are related to each other. When you hear or read the names of these concepts, you probably generate a mental image that helps you understand what they mean. However, they are still a bit too abstract to be useful from a measurement perspective.

Measurement Issues

What we need to do is move from abstract concepts to **operational variables**—*concrete, tangible, measurable representations of the concepts*. How are these concepts going to be represented when doing the research? There are a number of ways to think about measuring these concepts (see Table 1.2). If we take "economic prosperity" to mean something like how well the economy is doing, we might decide to use some broad-based measure, such as percentage change in the gross domestic product (GDP), percentage change in personal income, or perhaps the unemployment rate at the national level. We could also opt for a measure of *perceptions* of the state of the economy, relying on survey questions that ask individuals to evaluate the state of the economy, and there are probably many other ways you can think of to measure economic prosperity. Even after deciding which measure or measures to use, there are still decisions to be made. For instance, let's assume we decide to use change in GPD as a measure of prosperity. We still need to decide over what time period we need to measure change. GDP growth since the beginning of the presidential term? Over the past year? The most recent quarter? It's possible to make good arguments for any of these choices.

TABLE 1.2 ■ Possible Operational Measures of Key Concepts

Concept	*Operational Variables*
Economic prosperity	GDP change, income change, economic perceptions, unemployment, etc.
Incumbent support	Approval rating, presidential elections, congressional elections, etc.

The same decisions have to be made regarding a measure of incumbent support. In this case, we might use polling data on presidential approval, presidential election results, or perhaps support for the president's party in congressional elections. The point is that before you can begin gathering relevant data, you need to know how the key variables are being measured. In the example used here, it doesn't require much effort to figure out how to operationalize the concepts that come from the hypotheses, but that is not always the case (think about measuring concepts such as power, justice, equality, fairness).

There are two very important concerns at the measurement stage: **validity** and **reliability**. The primary concern with validity is to make sure you are measuring what you think you are measuring. You need to make sure that the operational variables are good representations of the concepts. It's tough to be certain of this, but one important, albeit imprecise, way to assess the validity of a measure is through its **face validity**. By this we mean, *on its face, does this operational variable make sense as a measure of the underlying concept*? If you have to work hard to convince yourself and others that you have a valid measure, then you probably don't.

For reliability, the concern is with consistency. Here the question is whether you would get the same (or nearly the same) results if you measured the concept at different points in time or across different (but similarly drawn) samples. So, for instance, in the case of measuring presidential approval, you would expect that outcomes of polls used do not vary widely from day to day and that most polls taken at a given point in time would produce similar results.

Data Gathering

Once a researcher has determined how they intend to measure the key concepts, they must find the data. Sometimes, a researcher might find that someone else has already gathered data they can use for their project. For instance, researchers frequently rely upon regularly occurring, large-scale surveys of public opinion that have been gathered for extended periods of time, such as the American National Election Study (ANES), the General Social Survey (GSS), or the Cooperative Election Study (CES). These surveys are based on large, scientifically drawn samples and include hundreds of questions on topics of interest to social scientists. Using data sources such as these is referred to as *secondary data analysis*.

Even when researchers are putting together their own dataset, they frequently use secondary data. For instance, to test the hypotheses we previously discussed, a researcher may want to track election results and some measure of economic activity, economic growth. These data do not magically appear. Instead, the researcher has to put on their thinking cap and figure out where they can find sources for these data. As it happens, election results can be found at David Leip's Election Atlas (https://uselectionatlas.org), and the economic data can be found at the Federal Reserve Economic Data website (https://fred.stlouisfed.org) and other government sites, though it takes a bit of poking around to actually find the right information.

Even after figuring out where to get their data, researchers still have several important decisions to make. Sticking with the retrospective voting hypothesis, if the focus is on national outcomes of U.S. presidential elections, there are a number of questions that need to be answered. In what time period are we interested? All elections? Post-WWII elections?

How shall incumbent support be measured? Incumbent party percentage of the total vote or percentage of the two-party vote? If using the growth rate in GDP, over what period of time should growth be measured? Researchers need to think about these types of questions before gathering data.

As you can tell, measurement decisions are important and can be complicated to resolve. Time spent here, however, is almost guaranteed to deliver substantial payoffs further down the research process.

Data Analysis and Interpretation

Assuming a researcher has gathered appropriate data for testing their hypotheses and that the data have been coded in such a way that they are suitable to the task (more on this in the next chapter), the researcher can now subject the hypothesis to empirical scrutiny. By this, I mean that they can compare the state of the world as suggested by the hypothesis to the actual state of the world, as represented by the data gathered by the researcher. Generally, to the extent that the relationship between variables stated in the hypothesis resembles the relationship between the operational variables in the real world, then there is support for the hypothesis. If there is a significant disconnect between expectations from the hypothesis and the findings in the data, then there is less support for the hypothesis. Hypothesis testing is a lot more complicated than this, as you will see in later chapters, but for now, let's just think about it as comparing the hypothetical expectations to patterns in data.

Most of the rest of this book explores many different methods you can use to evaluate your hypotheses, some more appropriate than others, depending on the your research goals and the type of data you are using. Researchers are typically interested in two things, the **strength** of the relationship and the level of **confidence** in the findings. By "strength of the relationship," we mean how closely outcomes on the dependent and independent variables track with each other. For instance, if there is a clearer and more consistent tendency for incumbent presidents to do better in elections when the economy is doing well than when it is in a slump, then the relationship is probably pretty strong. If there is a slightly greater tendency for incumbent presidents to do better in elections when the economy is doing well than when it is in a slump, then the relationship is probably weak.

Figure 1.2 provides a hypothetical example of what weak and strong relationships might look like, using generic independent and dependent variables. The scatterplot (you'll learn much more about these in Chapter 14) on the left side illustrates a weak relationship. The first thing to note is that the pattern is not very clear; there is a lot of randomness to it, and, without the solid line in the data points, which summarizes the trend in the data and tilts upward slightly, it would be hard to discern that there is weak positive relationship in the graph. The best way to appreciate how weak the pattern is on the left side is to compare it with the pattern on the right side, where you don't have to look very hard to notice a stronger trend in the data. In this case, there is a clear tendency for high values on the independent variable to be associated with high values on the dependent variable, indicating a strong, positive relationship.

FIGURE 1.2 ■ Simulated Examples of Strong and Weak Relationships

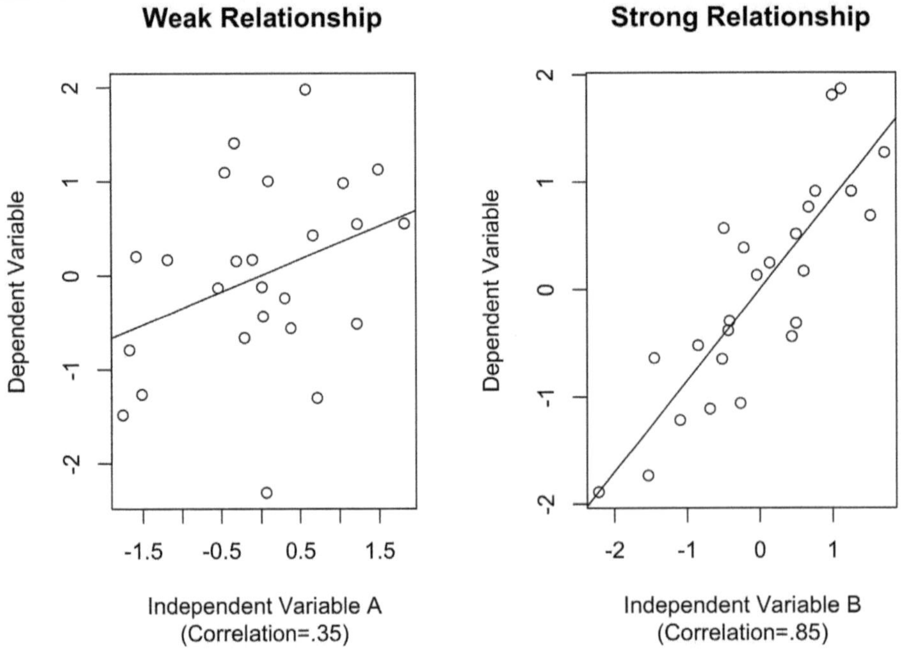

Figure 1.2 provides a good example of **data visualization**, a form of presentation that can be a very important part of communicating research results. The idea behind data visualization is to display research findings graphically, primarily to help consumers of research contextualize and understand the findings.

To appreciate the importance of visualization, suppose you do not have the scatterplots shown in Figure 1.2 but are instead presented with the correlations in the figure (.35 and .85). These correlations are statistics that summarize how strong the relationships are, with values close to 0 meaning there is not much of a relationship and values closer to 1 indicating strong relationships (much more on this and scatterplots in Chapter 14). If you had these statistics but no scatterplots, you would understand that Independent Variable *B* is more strongly related to the dependent variable than Independent Variable *A* is, but you might not fully appreciate what this means in terms of the predictive capacity of the two independent variables. The scatterplots help with this a lot.

At the same time, while the information in scatterplots gives you a clear intuitive impression of the differences in the two relationships, you can't be very specific about how much stronger the relationship is for Independent Variable *B* without more precise information, such as the correlation coefficients. Most often, the winning combination for communicating research results is some mix of statistical findings and data visualization.

In addition to measuring the strength of relationships, researchers also focus on their **level of confidence** in the findings. This is a key part of hypothesis testing and will be covered in much greater detail later in the book. The basic idea is that we want to know if the evidence of a relationship is strong enough that we can rule out the possibility that it occurred due to chance or perhaps to measurement issues. Usually, especially with large samples, researchers can have a high level of confidence in strong relationships. However, weak relationships, especially those based on a small number of cases, do not inspire confidence (for instance, Independent Variable *A* in Figure 1.2). This may be a bit confusing at this point, but good research will distinguish between confidence and strength when communicating results. This point is emphasized later, beginning in Chapter 10.

One of the most important parts of this stage of the research process is the **interpretation** of the results. The key point to get here is that the statistics and visualizations do not speak for themselves. It is important to understand that knowing how to type in computer commands and get statistical results is not very helpful if you can't also provide a coherent, substantive explanation of the results. Bottom line: Use words!

Typically, interpretations of statistical results focus on how well the findings comport with the expectations laid out in the hypotheses, paying special attention to both the strength of the relationships and the level of confidence in the findings, as previously discussed. A good discussion of research findings will also acknowledge potential limitations to the research, whatever those may be.

By way of example, let's look at a quick analysis of the relationship between economic growth and presidential support. To analyze this relationship, I gathered data on the percentage change in real GDP per capita during the first three quarters of the election year and the incumbent presidential party candidate's percentage of the two-party national popular vote in U.S. elections from 1948 to 2020. Note that these variables are concrete measures of abstract concepts (economic prosperity and presidential support) that flow from the theory of retrospective voting discussed earlier in this chapter.

An important linkage between data gathering and data analysis is organizing the data into a usable format. You will learn more about this in Chapter 2, but in the meantime, Table 1.3 provides you with a look at how the data are organized for this analysis.

TABLE 1.3 ■ Data for Testing the Retrospective Voting Hypothesis

Year	Vote	Gdp Change
1948	52.4	1.03
1952	44.6	2.93
1956	57.8	1.00
1960	49.9	-2.80
1964	61.3	1.93
1968	49.6	2.07
1972	61.8	4.17
1976	48.9	1.27

(Continued)

TABLE 1.3 ■ Data for Testing the Retrospective Voting Hypothesis (*Continued*)

Year	Vote	Gdp Change
1980	44.7	−1.19
1984	59.2	2.84
1988	53.9	2.53
1992	46.5	2.07
1996	54.6	2.69
2000	50.2	1.80
2004	51.2	2.00
2008	46.7	−2.87
2012	52	0.14
2016	51.1	0.96
2020	47.7	−1.52

The relationship between change in GDP and vote share for the incumbent party is shown in Figure 1.3. Note that in the scatterplot, the circles represent each outcome and the year labels have been added to make it easier for the reader to relate to and understand the pattern in the data (you will learn how to create graphs like this later in the book).

FIGURE 1.3 ■ A Simple Test of the Retrospective Voting Hypothesis

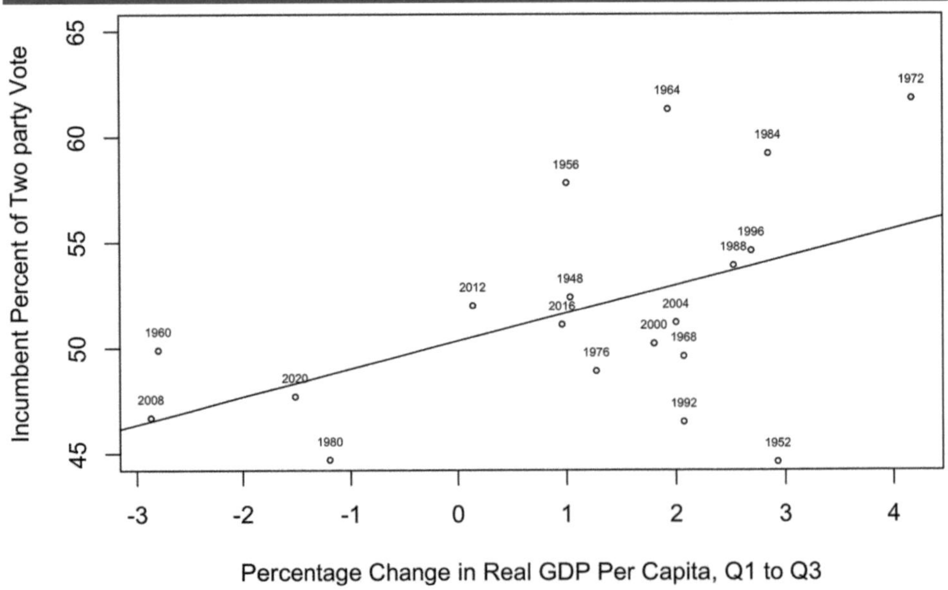

Here is an example of the type of interpretation, based on these results, that makes it easier for the research consumer to understand the results of the analysis:

> The results of the analysis provide some support for the retrospective voting hypothesis. The scatterplot shows that there is a general tendency for the incumbent party to struggle at the polls when the economy is relatively weak and to have success at the polls when the economy is strong. However, while there is a positive relationship between GDP growth and incumbent vote share, it is not a strong relationship. This can be seen in the variation in outcomes around the line of prediction, where we see a number of outcomes (1952, 1956, 1972, 1984, and 1992) that deviate quite a bit from the anticipated pattern. The correlation between these two variables (.49) confirms that there is a moderate, positive relationship between GDP growth and vote share for the incumbent presidential party. Clearly, there are other factors that help explain incumbent party electoral success, but this evidence shows that the state of the economy does play a role.

Feedback

Although it is generally accepted that theories should not be driven by what the data say (after all, the data are supposed to test the theory!), it would be foolish to ignore the results of the analysis and not allow for some feedback into the research process and reformulation of expectations. In other words, it is possible that you will discover something in the analysis that leads you to modify your theory or at least change the way you think about things. In the real world of social science data analysis, there is a lot of back-and-forth between theory, hypothesis formation, and research findings. Typically, researchers have an idea of what they want to test, perhaps grounded in some form of theory or maybe something closer to a solid rationale; they then gather data and conduct some analyses, sometimes finding interesting patterns that influence how they think about their research topic, even if they had not considered those things at the outset.

Let's consider the somewhat modest relationship between change in GDP and votes for the incumbent party, as reported in Figure 1.3. Based on these findings, you could conclude that there is a tendency for the electorate to punish the incumbent party for economic downturns and reward it for economic upturns, but the trend is not strong. Alternatively, you could think about these results and ask yourself if you are missing something. For instance, if the point of retrospective voting is to reward or punish the incumbent president for outcomes that occur during their presidency, then it should be easier to assign responsibility in years in which the president is running for another term. So maybe we need to examine the two sets of elections (incumbent running vs. open seat) separately before concluding that the retrospective model is only somewhat supported by the data.

Figure 1.4 illustrates how important it can be to allow the results of the initial data analysis to provide feedback into the research process. On the left side, there is a fairly strong, positive relationship between changes in GDP in the first three quarters of the year and the incumbent party's share of the two-party vote *when the incumbent is running*. There are a couple of years that deviate from the trend, but the overall pattern is much stronger here than it was in Figure 1.3, which included data from all elections. In addition, the scatterplot for

FIGURE 1.4 ■ Testing the Retrospective Voting Hypothesis in Two Different Contexts

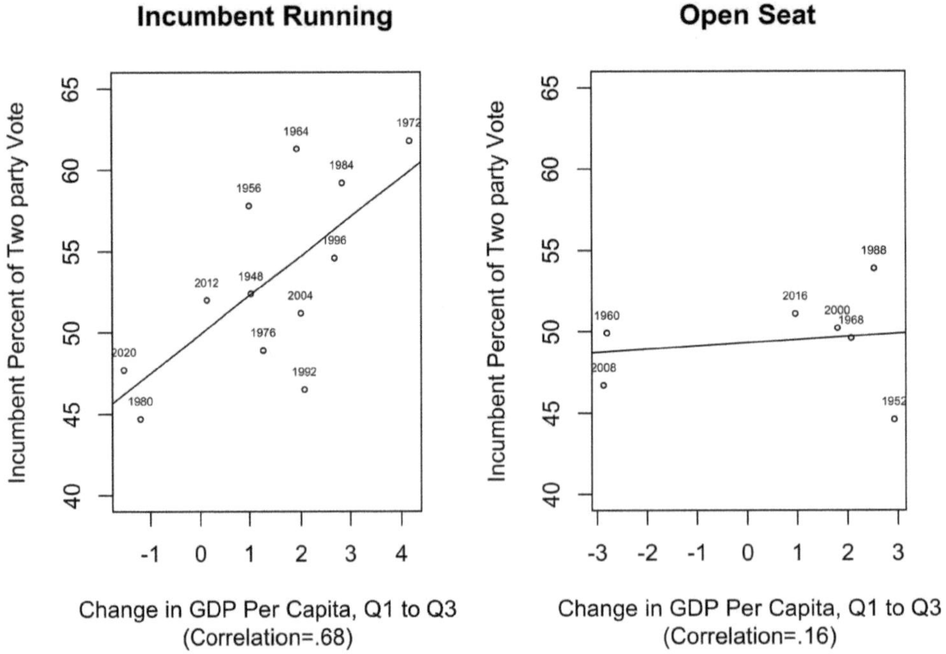

open-seat contests (right side) shows that *when the incumbent president is not running*, there is virtually no relationship between the state of the economy and the incumbent party share of the two-party vote. These interpretations of the scatterplot patterns are further supported by the correlation coefficients, .68 for incumbent races and a meager .16 for open-seat contests.

The lesson here is that it can be very useful to allow for some fluidity between the different components of the research process. When theoretically interesting possibilities present themselves during the data analysis stage, they should be given due consideration. Still, with this newfound insight, it is necessary to exercise caution and not be over-confident in the results, in large part because they are based on only 19 elections. With such a small number of cases, the next two or three elections could alter the relationship in important ways, if they do not fit the pattern of outcomes in Figure 1.4. This would not be as concerning if the findings were based on a larger sample of elections.

OTHER DATA-RELATED ISSUES

There are a number of important data characteristics that determine both what researchers can and cannot do at the data analysis stage of the research process and how they should discuss and interpret their research findings. Of particular importance are levels of measurement, level of analysis, and whether researchers are using observation or experimental data.

Levels of Measurement

One way in which data may differ from variable to variable is in terms of **level of measurement**. Essentially, the *level of measurement* of a variable describes how quantitative the variable is. This is a very important concept because making the appropriate choice of statistical method to use for a particular problem depends upon the level of measurement for the variables under examination. Generally, variables are classified along three different categories of level of measurement:

1. Nominal-level variables have categories or characteristics that differ in kind or quality only. There are *qualitative* differences between categories but not quantitative differences. This is sometimes hard to grasp, but I think the following example should help. Suppose we are interested in studying different aspects of religious attachment. For instance, we might ask survey respondents for their religious affiliation and end up collapsing their responses into the following broad categories:

Protestant
Catholic
Other Christian
Jewish
Other religion
No religion

Of course, we are interested in more than these six categories, but we'll leave it like this for now. The key thing is that as you move from one category to the next, you find different *types* of religion but not any sort of quantifiable difference in the labels used. For instance, "Protestant" is the first category and "Catholic" is the second, but we wouldn't say "Catholic" is twice as much as "Protestant" or one more unit of religion than "Protestant." Nor would we say that "Other religion" (the fifth category listed) is one unit more of religion than "Jewish" or one unit less than "No religion." These sorts of statements don't make sense, given the nature of this variable.

One way to appreciate the non-quantitative essence of these types of variables is to note that the information conveyed in this variable would not change and would be just as easy to understand if I listed the categories in a different order. Suppose "Catholic" switched places with "Protestant," and "Jewish" with "Other Christian," as shown in the following list. Doing so does not really affect how we react to the information we get from this variable.

Catholic
Protestant
Jewish
Other Christian
Other religion
No religion

A few other politically relevant examples of nominal-level variables are the following: region, marital status, race and ethnicity, and place of residence (urban/suburban/rural). Can you think of other examples?

2. Ordinal-level variables have categories or values that can be arranged in a meaningful order (the categories can be ranked) and for which it is possible to make greater/less than or magnitude-type statements—but without a lot of specificity. For instance, sticking with the example of measuring different aspects of religious attachment, you might be interested in ascertaining the level of religiosity (how religious someone is) of survey respondents. You could ask a question, something along the line of, "In your day-to-day life, how important is religion to you?" offering the following response categories:

Not at all important
Only slightly important
Somewhat important
Very important

A few things to note about this variable. First, the categories have some, though still limited, quantitative meaning. In terms of the thing being measured—the importance of religion—the level of importance increases as you move from the first to the last category. The categories are *ordered* from lowest to highest levels of importance. This is why variables like these are referred to as *ordinal* or *ordered* variables. You can appreciate the ordered nature of this variable by seeing what happens when the categories are mixed up:

Very important
Not at all important
Somewhat important
Only slightly important

In this configuration, the response categories don't seem to make as much sense as they did when they were ordered by magnitude. Moving from one category to the next, there is no consistently increasing or decreasing level of importance of religion.

Ordered variables such as these still have limited quantitative content, primarily because equal differences between ordinal categories do not have equal quantitative meaning. We still can't really say the response in the second category or the original variable ("only slightly important") is twice as important as the response in the first category ("not at all important"). We can't even say that the substantive difference between the first and second categories is the same as the substantive difference between the second and third categories. This is because the categories only represent differences in ranking of a trait from lowest to highest, not numeric distances.

Sometimes, ordinal variables may be hard to identify because the categories do not appear to range from "low" to "high" values. Take party identification, for instance, or political ideology, as presented next. In the case of party identification, you can think of the categories as growing more Republican (and less Democratic) as you move from "Democrat" to "Independent" to "Republican." Likewise, for ideology, categories grow more conservative (and less liberal) as you move from "Liberal" to "Moderate" to "Conservative."

Party ID	**Ideology**
Democrat	Liberal
Independent	Moderate
Republican	Conservative

Both nominal and ordinal variables are also referred to as **categorical variables**, emphasizing the role of labeled categories rather than numeric outcomes.

3. Interval- and ratio-level variables are the most quantitative in nature and have numeric values rather than category labels. This means that the outcomes can be treated as representing objective quantitative values, and equal numeric differences between categories have equal quantitative meaning. A true *interval scale* has an arbitrary zero point; in other words, zero does not mean "none" of whatever is being measured. The Fahrenheit thermometer is an example of this (zero degrees does not mean there is no temperature).[2] Due to the arbitrary zero point, interval variables should not be used to make ratio statements. For instance, it doesn't make sense to say that a temperature of 40 degrees is twice as warm as that of 20 degrees! But it is 20 degrees warmer, and that 20 degree difference has the same quantitative meaning as the difference between 40 and 60 degrees. *Ratio-level* variables differ from interval-level variables in that they have a genuine zero point; zero means none. Because of this, ratio statements can be made about ratio-level variables. For instance, 20% of the vote is half as much as 40% of the vote. For all practical purposes, other than making ratio statements, we can lump ratio and interval data together. Interval and ratio variables are also referred to as **numeric** variables.

To continue with the example of measuring religiosity, you might opt to ask survey respondents how many days a week they usually say at least one prayer. In this case, the response would range from 0 to 7 days, giving us a ratio-level measure of religiosity. Notice the type of language you can use when talking about this variable that you couldn't use when talking about nominal and ordinal variables. People who pray three days a week pray two more days a week than those who pray one day a week and half as many days a week as someone who prays six days a week.

Divisibility of Data

It is also possible to distinguish between variables based on their level of divisibility. A variable whose values are **finite** and cannot be subdivided is a **discrete** variable. Nominal and ordinal variables are always discrete, and some interval/ratio variables are also discrete (number of siblings, number of political science courses, etc.). A variable whose values can be **infinitely subdivided** is a continuous variable (time, weight, height, temperature, percentage voter turnout, percentage Republican vote, etc.). Only interval/ratio variables can be continuous, though they are not always. The table that follows (Table 1.4) helps organize information on levels of measurement and divisibility.

TABLE 1.4 ■ Levels of Measurement and Divisibility of Data

	Divisibility	
Level of Measurement	Discrete	Continuous
Nominal	Yes	No
Ordinal	Yes	No
Interval	Yes	Yes
Ratio	Yes	Yes

Level of Analysis

One other important characteristic of data is the level of analysis. Researchers in the social sciences typically study outcomes at the individual or aggregate levels. Usually, we consider *individual-level* data as those that represent characteristics of individual people (or some other basic level, such as firms or businesses). These types of data sometimes are also referred to as *micro* data. For instance, you might be interested in studying the political attitudes of individuals with different racial and ethnic backgrounds. For this, you could use a public opinion survey based on a random sample of individuals, and the survey would include questions about political attitudes and the racial and ethnic background characteristics of the respondents.

Aggregate data are usually based on aggregations of lower-level (individual/micro) data to some higher level. These types of data sometimes are also referred to as *macro* data. In political science, the aggregate levels are usually something like cities, counties, states, or countries. For instance, instead of focusing on individual differences in political attitudes on the basis of race and ethnicity, you might be interested in looking at the impact of the racial composition of states on state-level outcomes in presidential elections.

It is important to be aware of the level of analysis because this affects the types of valid inferences and conclusions you can make. If your analysis is based on individual-level data, then the inferences you make should be limited to individuals; and if your analysis is based on aggregate data, then the inferences you make should be limited to the level of aggregation you are studying. Inferring behavior at one level of analysis based on data from another level can be fraught with error. For instance, when using individual-level data, African American voters are the strongest Biden supporters in the 2020 elections (national exit polls show that 87% of Black voters supported Biden, compared to 65% of Latino voters, 61% of Asian American voters, and 41% of white voters). Based on this strong pattern *among individuals*, one might expect to finding a similar pattern between the size of the Black population and support for Biden *among the states*. However, this inference is completely at odds with the state-level evidence: There is no relationship between the percentage African American and the Biden percentage of the two-party vote among the states (the correlation is .009), largely because the greatest concentration of Black voters is in conservative southern states. It is also possible that you could start with the state-level finding and erroneously conclude that African Americans were no more or less likely than others to have voted for Biden in 2020, even though the individual-level data show high levels of support for Biden among African American voters.

This type of error is usually referred to as an error resulting from the **ecological fallacy**, which can occur when making assumptions about behavior at one level of analysis based on findings from another level. The key point here is to be careful of the language you use when interpreting your research findings.

Observational Versus Experimental Data

Ultimately, when testing hypotheses about how two variables are related to each other, we are saying that we think outcomes on the independent variable help shape outcomes on the dependent variable. In other words, we are interested in making causal statements. Causation, however, is very difficult to establish, especially when working with *observational data*, which is the type of data used in this book. You can think of observational data as *measures of outcomes that have already occurred*. As the researcher, you are interested in how *X* influences *Y*, so you gather data on already existing values of *X* and *Y* to see if there is a relationship between the two variables. A major problem is that there are multiple other factors that might produce outcomes on *X* and *Y* and, try as we might, it is very difficult to take all of those things into account when assessing how *X* might influence *Y*.

Experimental data, on the other hand, are data produced by the researcher, and the researcher is able to manipulate the values of the independent variable completely independent of other potential influences. Suppose, for instance, that we wanted to do an experimental study of retrospective voting in mayoral elections. We could structure the experiment in such a way that all participants are given the same basic information (background characteristics, policy positions, etc.) about both candidates (the incumbent seeking reelection and their challenger), but one-third of the participants (Group A) would be randomly assigned to receive information about positive outcomes during the mayor's term (reduced crime rate, increased property values, etc.), while another third (Group B) would receive information about negative outcomes (increased crime rate, decreased property values, etc.) during the mayor's term, and the remaining third of the respondents (Group C) would not receive any information about local conditions during the mayor's term. If, after receiving all of the information treatments, participants in Group A (positive information) give the mayor higher marks and are generally more supportive than participants in Groups B (negative information) and C (no information) and members of Group C are more supportive of the mayor than members of Group B, we could conclude that information about city conditions during the mayor's term caused these evaluations because the only difference between the three groups is whether they got the positive, negative, or no information on local conditions. In this example, the researcher is able to manipulate the information about conditions in the city *independent* of all other possible influences on the dependent variable.

This is not to say that experimental data do not have serious drawbacks, especially when it comes to connecting the experimental evidence to real-world politics. Consider, for instance, that the experimental scenario described here bears very little resemblance to the way voters encounter candidates and campaigns in real-world elections. So it is difficult to say if the results of the experiment are relevant to real-world mayoral elections. However, within the confines of the experiment itself, any differences in outcomes between Group A and Group B can be attributed to the difference in how city-level conditions were presented to the two groups.

CAUSAL LANGUAGE

One important thing to remember when discussing linkages between variables is to be careful about the language you use to describe those relationships. You should understand the limits of what you can say regarding the causal mechanisms while at the same time speaking confidently about what you think is going on in the data.

One way you can gain confidence in the causal nexus between variables, especially if you are using observational data, is by thinking about the **necessary conditions** for causality. By "necessary" conditions, I mean those conditions that must be met if there is a causal relationship. These should not be taken as sufficient conditions, however. If one of the conditions is not met, then there is no causal relationship. If they are all met, then you are on surer footing but still don't know that one variable "causes" the other. Meeting these conditions means that a causal relationship is possible.

Time order: Given our current understanding of how the universe operates, if *X* causes *Y*, then *X* must occur in time before *Y*.

Covariation: There must be an empirically discernible relationship between *X* and *Y*. If differences in *X* are not related to differences in *Y*, then it is hard to argue that outcomes on *X* play a role in shaping outcomes on *Y*.

Non-spurious: A spurious relationship between two variables is one that is produced by a third variable that is related to both the independent and dependent variables. In other words, while there may be a statistical relationship between *X* and *Y*, the relationship could reflect a third (confounding) variable that "causes" both *X* and *Y*. Remember, this is a problem with observational data but not experimental data. If you have heard researchers say something along the lines of "I have controlled for other influences," that means they have tried to address this issue. This problem is covered more extensively in Chapters 14, 16, and 17.

Theoretical grounding: Are there strong theoretical reasons for believing that *X* causes *Y*? This takes us back to the earlier discussion of the importance of having clear expectations and a sound rationale for pursuing your line of research. This is important because variables are sometimes related to each other *coincidentally* and may satisfy the time-order criterion and the relationship may persist when controlling for other variables. But if the relationship is nonsensical or at least seems like a real theoretical stretch, then it should not be assigned any causal significance.

Even if all of these conditions are met, it is important to remember that these are only *necessary* conditions. Satisfying these conditions is not a *sufficient* basis for making causal claims. Causal inferences must be made very cautiously, especially in a nonexperimental setting.

NEXT STEPS

This chapter reviewed a few of the topics and ideas that are important for understanding what data analysis is and what role it plays in the research process. Some of these things may still be hard for you to relate to at this point, especially if you have not been involved in a quantitative research project. Much of what is covered here will come up again in subsequent parts of the book, hopefully solidifying your grasp of the material.

The next couple of chapters cover other foundational material. Chapter 2 focuses on how to access R and use it for some simple tasks, such as examining imported datasets, and Chapter 3 introduces you to some basic statistical tables and graphs. As you read these chapters, it is very important that you follow the R demonstrations closely. In fact, I encourage you to make sure you can access R (download it or, preferably, use RStudio via Posit.cloud) and follow along, running the same R code you see in the book so you don't have to wait for an assignment to get your first hands on experience.

A quick word of warning, there will be errors. You should reconcile yourself now to getting errors while working in R, especially in the first few chapters. In fact, I made several errors trying to get things to run correctly so I could present the results you will see in the next couple of chapters. The great thing about this is that I learned a little bit from every error I made. You will have a much better experience if you can learn to chill about error messages. Between this book and help from your instructor, you will figure it out. Forge ahead, make mistakes, and learn!

EXERCISES

Concepts and Calculations

1. Identify the level of measurement (nominal, ordinal, interval/ratio) and divisibility (discrete or continuous) for each of the following variables.
   - Course letter grade
   - Voter turnout rate (%)
   - Marital status (married, divorced, single, etc.)
   - Occupation (Professor, cook, mechanic, etc.)
   - Body weight
   - Total number of votes cast
   - #Years of education
   - Subjective social class (poor, working class, middle class, etc.)
   - % below poverty-level income
   - Racial or ethnic group identification
2. For each of the pairs of variables listed, designate which one you think should be the dependent variable and which is the independent variable. Give a brief explanation.
   - Poverty rate/Voter turnout
   - Annual income/Years of education

- Racial group/Vote choice
- Study habits/Course grade
- Average life expectancy/Gross domestic product
- Social class/Happiness

3. Assume that the topics listed represent different research topics and classify each of them each of them as either a "political" or "social" topic. Justify your classification. If you think a topic could be classified either way, explain why.
- Marital satisfaction
- Racial inequality
- Campaign spending
- Welfare policy
- Democratization
- Attitudes toward abortion
- Teen pregnancy

4. For each of the pairs of terms listed, identify which one represents a concept and which one is an operational variable.
- Political participation/Voter turnout
- Annual income/Wealth
- Restrictive COVID-19 rules/Mandatory masking policy
- Economic development/Per capita GDP

5. What is a broad social or political subject area that interests you? Within that area, what is a narrower topic of interest? Given that topic of interest, what is a research question you think would be interesting to pursue? Finally, share a hypothesis related to the research question that you think would be interesting to test.

6. A researcher asked people taking a survey if they were generally happy or unhappy with the way life was going for them. They repeated this question in a follow-up survey two weeks later and found that 85% of respondents provided the same response. Is this a demonstration of *validity* or *reliability*? Why?

7. In an introductory political science class, there is a very strong relationship between accessing online course materials and course grade at the end of the semester: Generally, students who access course material frequently tend to do well, while those who don't access course material regularly tend to do poorly. This leads to the conclusion that accessing course material has a *causal* impact on course performance. What do you think? Does this relationship satisfy the necessary conditions for establishing causality? Address this question using all four conditions.

8. A scholar is interested in examining the impact of electoral systems on levels of voter turnout using data from a sample of 75 countries. The primary hypothesis is that voter turnout (% of eligible voters who vote) is higher in electoral systems that use proportional representation than in majoritarian/plurality systems.

- Is this an experimental study or an observational study?
- What is the dependent variable, and what is its level of measurement?
- What is the independent variable, and what is its level of measurement?
- What is the level of analysis for this study?

2 USING R TO DO DATA ANALYSIS

An important goal of this book is to introduce you to R, a programming language that facilitates data analysis. Everything you might want to do with data analysis requires that you use some sort of computer-based program to organize and evaluate the data. I think R is the right program for you to use, for a number of reasons. First, R is available for free. As long as you are able to access R remotely or can download it to your personal computing device, you can use it without paying. Among other things, this means that when you leave school, where you may have free or low-cost access to other programs, you can take R with you. In addition to being free, using R helps you develop some basic programming skills, which could be useful on the job market. Finally, by using R, you are more likely to understand how you are connected to the data you are using and to think a bit more systematically about the problems you are trying to solve.

One reason sometimes given for *not* using R is that there is a steep learning curve. Indeed, there is a bit of a learning curve with R, but the same is true for most statistical programs used in courses on data analysis, especially if students don't have prior experience with them. Whether using R, SPSS, Stata, SAS, or even Excel, students with no prior experience tend to struggle for the first few weeks of the semester. Also, some of the things that make R a bit more difficult to learn at the outset are also the things that provide added value from an educational perspective. Finally, it's all right for you to be challenged with something that is a bit different from what you are used to. That's one of the more valuable things about higher learning!

R is used throughout this book to generate all statistical results and graphics. In addition, as an instructional aid, in most cases, the code used to generate graphics and statistical results is provided in the text associated with those results. You can use the sample code to help you solve the R-based problems at the end of the chapters or as a guide to other problems and homework that might be assigned.

ACCESSING R

RStudio and Posit Cloud

One important tool that is particularly useful for new users of R is **RStudio**, a graphical user interface (GUI) that facilitates user-to-R interactions. I strongly recommend using the cloud-based version of RStudio, found at posit.cloud. This is a terrific online option for doing your work in the RStudio environment without having to download R and RStudio to your personal device. This is an especially attractive alternative if your personal computer is a bit out of date or underpowered. As long as you have reliable internet access, posit Cloud is the way to go, especially since, in doing so, you can avoid downloading and installing R and RStudio, a process that gives some students problems. Posit Cloud can be used by individuals for free or at a very low cost, and institutions can choose from a number of low-cost options to set up group workspaces where students can do their work. In addition, instructors who set up a class account on RStudio Cloud can take advantage of a number of options that facilitate working with their students on assignments.

You don't need to download anything to use the cloud-based version of RStudio. Just go to posit Cloud and click on "Sign Up" if you do not have an account or "Log In" if you already have an account. Signing up for the "Cloud Free" account should provide you with the space and resources you need to do the work associated with this book. If you end up needing more resources, the "Cloud Student" option is available for just five dollars per month. If your instructor is able to set up "Cloud Instructor" plan, then you can sign up for the free plan and join a work space set up for your class.

Downloading R and RStudio

If using the cloud-based version of RStudio is not optimal for you, there are other options. If you have reliable access to computer labs on your college or university campus, there's a good chance that R and RStudio are installed and available for you to use. If so and if you don't mind having access only in the labs, then great, there's no need to download anything. That said, one of the nice things about using R is that you can download it (and RStudio) for free and have instant, anytime access on your own computer.

Downloading and installing R and RStudio is at once pretty straightforward and at the same time, a point in the process where new users encounter a few problems. First, go to the Posit.co website and click on the DOWNLOAD RSTUDIO button in the upper-right corner. This will take you to a new page, where you should scroll down to the RStudio Desktop section. Click on the DOWNLOAD RSTUDIO button again. This should take you to a page that looks like Figure 2.1.

FIGURE 2.1 ■ RStudio Download Links

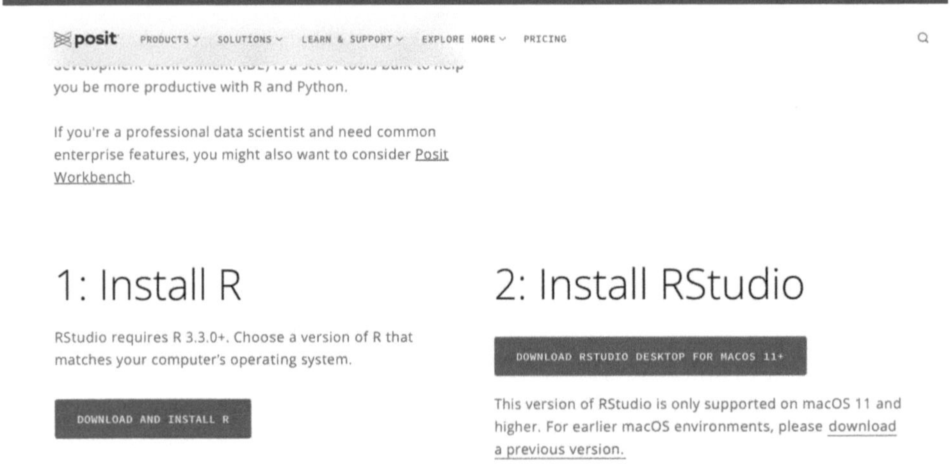

You need to install R before you can use RStudio, so click on the DOWNLOAD AND INSTALL R button. This will take you to a page on the Comprehensive R Archive Network (CRAN) that lists a collection of local links to CRAN mirrors (places from which to download R), arranged by country. You should select a CRAN that is geographically close to you. (Step 1 in Figure 2.2).

FIGURE 2.2 ■ Steps in Downloading R

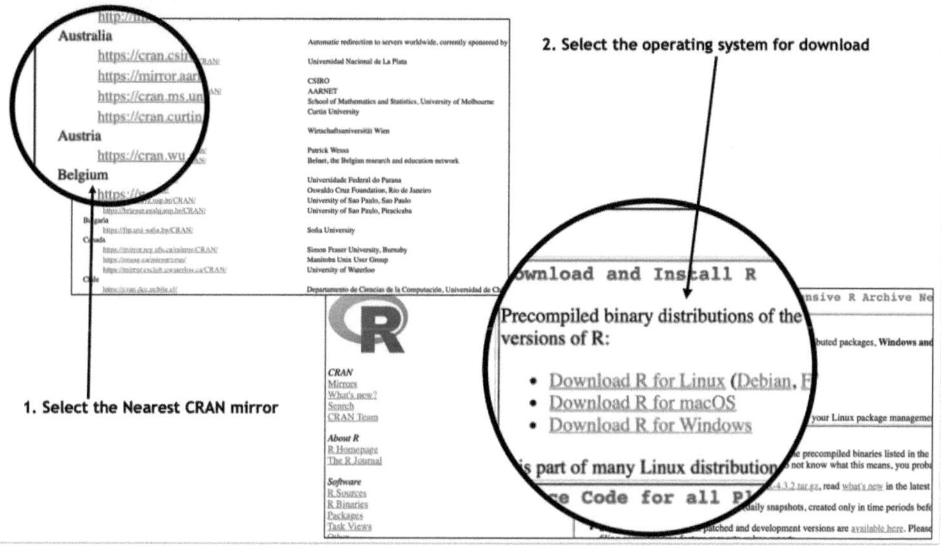

Once you choose a local CRAN, you will go to a page with separate links for the Linux, macOS, and Windows operating systems. Click on the link that matches the operating system on your computer (Step 2 in Figure 2.2), and you will be sent to a new page. If you are using a Windows operating system, you then click on Base, and you will be sent to a download link; click the download link and install the program. If you are using macOS, you will be presented with download choices based on your specific version of macOS; choose the download option whose system requirements match your operating system, and download as you would any other program from the web. For Linux, choose the distribution that matches your system and follow the installation instructions.

Downloading RStudio is much easier. Just click on the DOWNLOAD RSTUDIO DESKSTOP button (Step 2 in Figure 2.1, and RStudio will download directly to your desktop. Depending on the type of operating system you have, you may have an extra step of two left to install the program.

OPENING RSTUDIO

If you are using posit Cloud, when you log into your account, you will be sent to your posit Workspace, and that screen should look like the one in Figure 2.3. Click on the New Project tab and then on the New RStudio Project tab from the pull-down menu.

FIGURE 2.3 ■ Posit Cloud Screen After Login

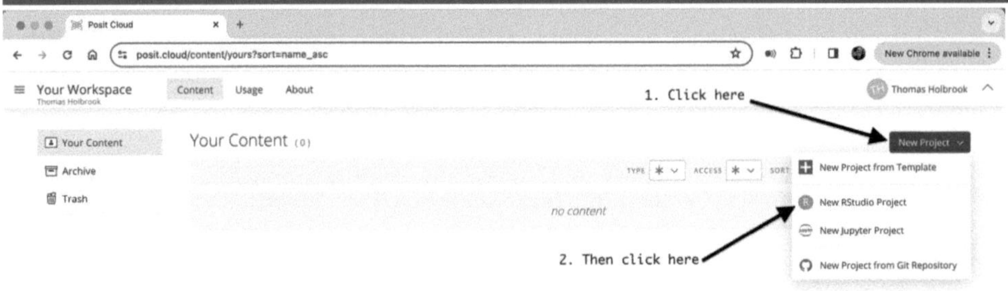

This will create a new unnamed project and will open the RStudio window. A project is sort of like a self-contained folder that contains things you use (data files, R packages, etc.) and things you create and save (script files, graphs, R output) while working in the project. Once you are in the RStudio window, you should replace "Untitled Project" with a project title, perhaps something like the course code for you class (e.g., PolSci390) or something else that is short, simple, and meaningful. Students tend to go a little overboard, creating new projects almost every week. I strongly suggest that you create just one project for the entire semester. The next time you open posit Cloud, your newly created project will be listed under "Your Content," and you just have to click on the project name to get back to your work.

If you've downloaded RStudio to your personal device, open it by clicking on the appropriate icon or file name. Initially, you should see three different RStudio windows that look something like the image in the first illustration that follows. It's possible that your screen will look a bit different due to differences in operating systems, the version of RStudio you downloaded, or some other reason, but generally, it should look like Figure 2.4.

The first thing you should do is click on the boxes in the upper-right corner of the **Console** window on the left side. This will open up an additional window, the **Source** window. Once you do this, your computer display should look like the image in Figure 2.5 (except for the annotations). Again, there may be some slight differences but not enough to matter for learning about the RStudio environment.

You will use the Source window a lot. In fact, almost everything you do will originate in the Source window. There is also a lot going on in the other windows but, generally, everything that happens (well, okay, *most* things) in the other windows happens because of what you do in the Source window. This is the window where you write your commands (the code that tells R what to do). Some people refer to this as the *Script* window because this is where you create the "script" file that contains your commands. Script files are very useful. They help you stay organized and can save you a lot of time and effort over the course of a semester or while you are working on a given research project.

FIGURE 2.4 ■ An Annotated Display of Initial RStudio Windows

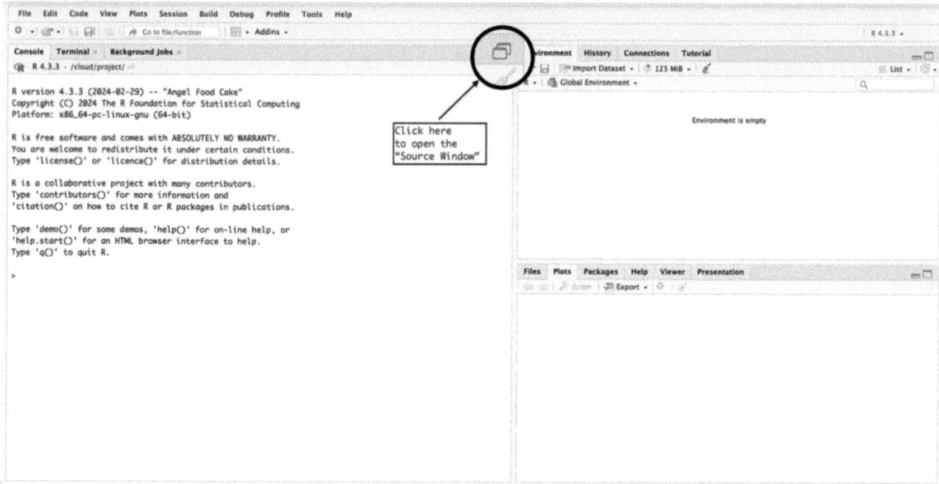

FIGURE 2.5 ■ An Annotated Display of RStudio Windows

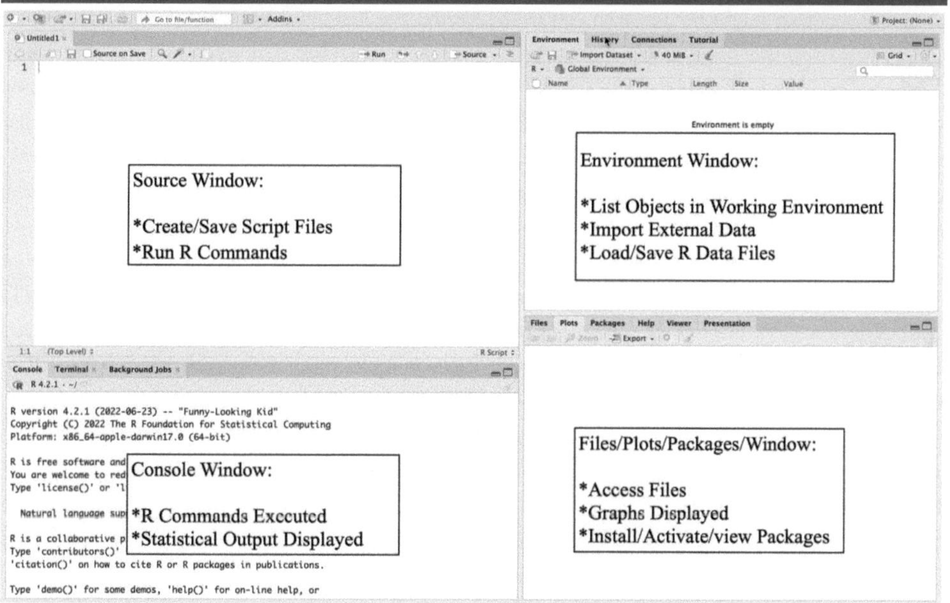

Here's how the script file works. You start by writing a command at the cursor. You then leave the cursor in the command line and click on the "Run" icon (upper-right part of the window). Alternatively, you can leave the cursor in the command line and press `control(CNTRL) + return(Enter)` on your keyboard. When you do this, you will see that the command is copied and processed at the > prompt in the Console window. As indicated, what you do in the Source window controls what happens in the other windows. You control what is executed in the Console window by writing code in the Source window and telling R to run the code. You can also just write the command at the prompt in the Console window, but this tends to get messy, and it is easier to write, edit, and save your commands in the Source window.

Once you execute the command, the results usually appear in one of two places, either the Console window (lower left) or the Plots tab in the lower right window. If you are running a statistical function, the results will appear in the Console window. If you are creating a graph, it will appear in the Plots window. If your command has an error in it, the error message will appear in the Console window.

Anytime you get an error message, you should edit the command in the Source window until the error is corrected and make sure the part of the command that created the error message is deleted. Once you are finished running your commands, you can save the contents of the Source window as a *script* file, using the Save icon in the upper-left corner of the window. When you want to open the script file again, you can use the Open-Folder icon in the toolbar just above the Source window to see a list of your files. It is a good idea to give the script file a substantively meaningful name when saving it so you know which file to access for later uses.

Creating and saving script files is one of the most important things you can do to keep your work organized. Doing so allows you to start an assignment, save your work, and come back to it some other time without having to start over. It's also a good way to have access to any previously used commands or techniques that might be useful to you in the future. For instance, you might be working on an assignment from this book several weeks down the road and realize that you could use something from the first couple of weeks of the semester to help you with the assignment. You can then open your earlier script files to find an example of the technique you need to use. It is probably the case that most people who use R rely on this type of recovery process—trying to remember how to do something, then looking back through earlier script files to find a useful example—at least until they have acquired a high level of expertise. In fact, the writing of this textbook relied on this approach a lot!

The upper-right window in RStudio includes a number of different tabs, the most important of which for our purposes is the Environment tab, which provides a list of all objects you are using in your current session. For instance, if you import a new dataset, it will be listed here. Similarly, if you store the results of some analysis in a new object (this will make more sense later), that object will be listed here. There are a couple of options here that can save you from having to type in commands. First, the open-folder icon opens your working directory (the place where R saves and retrieves files) so you can quickly add any preexisting R-formatted datasets to your environment. You can also use the Import Dataset tab to import datasets from other formats (SPSS, Stata, SAS, Excel) and convert them to datasets that can be used in R. We will do this later in this chapter. The History tab keeps a record of all commands you have used in your R session. This is similar to what you will find in the script file you create, except that it tends to be a lot messier and includes lines of code that you don't need to keep, such as those that contain errors or multiple copies of commands that you end up using several times.

The lower-right window includes a lot of useful options. The Plots tab is probably the most often used tab in the lower-right window. This is where graphs appear when you run commands to create them. This tab also provides options for exporting the graphs. The Files tab includes a list of all files in your working directory. This is where your script file can be found after you've saved it, as well as any R-generated files (e.g., datasets, graphs) you've saved. If you click on any of the files, they will open in the appropriate window. The Packages tab lists all installed packages (usually bundles of R commands and datasets you can use–more on this later) and can be used to install new packages and activate existing packages. The Help tab is used to search for help with R commands and packages. It is worth noting that there is a lot of additional online help to be had with simple web searches. The Viewer tab is unlikely to be of much use to new R users.

This overview of the RStudio setup is going to be most useful if you jump right in and get started. As with learning anything new, the best way to learn is to practice, make mistakes, and practice some more. There will be errors, and you can learn a lot by figuring out how to correct them.

UNDERSTANDING WHERE R (OR ANY PROGRAM) FITS IN

Before getting into the specific aspects of exactly how you tell R to do something, it is important to understand the relationship between you (the researcher), the data, and R. These connections may seem obvious to professors and experienced researchers or to students who have had a couple of research-based courses, but they are not so obvious to a lot of students, especially those with relatively little experience. For many students, there is a *black box* element to the research process—they're given some data and told what commands to use and how the results should be interpreted but without really understanding the process that ties things together. Understanding how things fit together is important because it helps reduce user anxiety and facilitates somewhat deeper learning.

The five panes that follow in the *R and Amazing College Student* comic strip (below) are a summary of how the researcher is connected to both R and the data and how these connections produce results. The first thing to realize is that R does things (performs operations) because you tell it to. Without your ideas and input R does nothing, so *you are in control*. It all starts with you, the researcher. Keep this in mind as you progress through this book.

PANE 1 ■ You Are in Charge

You, the college student, are sitting at your desk with your laptop, thinking about things that interest you (first pane). Maybe you have an idea about how two variables are connected to each other, or maybe you are just curious about some sort of statistical pattern. In this scenario, let's suppose that you are interested in how the states differ from each other with respect to presidential approval in the first few months of the Biden administration. No doubt, President Biden is more popular in some states than in others, and it might be interesting to look at how states differ from each other. In actuality, you are probably interested in discovering patterns of support and testing some theory about *why* some states have high approval levels and some states have low approval levels. For now, though, let's just focus on the differences in approval.

Odds are that you don't have this information at your finger tips, so you need to go find it somewhere. You're still sitting with your laptop, so you decide to search the internet for this information (second comic pane).

PANE 2 ■ Get Some Data

You find the data at civiqs.com, an online public opinion polling firm. After digging around a bit on the civiqs.com site, you find a page that has information that looks something like Figure 2.6.

FIGURE 2.6 ■ Presidential Approval Data From civiqs.com

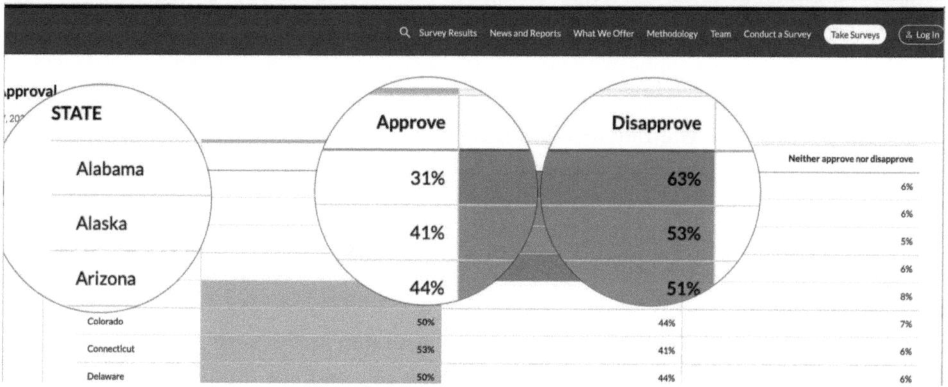

For each state, this site provides an estimate of the percentage who approve, disapprove, and who neither approve nor disapprove of the way President Biden is handling his job as president, based on surveys taken from January to June of 2021. This is a really nice table, but it's still a little bit hard to make sense of the data. Suppose you want to know which states have relatively high or low levels of approval? You could hunt around through this alphabetical listing and try to keep track of the highest and lowest states, or you could use a program like R to organize and process data for you. To do this, you need to download the data to your laptop and put it into a format that R can use (second comic pane). In this case, that means copying the data into a spreadsheet that can be imported to R (this has been done for you, saved as "Approve21.xlsx") or using more advanced methods to have R go right to the website and grab the relevant data (beyond the scope of what we are doing). Once the data are downloaded, you are ready to use R.

TIME TO USE R

Downloading and Importing Data

The first thing you should do is create a new folder on your personal computer, making sure that it is not buried very deep in the tangle of directories we all have on our computers. It might be helpful to put this folder on your desktop and give it short, meaningful name. Now, you need to **Download the Textbook Data Sets** to your new folder. Go to this book's page on Sagepub.com and look up the student resources, among which you will find the datasets (a collection of files with.rda and.xlsx extensions), and download all of these files into the folder you created. If you are working in posit Cloud, click on the Files tab in the lower-right window, then on the Upload tab, and go to the folder you created on your computer and upload all of the course files into posit Cloud. You will be using these files throughout the remaining chapters of this book.

From this point on, the process is all about you, the researcher, requesting things from R, with R providing responses. Hopefully, R will give you what you asked for, but it might also tell you that your request created an error. That's part of the process. The first step is to make sure the Source window in RStudio is open, so you can easily edit and save your commands from this session. If it is not open, you can reduce the size of the Console window, as instructed, or click on the Plus icon in the upper left corner and select "R Script."

Now, let's see how we can use R to take a closer look at the presidential approval data, which has motivated the student in Comic Pane 1. First things first, tell R to get the data. For demonstration purposes, we will import the Biden approval data we got from civiqs.com, stored as "Approve21.xlsx." In RStudio, you can do this by clicking the `Import Dataset` tab in the Environment/History window, then choose "From Excel," then click on "Browse" to find the file you want to open ("Approve21.xlsx"), and click on "Import."

If you do this, the following commands will be executed. Note that you may be prompted to install a missing package (`readxl`). If so, click "yes."

In segments like the following one, **italicized text that starts with # are comments or instructions to help you understand what's going on, and the boldface lines are commands that tell R what to do.**

```
#Tell R to install the `readxl` package (if you haven't already done so)
install.packages("readxl")

#Tell R to make "readxl" available for use
library(readxl)

#Read the Excel file into R, call it Approve21
Approve21 <- read_excel("Approve21.xlsx")

#Show the data set as a spreadsheet
View(Approve21)
```

You can also copy these commands into a script file and run them. If you do this and get an error message ("Error: 'path' does not exist: 'Approve21.xlsx'"), there is a problem with the file location. This topic is taken up at the end of the chapter, so for right now, just access the file using the `Import` tab as described above. If you still have a problem, check in with your instructor.

These lines of code use the `read_excel` command to convert the Excel file into a dataset that R can use. This new dataset is named `Approve21`. The first thing you will see after importing the data is a spreadsheet view of the dataset, which pops up in the Source window. Exit this window by clicking on the x on the right side of the Spreadsheet View tab. Some of the information we get with R commands that follow can also be obtained by inspecting the spreadsheet view of the dataset, but that option doesn't work very well for larger datasets.

At this point, in RStudio, you should see that the import commands have been executed in the Console (lower-left) window. Also, if you look at the Environment/History (upper-right) window, you should see that a dataset named `Approve21` (the civiqs.com data) has been added. This means that this dataset is now available for you to use. If you don't see these things or if you get an error message, check in with your course instructor before proceeding.

In the remainder of this chapter, you will use a set of simple commands to examine the presidential approval data. R commands usually tell R to do something to the entire dataset or to a particular variable from the dataset. The first couple of commands used in the following section focus on the entire dataset.

Examine the Dataset

Usually, before jumping into producing graphs or statistics, we want to examine the dataset to become more familiar with it. This is also an opportunity to make sure there are no obvious problems with the data. First, let's check the dimensions of the dataset (how many rows and columns).

```
#Remember, `Approve21` is the name of the data set
#Get dimensions of the data set

dim(Approve21)
[1] 50 5
```

Ignoring "[1]," which just tells you that this is the first line of output, the first number produced by `dim()` is the number of rows in the dataset, and the second number is the number of columns. The number of rows is the number of cases or observations, and the number of columns is the number of variables in the dataset. According to these results, this dataset had 50 cases and five variables. (If you look in the Environment/History window, you should see this same information listed with the dataset as "50 obs. of 5 variables.")

Let's take a second to make sure you understand what's going on here. As illustrated in the comic pane that follows, you sent requests to R in RStudio ("Hey R, use `Approve21` as the

PANE 3 ■ Interact With R Through Rstudio

dataset, and tell me its dimensions"), and R sent information back to you. Check in with your instructor if this doesn't make sense to you.

Do 50 rows and 5 columns make sense, given what we know about the data? It makes sense that there are 50 rows, since we are working with data from the states, but if you compare this to the original civiqs.com data in Figure 2.6, it looks like there is an extra column in the new dataset. To verify that the variables and cases are what we think they are, we can tell R to show us some more information about the dataset. First, we can have R show the names of the variables (columns) in the dataset.

```
#Get the names of all variables in "Approve21"
names(Approve21)
[1]  "state"    "stateab"    "Approve"    "Disapprove"    "Neither"
```

Okay, this is helpful. Now we know that the data in the first and second columns contain some sort of state identifiers, and it looks like the last three columns contain data on the percentage of who approve, disapprove, or neither approve or disapprove of President Biden. If you compare this to the columns in Figure 2.6, you will note that what appears to be a state identifier, `stateab`, was added to the dataset after the data were downloaded from civiqs.com.

Now, suppose we want to get information about specific variables in the dataset. When using both the `dim()` and `names()` commands, we asked R to give us specific information about the dataset named `Approve21`. Generally, when asking R to do something with a specific variable (column) from a dataset, we follow the format `DatasetName$VariableName` to identify the variable, separating the dataset and variable name with a dollar sign. Following, we use the `class()` command to get information about the level of measurement of a couple of variables.

```
# Get the "class" of a few variables.
class(Approve21$state)
[1] "character"
class(Approve21$stateab)
[1] "character"
class(Approve21$Approve)
[1] "numeric"
```

This confirms that `state` and `stateab` are character (non-numeric) variables and `Approve` is numeric, as expected (go ahead and run this command for the remaining variables!). This is an important step because it can help flag certain types of errors. If there had been a coding error in which a stray character or symbol was entered in place of a neighboring number on the keyboard (e.g., 3o or 3p instead of 30) for `Approve`, then it would have shown up here as a character variable instead of the expected numeric variable.

We don't know from this output if the dataset uses a state abbreviation or the full state names to identify the cases, and we also don't know if `Approve`, `Disapprove`, and `Neither` are expressed as percentages or proportions. So, we need to take a closer look at the data. One useful pair of commands, `head()` and `tail()`, can be used to look at the first and last several rows in the dataset, respectively.

```
#List the first several rows of data
head(Approve21)

# A tibble: 6 × 5
  State      stateab Approve Disapprove Neither
  <chr>      <chr>     <dbl>      <dbl>   <dbl>
1 Alabama    AL           31         63       6
2 Alaska     AK           41         53       6
3 Arizona    AZ           44         51       5
4 Arkansas   AR           31         63       6
5 California CA           57         35       8
6 Colorado   CO           50         44       7
```

Here, you see the first six rows of the dataset and learn that `state` uses full state names, `stateab` is the state abbreviation, and the other variables are expressed as whole numbers. Note that this command also gives you information on the level of measurement for each variable: "chr" means the variable is a character variable, and "dbl" means that the variable is a numeric variable that could have a decimal point (though none of these variables do).

We can also look at the bottom six rows.

```
#List the last several rows of data
tail(Approve21)

# A tibble: 6 × 5
  State         stateab Approve Disapprove Neither
  <chr>         <chr>     <dbl>      <dbl>   <dbl>
1 Vermont       VT           58         36       7
2 Virginia      VA           48         46       6
3 Washington    WA           53         39       7
4 West Virginia WV           23         72       5
5 Wisconsin     WI           47         48       5
6 Wyoming       WY           26         68       5
```

If we pay attention to the values of `Approve` and `Disapprove` in these two short lists of states, we see quite a bit of variation in the levels of support for President Biden in his first six months in office. Not Surprisingly, the president was fairly unpopular in conservative places, such as Alabama, Arkansas, West Virginia, and Wyoming, and fairly popular in more liberal states, such as California, Vermont, and Washington. However, because the dataset is ordered alphabetically by state name, it is hard to get a more general sense of what types of states give President Biden high or low marks for how he is handling his job. You could "eyeball" the entire dataset and try to keep track of states that are relatively high or low on presidential approval, but that is not very efficient and is probably prone to some level of error.

To get a listing of the lowest and highest approval states, we still use the `head()` and `tail()` but add the `order()` command to list the states in order of approval rather than alphabetically, by state. The following command is a bit advanced, so I'll break it down. In broad terms, the right side of the arrow is telling R to take the Approve21 data and reorder the rows so they are listed in order of the values on `Approve`. The square brackets `[ ]` are usually used to subset a dataset, and here they are saying to use the `order` command to sort the `Approve21` dataset by the values of the `Approve` variable. Let's see how this changes the head and tail of the dataset (if you type this into your computer, make sure everything is exactly the same).

```
#Sort by "Approve" and copy over original data set
Approve21 <- Approve21[order(Approve21$Approve),]

#list the first several rows of data, now ordered by "Approve"
head(Approve21)
# A tibble: 6 × 5
  State          stateab Approve Disapprove Neither
  <chr>          <chr>     <dbl>      <dbl>   <dbl>
1 West Virginia  WV           23         72       5
2 Wyoming        WY           26         68       5
3 Oklahoma       OK           29         65       6
4 Idaho          ID           30         65       5
5 Alabama        AL           31         63       6
6 Arkansas       AR           31         63       6
```

This shows us the first six states in the dataset, those with the lowest levels of presidential approval. Our initial impression from the alphabetical list was verified: President Biden is given his lowest marks in states that are usually considered very conservative. There is also a bit of a regional pattern, as these are all Southern or Mountain West states.

Now, let's look at the states with the highest approval levels by looking at the last six states in the dataset. Notice that you don't have to sort the dataset again because the earlier sort changed the order of the observations until they are sorted again for some reason.

```
#list the last several rows of data, now ordered by "Approve"
tail(Approve21)

# A tibble: 6 × 5
  State          stateab Approve Disapprove Neither
  <chr>          <chr>     <dbl>      <dbl>   <dbl>
1 California     CA           57         35       8
2 Rhode Island   RI           57         37       7
```

```
3  Vermont        VT    58    36    7
4  Maryland       MD    59    33    8
5  Hawaii         HI    61    33    6
6  Massachusetts  MA    63    30    7
```

Again, our initial impression is confirmed. President Biden gets his highest marks in states that are typically considered fairly liberal states. There is also an East Coast and Pacific West flavor to this list of high approval states.

It is much neater and less distracting to show just the variables of interest, in this case `state` and `Approve`, instead of the entire dataset. To do this, you can modify the `head` and `tail` commands to designate that certain columns be displayed (shown below with `head`). The language within the brackets, `c('state', 'Approve')`, is used to tell R to combine the columns headed by `state` and `Approve` from the dataset `Approve21`.

```
#list the first several rows of data for "state" and "Approve"
head(Approve21[c('state', 'Approve')])

# A tibble: 6 × 2
   State          Approve
   <chr>            <dbl>
1  West Virginia       23
2  Wyoming             26
3  Oklahoma            29
4  Idaho               30
5  Alabama             31
6  Arkansas            31
```

Get a Graph

The preceding commands helped you get more familiar with the shape of the dataset, the nature of the variables, and the types of states at the highest and lowest end of the distribution; but it would be nice to have a more general sense of the variation in presidential approval among the states. For instance, how are states spread out between the two ends of the distribution? Are they clustered in the middle, or are they spread out more evenly. You will learn about a number of statistical techniques that provide information about the general tendency of the data, but graphing the data as a first step can also be very effective. **Histograms** are a particularly helpful type of graph for this purpose. To get a histogram, you need to tell R to get the dataset, use a particular variable, and summarize its values in the form of a graph (as in Pane 4).

PANE 4 ■ Tell R You Want a Histogram

The following very simple command will produce a histogram.

```
#This command tells R to use the data in the "Approve" column
# of the "Approve21" data set to make a histogram.

hist(Approve21$Approve)
```

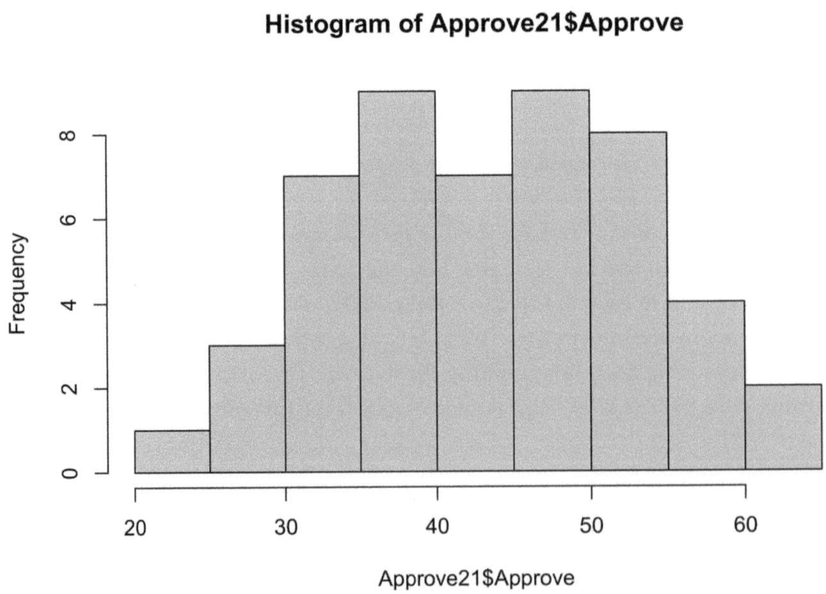

Here, the numbers on the horizontal axis represent approval levels, and the values on the vertical axis represent the number of states in each category. Each vertical bar represents a range of values on the variable of interest. You'll learn more about histograms in the next few chapters, but you can see in this graph that there is quite a bit of variation in presidential approval, ranging from 20%–25% at the low end to 60%–65% at the high end, with the most common outcomes falling between 35% and 55%.

One of the important things you can do in R is modify the appearance of graphs to make them look a bit nicer so they are easier to understand. This is especially important if the information is being shared with people who are not as familiar with the data as the researcher is. In this case, you modify the graph title (`main`), the horizontal (`xlab`), and the vertical (`ylab`) labels.

```
# Axis labels in quotes and commands separated by commas
hist(Approve21$Approve,
     main = "State-level Biden Approval January-June 2021", #Graph title
     ylab = "Number of States", #vertical axis label
     xlab = "% Approval") #Horizontal axis label
```

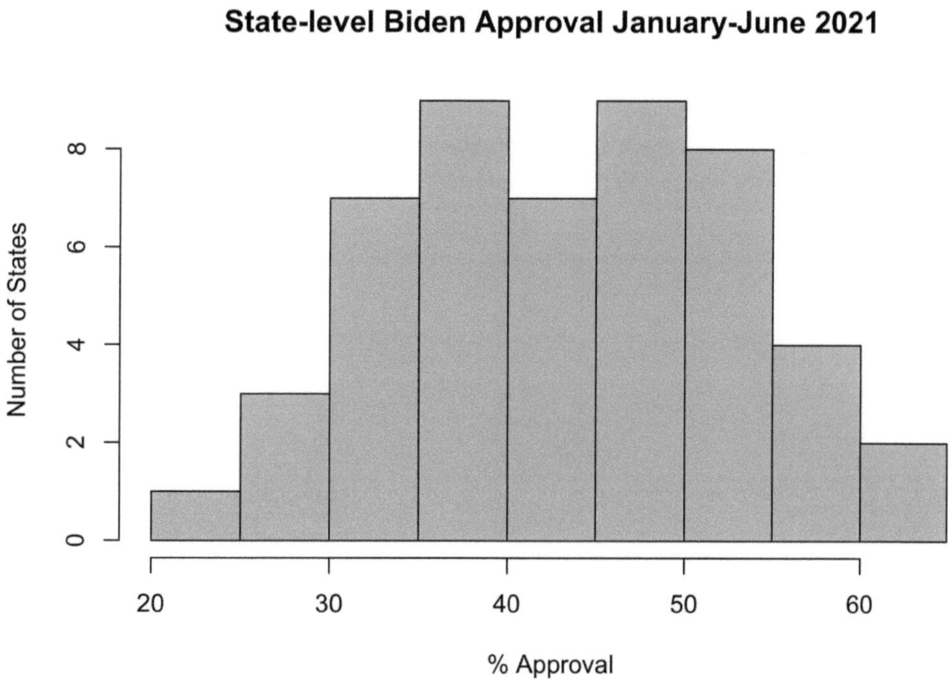

Note that all of the new labels are in quotes and the different parts of the command are separated by commas. You will get an error message if you do not use quotes and commas appropriately.

Making these slight changes to the graph improved its appearance and provided additional information to facilitate interpretation. The cartoon student asked R for a histogram of a specific variable from a specific dataset that they had previously told R to load, and R gave them this nice-looking graph. That, in a nutshell, is how this works. The cartoon student is happy that they have learned a lot, and they are ready to learn more!

PANE 5 ■ Making Progress

Before moving on, let's check in with RStudio to see what the various windows look like after running all of the commands listed previously (Figure 2.7). Here, you can see the script file

FIGURE 2.7 ■ The RStudio Windows After Running Several Commands

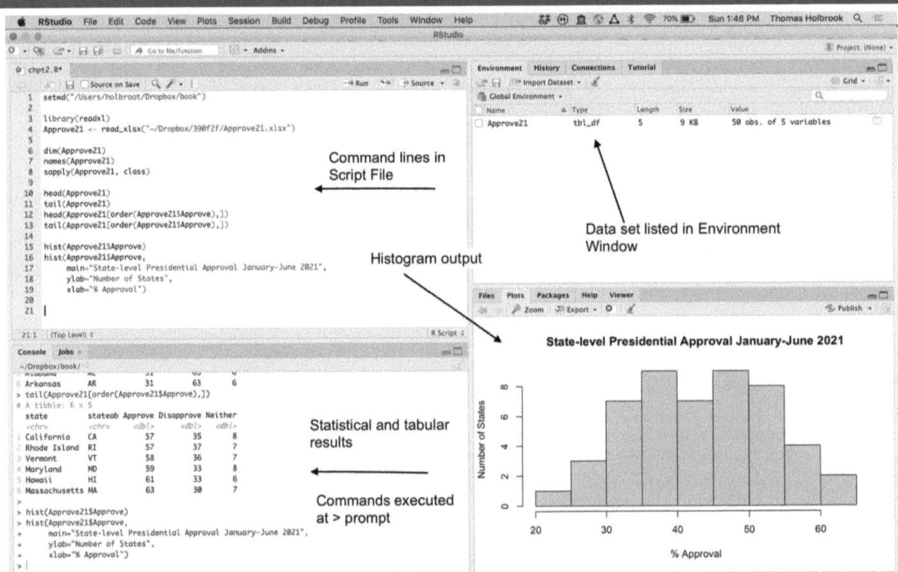

(named "chpt2.R") that includes all of the commands previously run. This file is now saved in the working directory and can be used later. The Environment window lists the `Approve21` dataset, along with a little bit of information about it. The Console window shows the commands that were executed, along with the results for the various data tables we produced. Finally, the Plot window in the lower-right corner shows the histogram produced by the histogram command.

SOME R TERMINOLOGY

Now that you have already used R a bit, we need to cover a bit of terminology that sheds light on how R operates. It is no accident that this discussion takes place after you have already been using R. From a learning perspective, it is important to have a bit of hands-on experience with the tools you will be using before getting into definitions and terminology related to those tools. Without walking through the process of using R to produce results, much of the following discussion would be too abstract and you might have trouble making the connections between terminology and actually using R. As you will see, we have already been using many of the terms discussed.

Data Frames

One of the most important things to understand is the concept of a **data frame**. A data frame is a collection of columns and rows of data, as depicted in Figure 2.8. Columns in a data frame represent the values of the variables we are interested in. The column headings are the variable names we use when telling R what to do. In Figure 2.8, the highlighted column (surrounded by a solid line) contains the state abbreviations. If you want R to use the state abbreviation, you will refer to the column as `stateab`. The rows represent the individual cases or observations. The highlighted row below (surrounded by the dashed line) includes information from all five columns. Although we know the second row includes outcomes for Alaska, we wouldn't normally refer to this as "row Alaska." Instead, from R's perspective, this is "row 2" (the top row in a data frame is not considered row 1, as it is where the column names are stored). This collection of rows and columns together constitute a data frame (surrounded by the solid double line). For our purposes, the terms data frame and dataset are used interchangeably.

FIGURE 2.8 ■ An Illustration of a Data Frame

state	stateab	Approve	Disapprove	Neither
Alabama	AL	31	63	6
Alaska	AK	41	53	6
Arizona	AZ	44	51	5
Arkansas	AR	31	63	6
California	CA	57	35	8
Colorado	CO	50	44	7

One of the examples shown earlier used the `names(Approve21)` command to get the names of all five columns (variables). In this case, we told R to do something with the entire dataset. However, when getting the class of selected variables, sorting the dataset, and producing the histogram, we told R to use just one column from the dataset. We specified that R should go to `Approve21` and get information from the column headed by `Approve`, and we communicated this as `Approve21$Approve`.

Sometimes, it can hard to connect with and understand the different elements of a data frame. To make these things a bit more concrete and hopefully more understandable, I assemble part of this dataset, using the `c()` function in R to combine data points into vectors (lists of values that can be stored together) representing individual variables, and then join all of those vectors using `data.frame` to create a dataset.

Let's use the first four rows of data from Figure 2.8 to demonstrate this. The first variable is "state," so we can begin by creating a vector named state that includes the first four state names in order.

```
#create vector of state names and store as "state"
state <- c("Alabama", "Alaska", "Arizona", "Arkansas")

#Show state names
state
[1] "Alabama"   "Alaska"   "Arizona"   "Arkansas"
```

As you see, the vector named `state` has the first four state names.

Now, let's do this for the first four observations of the remaining columns.

```
#create vector of state abbreviations and store as "stateab"
stateab <- c("AL", "AK", "AZ", "AR")

#Show state abbreviations
stateab
[1] "AL" "AK" "AZ" "AR"

#create vector of state abbreviations and store as "Approve"
Approve <- c(31, 41, 44, 31)

#Show approval values
Approve
[1] 31 41 44 31

#create vector of state abbreviations and store as "Approve"
Disapprove <- c(63,53,51,63)

#Print (show) disapproval values
Disapprove
[1] 63 53 51 63

#create vector of state abbreviations and store as "Neither"
Neither <- c(6,6,5,6)

#Print (show) neither values
Neither
[1] 6 6 5 6
```

As you can see, all of the information from the first four rows of the approval dataset are now represented by five different vectors. To create a data frame, we need to stitch these vectors together, using the `data.frame` function.

```
#Combine separate vectors into a single data frame
df <- data.frame(state, stateab, Approve, Disapprove, Neither)
#Show the data frame
df
      state  stateab  Approve  Disapprove  Neither
1   Alabama       AL       31          63        6
2    Alaska       AK       41          53        6
3   Arizona       AZ       44          51        5
4  Arkansas       AR       31          63        6
```

Now you have a small data frame named `df` with the same information found in the first few rows of data in Figure 2.8. Hopefully, this illustration helps you connect to the concept of a data frame. Fortunately, researchers don't usually have to input data in this manner, but it is a helpful tool to use here to illustrate the concept of a data frame.

As a quick aside, when we imported the `Approve21` dataset, R created a "tibble," which is a particular type of data frame, labeled by R as `tbl_df` (for tibble data frame). The classic R dataset is usually identified as a `data.frame` and has some features and characteristics that are different from a `tibble` dataset. We ended up with a `tbl_df` because the default for the Import Dataset tab in the Environment window is to create a tibble when importing data. This distinction is not something you need to worry about, and it is only mentioned here so you don't get confused if you come across a `tbl_df` instead of a `data.frame`.

Objects and Functions

Everything in R is an **object**. I know that doesn't sound very useful, but it is true. Everything you've seen in this chapter is an object; the data frames, the variables inside the data frames, the vectors we created, the histogram, and the commands we used to tell R what to do. These things are all objects. **Functions** are a particular type of object. You can think of functions as commands that tell R what to do. Anything R does, it does with a function, and those functions tell R to do something with an object.

As an example, the code we used to create the histogram also includes both a function and an object.

```
hist(Approve21$Approve,
     main = "State-level Biden Approval January-June 2021",
     xlab = "% Approval")
```

Here, `hist` is a function call (telling R to use a function called "hist"), and `Approve21$Approve` is an object. The function `hist` is performing operations on the object

Approve21$Approve. Other "action" parts of the function are main, xlab, and ylab, which are used to create the main title and the labels for the *x* axis, and the *y* axis.

We also just used two functions (c and data.frame) to create six objects: the five vectors of data (state, stateab, Approve, Disapprove, and Neither) and the data frame where they were joined (df).

Storing Results in New Objects

One of the really useful things you can do in R is create new objects, either to store the results of some operation or to add new information to existing datasets. For instance, at the very beginning of the data example in this chapter, we imported data from an Excel file and stored it in a *new object* named Approve21 using the following command.

```
Approve21 <- read_excel("Approve21.xlsx")
```

Here and in multiple places throughout this chapter, I used <- to tell R to put the results of some operation into a new object, usually a dataset or a variable within a dataset. Pay close attention to this, as you will use the <- a lot. This is a convention in R that I think is helpful because it is literally pointing at the new object and saying to take the result of the right side action and **assign** it to the object on the left. We also updated an object when we sorted Approve21 and replaced the original data with sorted data by level of approval, as shown next.

```
Approve21 <- Approve21[order(Approve21$Approve),]
```

We can also modify existing datasets by creating new variables based on **transformations of existing variables** (more on this in Chapter 4). For instance, suppose that you want to express presidential approval as a proportion instead of a percentage. You can do that by dividing Approve21$Approve by 100, but you don't want to replace the original variable, so the results are stored in a new variable, Approve21$Approve_prop, which is added to the original dataset.[1]

```
#Calculate and add "Approve_prop" to Approve21
Approve21$Approve_prop <- Approve21$Approve/100
```

You can check with the names function to see if the new variable is now part of the dataset.

```
names(Approve21)
[1]  "state"          "stateab"   "Approve"   "Disapprove"   "Neither"
[6]  "Approve_prop"
```

The new object is listed as the sixth variable in the dataset. You can also look at its values to make sure they are expressed as proportions rather than percentages. Typing just object name at the > prompt will generate a listing of all values for that object.

```
Approve21$Approve_prop
[1] 0.23 0.26 0.29 0.30 0.31 0.31 0.31 0.32 0.33 0.34 0.35 0.36 0.36 0.36 0.37
[16] 0.37 0.37 0.39 0.39 0.39 0.41 0.41 0.42 0.43 0.44 0.44 0.45 0.46 0.47 0.47
[31] 0.47 0.48 0.48 0.50 0.50 0.50 0.51 0.52 0.52 0.52 0.53 0.53 0.53 0.54 0.57
[46] 0.57 0.58 0.59 0.61 0.63
```

It looks like all of the values of `Approve21$Approve_prop` are indeed expressed as proportions. The bracketed numbers do not represent values of the object. Instead, they tell the order of the outcomes in the dataset. For instance, [1] is followed by .23, .26, and 29, indicating the value of the first observation is .23, the second is .26, and the third is .29. Likewise, [16] is followed by .37, .37, and .39, indicating the value for the sixteenth observation is .37, the seventeenth is .37, and the eighteenth is .39, and so on.

Packages and Libraries

Packages and libraries are very important objects in the R environment. **Packages** are collections of functions, objects (usually data sources), and code that enhance the base functions of R. For instance, the function `hist` (used to create a histogram) is part of a package called `graphics` that is automatically installed as part of the R base, and the function `read_xlsx` (used to import the Excel spreadsheet) is part of a package called `readxl`. Once you have installed a package, you should not have to reinstall it unless you start an R session on a different device.

If you need to use a function from a package that is not already installed, you use the following command.

```
install.packages("package_name")
```

You can also install packages using the Packages tab in the lower-right window in RStudio. Just go to the Install icon, search for the package name, and click on "Install." When you install a package, you will see a lot of stuff happening in the Console window. Once the > prompt returns without any error messages, the new package is installed. In addition to the core set of packages that most people use, we will rely on a number of other packages in this book.

Each chapter in this book begins with a list of the packages that will be used in the chapter, and you can install them as you work through the chapters, if you wish. Alternatively, you can copy, paste, and run the code that follows to install all of the packages used in this book in one fell swoop.

```
install.packages(c("dplyr", "descr", "DescTools", "effectsize",
                   "gplots", "Hmisc", "lmtest", "ppcor", "readxl",
                   "sandwich", "stargazer"))
```

Libraries are where the packages are stored for active use. Even if you have installed all of the required packages, you generally can't use them unless you have loaded the appropriate library.

```
#Note that you do not use quotation marks for package names here.
library(package_name)
```

You can also load a library by going to the Packages tab in the lower-right window and clicking on the box next to the already installed package. This command makes the package available for you to use. I used the command `library(readxl)` before using `read_xlsx()` to import the presidential approval data. This told R to make the functions in `readxl` available for me to use. If I had not done this, I would have gotten an error message something like "Error: could not find function `read_xlsx`." You will also get this error if you misspell the function name (happens a lot). Installing the packages and attaching the libraries allows us to access and use any functions or objects included in these packages.

MANAGING FILES AND OUTPUT

Working Directory

When you create files, ask R to use files, or import files to use in R, they are stored in your **working directory.** Organizationally, it is important to know where this working directory is or to change its location so you can find your work when you want to pick it up again. The best practice in R is to designate a specific folder on your computer as your working directory. If you opened a new folder and downloaded the book files there as I suggested earlier in this chapter, you should use that as your working directory. You can set this as your working directory, as follows.

```
# Here, the working directory is set to a folder on my desktop.
setwd("/Users/username/Desktop/PolSci390")
```

You can also check on the location of the working directory to make sure it is where you think it is with the `getwd()` command.

```
getwd()
[1] "/Users/username/Desktop/PolSci390"
```

The nice thing about keeping all of your work (data files, script files, etc.) in the working directory is that you do not have to include the sometimes cumbersome file path names when accessing the files So, instead of having to write something like `readxl("/Users/username/Desktop/PolSci390/Approve21.xlsx")`, you just have to write `readxl("Approve21.xlsx")`. If you need to use files that are not stored in your working directory and you don't remember their location, you can use the `file.choose()` command to locate them on your computer. The output from this command will be the correct path and file name, which you can then copy and paste into R.

If you are using **posit Cloud** and you have uploaded the course files, all of those files should be in the working directory in the cloud. You can check this by clicking on the Files tab in the lower-right window of RStudio. This should show you all of the uploaded files, and any work you save in the posit Cloud environment should be saved there.

Save Your Work

Before wrapping up the material in this chapter, you should save your work. This means not just saving the script file but also saving the `Approve21` dataset as an R data file. To save the script file, just go to the File icon at the top of the Source window, click the icon, and assign your file an intuitive name that you will be able to remember. I use "chpt2" as the name, which R saves as `chpt2.R`. This file will be saved to the working directory. Now, any time you need to use commands from this file, you can click on the folder icon at the top of the RStudio window, locate the file, and open it.

You should also save the `Approve21` dataset as an R data file so you won't need to import the Excel version of this dataset when you want to use it again. The command for saving an object as an R data file is `save(objectname, file = filename.rda)`. Here, the object is `Approve21`, and I want to keep that name as the file name, so I use the following command.

```
save(Approve21, file = "Approve21.rda")
```

This will save the file to your working directory. When you want to use this file in the future, you can use the `load()` command to access the dataset.

```
load("Approve21.rda")
```

Creating Documents

If you are working with R, odds are that you are working on something that you need to share with others. For most of you, this means turning in end-of-chapter assignments or perhaps a longer writing assignment. It's also possible that you need to submit a report of some sort for your job or you need to put together a presentation on some subject. Either way, you need to integrate the on-screen results you get using R into some sort of document. There are a couple of options for integrating R results into documents: using a standard word processing program or by creating a Quarto document.

Standard Word Processing Document

First, you can copy and paste the findings into another document, probably using Microsoft Word or some similar word-processing product. When you do this, you should make sure you format the pasted output using a *fixed-width font* ("Courier" and "Courier New" work well). Otherwise, the columns in the R output will wander all over the place, and it will be hard to evaluate your work. You should always check to make sure the columns are lining up on paper the same way they did on screen. You might also have to adjust the font size to accomplish this. If you save your document as a PDF, make sure to double-check the formatting in the final product.

To copy graphs, you can click "Export" in the graph window and save your graph as an image (make sure you keep track of the file location). Then, you can insert the graph into your document. Alternatively, you can also click "copy to clipboard" after clicking on "Export" and copy and paste the graph directly into your document. You might have to resize the graphs after importing, following the conventions used by your document program.

You may have to deal with other formatting issues, depending on what document software you are using. The main thing is to always take time to make sure your results are presented as neatly as possible. In most cases, it a good idea to include only your final results, unless instructed otherwise.

Quarto Document

Alternatively, you can create a Quarto document. Quarto is a posit Cloud product that allows you to integrate your code, the results of your analysis, and your discussion of the results into a single document, all while working in the RStudio environment. Generally, Quarto documents look much better than those created by copying and pasting results into a standard word-processing program. Another benefit to Quarto is that learning to use it gives you another important skill that you can add to your resume. Although your experience with Quarto may be a bit bumpy at the beginning, it is really a breeze once you've created a couple of documents. You can open a Quarto document by clicking on the New File tab, where you open your script file. If you decide to use Quarto or if your instructor requires it, you will find a brief tutorial in Appendix B.

NEXT STEPS

Everything in this chapter will help you get started on your journey to using R for political and social data analysis. That said, this information is really just the very small piece of the tip of a very large iceberg of information. There is a lot more to learn. I say this not to intimidate or cause concern but because I think this is a reason for excitement. For most students, doing data analysis and working with R are both new experiences, and that's exactly what makes this an exciting opportunity. This book is designed to make it possible for any student to learn and improve their skill level, as long as they are willing to commit to putting in the time and effort required.

You will learn more about data analysis and how to use additional R commands in the next several chapters, and you probably will suffer through some frustration as R gives you the occasional error message, but if you stay on top of things (on schedule or ahead of schedule) and ask for help before it's too late, you will be fine.

EXERCISES

Concepts and Calculations

1. The following lines of R code were used earlier in the chapter. For each one, identify the object(s) and function(s). Explain your answer.
   - `class(Approve21$Disapprove)`
   - `save(Approve21, file = "Approve21.rda")`
   - `Approve21$Approve_prop <- Approve21$Approve/100`
   - `Approve21 <- read_excel("Approve21.xlsx")`

2. In your own words, describe what a script file is and what its benefits are.

3. The following command saves a copy of the `states20` dataset: `save(states20, file = "states20.rda")`. Is this file being saved to the working directory or to another directory? How can you tell?

4. After using `head(Approve21)` to look at the first few rows of the Approve21 dataset, I pasted the screen output to my MS Word document and it looked like this:

   state stateab Approve Disapprove Neither

   Alabama AL 31 63 6
   Alaska AK 41 53 6
   Arizona AZ 44 51 5
   Arkansas AR 31 63 6
   California CA 57 35 8
   Colorado CO 50 44 7

Why did I end up with these crooked columns that don't align with the variable names? How can I fix it?

R Problems

1. I tried to load the `anes20.rda` data file using the following command: `load(Anes20.rda)`. It didn't work, and I got the following message. **Error in load(Anes20.rda): object 'Anes20.rda' not found**. What did I do wrong? (Hint: there are two errors.)

2. Load the `countries2` dataset and get the names of all of the variables included in it. Based just on what you can tell from the variable names, what sorts of variables are in this dataset? Identify one variable that looks like it might represent something interesting to study (a potential dependent variable), and then identify another variable that you think might be related to the first variable you chose.

3. Use the `dim` function to tell how many variables and how many countries are in the dataset.

4. Use the `class()` function to identify the level of measurement of `ccode` and `matmort` from the `countries2` dataset.

5. Use the `Approve21` dataset and create a new object, `Approve21$net_approve`, which is calculated as the percentage in the state who approve of the Biden's performance MINUS the percentage in the state who disapprove of Biden's performance. Sort the dataset by net approve using the following command:
`Approve21 <- Approve21[order(Approve21$net_approve), ]`, and list the six highest and lowest states. Say a few words about the types of states in these two lists.

6. Produce a histogram of `Approve21$net_approve` and describe what you see. Be sure to provide substantively meaningful labels for the histogram.

3 FREQUENCIES AND BASIC GRAPHS

GET READY

In this chapter, we explore how to examine the distribution of variables using some basic data tables and graphs. In order to follow along in R, you should load the `anes20.rda` and `states20.rda` data files, as shown. If you have not already downloaded the datasets, see the instructions at the end of the "Accessing R" section of Chapter 2.

```
#If files are in your working directory, just:
load("anes20.rda")
load("states20.rda")
```

If you get errors at this point, check to make sure the files are in your working directory and make sure you spelled the file names correctly and enclosed the file names in quotes.

In addition, you should also load the libraries for `descr` and `Desctools`, two packages that provide many of the functions we will be using. You may have to install the packages, if you have not done so already.

```
library(descr)
library(DescTools)
```

INTRODUCTION

Sometimes, very simple descriptive statistics or graphs can convey a lot of information and play an important role in the presentation of data analysis findings. Advanced statistical and graphing techniques are normally required to speak with confidence about data-based findings, but it is almost always important to start your analysis with some basic tools, for two reasons. First, these tools can be used to alert you to potential problems with the data. If there are issues with the way the data categories are coded or perhaps with missing data, those problems are relatively easy to spot with some of the simple methods discussed in this chapter. Second, the distribution of values (how spread out they are or where they tend to cluster) on key variables can be an important part of the story told by the data, and some of this information can be hard to grasp when using more advanced statistics.

Two data sources are used to provide examples in this chapter: a dataset comprised of selected variables from the 2020 American National Election Study,[1] a large-scale academic survey of public opinion in the months just before and after the 2020 U.S. presidential election (saved as `anes20.rda`), and a state-level dataset containing information on dozens of political, social, and demographic variables from the fifty states (saved as `states20.rda`). In the `anes20` dataset, most variable names follow the format "V20####," while in the `states20` dataset the variables are given descriptive names that reflect the content of the variables. Codebooks for these datasets are included in the appendix to this book.

FREQUENCIES

One of the most basic things we can do is count the number of times different outcomes for a variable occur. Usually, this sort of counting is referred to as getting the **frequencies** for a variable. There are a couple of ways to go about this. Let's take a look at a variable from `anes20` that measures the extent to which people want to see federal government spending on "aid to the poor" increased or decreased (`anes20$V201320x`).

One of the first things to do is familiarize yourself with this variable's categories. You can do this with the `levels()` command.

```
#Show the value labels for V201320x, from the anes20 data set.
#If you get an error message, make sure you have loaded the data set.

levels(anes20$V201320x)
[1] "1. Increased a lot"    "2. Increased a little" "3. Kept the same"
[4] "4. Decreased a little" "5. Decreasaed a lot"
```

Here, you can see the category labels, which are ordered from "1. Increased a lot" at one end to "5. Decreased a lot" at the other. This is useful information, but what we really want to know is how many survey respondents chose each of these categories. Does one outcome seem like it occurs a lot more than all of the others? Are there some outcomes that hardly ever occur? Do the outcomes generally lean toward the "Increase" or "Decrease" side of the scale?

You need to create a table that summarizes the distribution of responses. This table is usually known as a frequency distribution or a frequency table. The base R package includes a couple of commands that can be used for this purpose. First, you can use `table()` to get simple frequencies.

```
#Create a table showing the how often each outcome occurs
table(anes20$V201320x)

   1. Increased a lot  2. Increased a little  3. Kept the same
                 2560                   1617              3213
 4. Decreased a little    5. Decreasaed a lot
                   446                    389
```

Now you can see not just the category labels but also the number of survey respondents in each category. From this, you can see that there is more support for increasing spending on the poor than for decreasing it, and it is clear that the most common choice is to keep spending the same.

Okay, this is useful information and certainly an improvement over just listing all of the data and trying to make sense out of them that way. However, this information could be more useful if we expressed it in relative rather than absolute terms. As useful as the simple raw frequencies are, the drawback is that they are a bit hard to interpret on their own (at least without doing a bit of math in your head). Let's take the 2560 "Increased a lot" responses as an example. This seems like a lot of support for this position, and we can tell that it is compared to the number of responses in most other categories; but 2560 responses can mean different things from one sample to another, depending on the sample size (2560 out of how many?). Certainly, the magnitude of this number means something different if the total sample is 4000 rather than if it is 10,000. Since R can be used as an overpowered calculator, we can add up the frequencies from all categories to figure out the total sample size.

```
#Add up the category frequencies and store them in "sample_size"
sample_size <- 2560 + 1617 + 3213 + 446 + 389
#Print the value of "sample_size"
sample_size
[1] 8225
```

So now we know that there were 8225 total valid responses to the question on aid to the poor, and 2560 of them favored increasing spending a lot. Now we can start thinking about whether that seems like a lot of support, *relative to the sample size.* What we need to express is the frequency of the category outcomes *relative* to the total number of outcomes. These relative frequencies are usually expressed as *percentages* or *proportions.*

Percentages express the relative occurrence of each value of x. For any given category, this is calculated as the number of observations in the category, divided by the total number of valid observations across all categories, multiplied times 100:

$$\text{Category Percent} = \frac{\text{Total cases in category}}{\text{Total valid cases}} * 100$$

Or if you are really itching for something that looks a bit more complicated (but isn't),

$$\text{Category Percent} = \frac{f_k}{n} * 100$$

Where

f_k = frequency, or number of cases in any given category

n = the number of valid cases from all categories

This simple statistic is very important for making relative comparisons. *Percent* literally means *per one hundred*, so regardless of overall sample size, we can look at the category percentages and get a quick, standardized sense of the relative number of outcomes in each category. This is why percentages are also referred to as relative frequencies. In frequency tables, the percentages can range from 0 to 100 and should always sum to 100.

Proportions are calculated pretty much the same way, except without multiplying times 100:

$$\text{Category Proportion} = \frac{\text{Total cases in category}}{\text{Total valid cases}}$$

The main difference here is that proportions range in value from 0 to 1 rather than 0 to 100. It's pretty straightforward to calculate both the percentage and the proportion of respondents who thought that federal spending to aid the poor should be increased a lot.

```
#Calculate Percent in "Increased a lot" category
(2560/sample_size)*100
[1] 31.12462

#Calculate Proportion in "Increased a lot" category
2560/sample_size
[1] 0.3112462
```

So we see that about 31% of all responses are in this category. What's nice about percentages and proportions is that, for all practical purposes, the values have the same meaning from one sample to another. In this case, 31.1% (or .311) means that slightly less than one-third of all responses are in this category, regardless of the sample size. Of course, we can also look at the percentages for the other response categories to gain a more complete understanding of the relative popularity of the response choices.

Fortunately, we do not make this calculation manually for every category. Instead, we can use the `prop.table` function to get the proportions for all five categories. We need to *store the results of the raw frequency table in a new object* and then have `prop.table` use that object to calculate the proportions. Note that I use the extension ".tbl" when naming the new object. This serves as a reminder that this particular object is a table. When you execute commands such as this, you should see the new object appear in the Global Environment window.

```
#Store the frequency table in a new object called "poorAid.tbl"
poorAid.tbl <- table(anes20$V201320x)

#Create a proportion table using the contents of "poorAid.tbl"
prop.table(poorAid.tbl)

   1. Increased a lot  2. Increased a little  3. Kept the same
          0.31124620             0.19659574        0.39063830
4. Decreased a little     5. Decreasaed a lot
          0.05422492             0.04729483
```

There are a couple of key takeaway points from this table. First, there is very little support for decreasing spending on federal programs to aid the poor and a lot of support for increasing spending. Only about 10% of all respondents (combining the two "Decreased" categories) favor cutting spending on these programs, compared to just over 50% (combining the two "Increased" categories) who favor increasing spending. Second, the single most popular response is to leave spending levels as they are (39%). Bottom line, there is not much support in these data for cutting spending on programs for the poor.

There are some things to notice about this interpretation. First, I didn't get too bogged down in comparing all of the reported proportions. If you are presenting information like this, the audience (e.g., your professor, classmates, boss, or client) is less interested in the minutia of the table than the general pattern. Second, while focusing on the general patterns, I did provide a few specifics. For instance, instead of just saying "There is very little support for cutting spending," I included specific information about the percentage who favored and opposed spending and who wanted it kept the same but without getting too bogged down in details. Finally, I referred to percentages rather than proportions, even though the table reports proportions. This is really just a personal preference, and in most cases, it is okay to do this. Just make sure you are consistent within a given discussion.

Okay, so we have the raw frequencies and the proportions, but note that we have to use two different tables to get this information, and those tables are not exactly "presentation ready." Fear not, for there are a couple of alternatives that save steps in the process and still provide all the information you need in a single table. The first alternative is the `freq` command, which is provided in the `descr` package. Here's what you need to do.

```
#Provide a frequency table, but not a graph
freq(anes20$V201320x, plot = F)

PRE: SUMMARY: Federal Budget Spending: aid to the poor
                        Frequency   Percent Valid  Percent
1. Increased a lot           2560   30.9179          31.125
2. Increased a little        1617   19.5290          19.660
3. Kept the same             3213   38.8043          39.064
4. Decreased a little         446    5.3865           5.422
5. Decreasaed a lot           389    4.6981           4.729
NA's                           55    0.6643
Total                        8280  100.0000         100.000
```

As you can see, we get all of the information provided in the earlier tables, plus some additional information, and the information is somewhat better organized. The first column of data shows the raw frequencies, the second shows the total percentages, and the final column is the valid percentages. The valid percentages match up with the proportions reported earlier, while the "Percent" column reports slightly different percentages based on 8280 responses (the 8225 valid responses and 55 survey respondents who did not provide a valid response). When conducting surveys, some respondents refuse to answer some questions or may not have an opinion or might be skipped for some reason. These 55 responses in the preceding table are considered *missing data* and are denoted as `NA` in R. It is important to be aware of the level of missing data and usually a good idea to have a sense of why they are missing. Sometimes, this requires going back to the original codebooks or questionnaires (if using survey data) for more information about the variable. Generally, researchers present the valid percentage when reporting results.

So what about the `table` function, which we originally used to get frequencies? Is it still of any use? You bet it is! In fact, many other functions make use of the information from the `table` command to create graphics and other statistics, as you will see shortly.

Besides providing information on the distribution of single variables, it can also be useful to compare the distributions of multiple variables if there are sound theoretical reasons for doing so. For instance, in the previous example, the data showed widespread support for federal spending on aid to the poor. It is interesting to ask, though, about how supportive people are when we refer to spending not as "aid to the poor" but as "welfare programs," which technically are programs to aid the poor. The term "welfare" is viewed by many as a "race-coded" term, one that people associate with programs that primarily benefit racial minorities (mostly African Americans), which leads to lower levels of support, especially among whites. As it happens, the 2020 ANES asked the identical spending question but substituted "welfare programs" for "aid to the poor," shown next.

```
freq(anes20$V201314x, plot = FALSE)

PRE: SUMMARY: Federal Budget Spending: welfare programs
                          Frequency  Percent Valid  Percent
1. Increased a lot             1289   15.5676           15.69
2. Increased a little          1089   13.1522           13.26
3. Kept the same               3522   42.5362           42.88
4. Decreased a little          1008   12.1739           12.27
5. Decreasaed a lot            1305   15.7609           15.89
NA's                             67    0.8092
Total                          8280  100.0000          100.00
```

On balance, there is much less support for increasing government spending on programs when framed as "welfare" than as "aid to the poor." Almost 51% favored increasing spending on aid to the poor; only 29% favored increased spending on welfare programs. And, while only 10% favored decreasing spending on aid to the poor, 28% favored decreasing funding for welfare programs. Clearly, in their heads, respondents see these two policy areas as different, even though the primary purpose of "welfare" programs is to provide aid to the poor.

This single comparison is a nice illustration of how even very simple statistics can reveal substantively interesting patterns in the data.

The Limits of Frequency Tables

As useful and accessible as frequency tables can be, they are not always as straightforward and easy to interpret as the ones I just presented, especially for many numeric variables. The basic problem is that once you get beyond 7–10 categories, it can be difficult to see the patterns in the frequencies and percentages. Sometimes, there is just too much information to sort through effectively. Consider the case of presidential election outcomes in the states in 2020. Here, we will use the `states20` dataset mentioned earlier in the chapter. The variable of interest is `d2pty20`, Biden's percentage of the two-party vote in the states.

```
# Make sure you've loaded the 'states20' data set.
#Frequency table for "d2pty20" in the states
freq(states20$d2pty20, plot = FALSE)

states20$d2pty20
          Frequency  Percent
27.52             1        2
30.2              1        2
32.78             1        2
33.06             1        2
34.12             1        2
35.79             1        2
36.57             1        2
36.8              1        2
37.09             1        2
38.17             1        2
39.31             1        2
40.22             1        2
40.54             1        2
41.6              1        2
41.62             1        2
41.8              1        2
42.17             1        2
42.51             1        2
44.07             1        2
44.74             1        2
45.82             1        2
45.92             1        2
47.17             1        2
48.31             1        2
49.32             1        2
50.13             1        2
50.16             1        2
50.32             1        2
50.6              1        2
51.22             1        2
51.41             1        2
53.64             1        2
```

53.75	1	2
54.67	1	2
55.15	1	2
55.52	1	2
56.94	1	2
58.07	1	2
58.31	1	2
58.67	1	2
59.63	1	2
59.93	1	2
60.17	1	2
60.6	1	2
61.72	1	2
64.91	1	2
65.03	1	2
67.03	1	2
67.12	1	2
68.3	1	2
Total	50	100

There is not much that is useful here, other than that vote share ranges from 27.5 to 68.3. Beyond that, this frequency table includes too much information to absorb in a meaningful way. There are fifty different values, and it is really hard to get a sense of the general pattern in the data.

In cases like this, it is useful to collapse the data into fewer categories that represent ranges of outcomes. Fortunately, the `Freq` function does this automatically for numeric variables (note the uppercase F, as R is case-sensitive).

```
Freq(states20$d2pty20)
```

	Level	freq	perc	cumfreq	cumperc
1	[25,30]	1	2.0%	1	2.0%
2	(30,35]	4	8.0%	5	10.0%
3	(35,40]	6	12.0%	11	22.0%
4	(40,45]	9	18.0%	20	40.0%
5	(45,50]	5	10.0%	25	50.0%
6	(50,55]	9	18.0%	34	68.0%
7	(55,60]	8	16.0%	42	84.0%
8	(60,65]	4	8.0%	46	92.0%
9	(65,70]	4	8.0%	50	100.0%

Here, you get the raw frequencies, the valid percentage (note that there are no NAs listed), the cumulative frequencies, and the cumulative percentage. When data are collapsed into ranges like this, the groupings are usually referred to as *intervals, classes*, or *bins* and are labeled with the upper and lower *limits* of the category. This function uses what are called **right closed** intervals (indicated by] on the right side of the interval), so the first bin (also closed on the left, `[25, 30]`) includes all values of presidential approval ranging from 25 to 30, the second bin (`(30, 35]`) includes all values ranging from just more than 30 to 35, and so on. In this instance, binning the data makes a big difference. Now it is much easier to see that there are relatively few states with very high or very low levels of Biden support, and most states are in the middle of the distribution.

The last two columns provide useful information that was not provided in the earlier frequency table, which used the `freq` (lower-case) function. Of particular use is the **cumulative percentage**, which is the percentage of observations in or below (in a numeric or ranking sense) a given category. You calculate the cumulative percentage for a given ordered or numeric value by summing the percentage with that value and the percentage in all lower-ranked values. In the preceding table, the cumulative percentage shows that then-candidate Biden received 50% or less of the two-party vote in exactly half of the states.

You might think that using nine categories is a bit too many; you can adjust the number of categories, as well as the limits of the bins, by using the `breaks` command. Following, I wanted to create five bins with a width of 10 ranging from 20% to 70% of the two-party vote.

```
#Create a frequency table with five user-specified groupings
Freq(states20$d2pty20, breaks = c(20, 30, 40, 50, 60, 70))

     level  freq    perc  cumfreq  cumperc
1  [20, 30]    1    2.0%        1     2.0%
2  (30, 40]   10   20.0%       11    22.0%
3  (40, 50]   14   28.0%       25    50.0%
4  (50, 60]   17   34.0%       42    84.0%
5  (60, 70]    8   16.0%       50   100.0%
```

We'll explore regrouping data like this in greater detail in the next chapter.

GRAPHING OUTCOMES

As useful as frequency tables are, basic univariate graphs complement this information and are sometimes much easier to interpret. As discussed in Chapter 1, data visualizations help *contextualize* the results, giving the research consumer an additional perspective on the statistical findings. In the graphs examined here, the information presented is *exactly the same* as some of the information presented in the frequencies previously discussed, albeit in a different format.

Bar Charts

Bar charts are simple graphs that summarize the occurrence of outcomes in categorical variables, providing some of the same information found in frequency tables. The category labels are on the horizontal axis, just below the vertical bars; ticks on the vertical axis denote the number (or percentage) of cases; and the height of each bar represents the number (or percentage) of cases for each category. It is important to understand that the horizontal axis represents categorical differences, not quantitative distances between categories.

The following code is used to generate the bar chart for `anes20$V201320x`, the variable measuring spending preferences on programs for the poor. Note here that the `barplot` command uses the initial *frequency table*, saved earlier as `poorAid.tbl`, as input rather than the name of the variable. This illustrates what I mentioned earlier, that even though the `table` command does not provide a lot of information, it can be used to help with other R commands. It also reinforces an important point: Bar charts are the graphic representation of the raw data from a frequency table.

```
#Plot the frequencies of anes20$V201320x
barplot(poorAid.tbl)
```

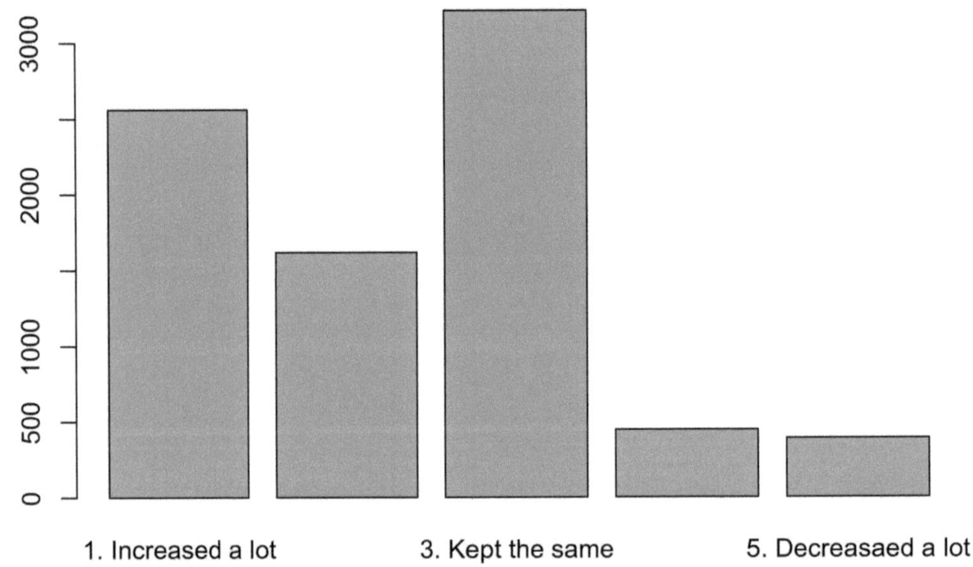

Sometimes you have to tinker a bit with graphs to get them to look as nice as they should. For instance, you might have noticed that not all of the category labels are printed in the preceding bar chart. This is because the labels themselves are a little bit long and clunky, and with five bars to display, some of them were dropped due to lack of room. We could add a command to reduce the size of the labels, but that can lead to labels that are too small to read easily (still, we will look at that command later). Instead, we can replace the original labels with shorter ones that still represent the meaning of the categories, using the `names.arg` command. Make sure to notice the quotation marks and commas in the command. Axis labels and a main title are also added to the graph to make it easier for your target audience to understand what's being presented.

```
#Same as above but with labels altered for clarity in "names.arg"
barplot(poorAid.tbl,
        names.arg = c("Increase/Lot", "Increase",
                      "Same", "Decrease", "Decrease/Lot"),
        xlab = "Increase or Decrease Spending on Aid to the Poor?",
        ylab = "Number of Cases",
        main = "Spending Preference")
```

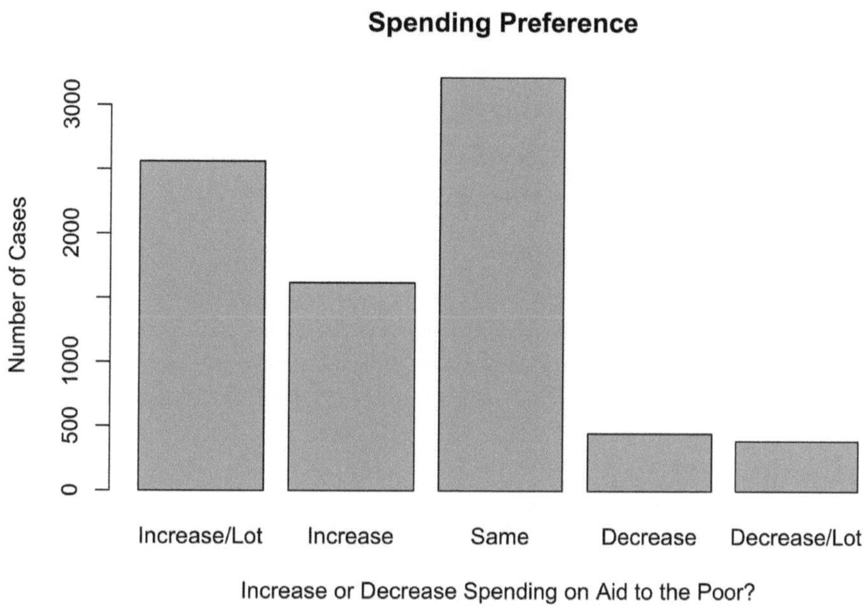

I think you'll agree that this looks a lot better than the first graph. By way of interpretation, you don't need to know the exact values of the frequencies or percentages to tell that there is very little support for decreasing spending and substantial support for keeping spending the same or increasing it. This is made clear simply from the visual impact of the differences in the height of the bars. Images like this often make quick and clear impressions on people.

Now let's compare this bar chart to one for the question that asked about spending on welfare programs. Since we did not save the contents of the original frequency table for this variable to a new object, we can insert `table(anes20$V201314x)` into the `barplot` command.

```
#Tell R to use the contents of "table(anes20$V201314x)" for graph
barplot(table(anes20$V201314x),
        names.arg = c("Increase/Lot", "Increase",
                      "Same", "Decrease", "Decrease/Lot"),
        xlab = "Increase or Decrease Spending on Welfare?",
        ylab = "Number of Cases",
        main = "Spending Preference")
```

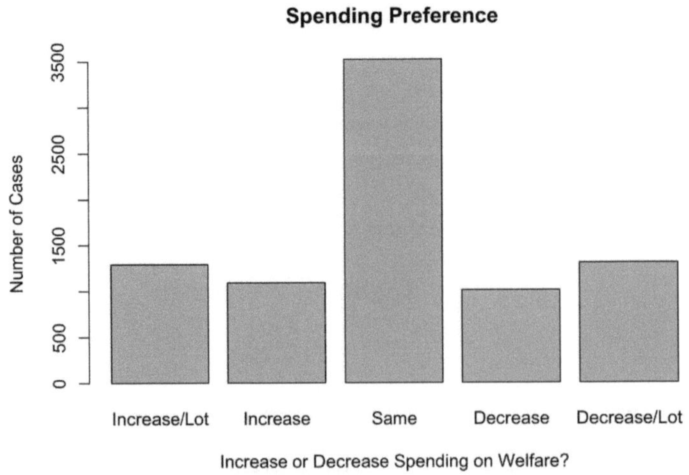

As was the case when we compared the frequency tables for these two variables, the biggest difference that jumps out is the lower level of support for increasing spending and the higher level of support for decreasing welfare spending, compared to preferences of spending on aid to the poor.

If you prefer to **plot the proportions** rather than the raw frequencies, you just have to specify that R should use the proportions table as input and change the *y*-axis label accordingly.

```
#Use "prop.table" as input
barplot(prop.table(poorAid.tbl),
        names.arg = c("Increase/Lot", "Increase",
                      "Same", "Decrease", "Decrease/Lot"),
        xlab = "Increase or Decrease Spending on Aid to the Poor?",
        ylab = "Proportion of Cases",
        main = "Spending Preference (Relative Frequencies)")
```

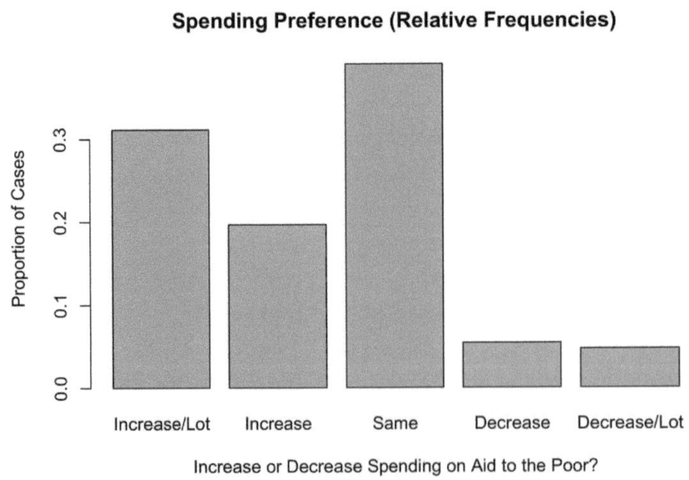

Getting Bar Charts With Frequencies

We have been using the `barplot` command to get bar charts, but these charts can also be created by modifying the `freq` command. Throughout the earlier discussion of frequencies, I used commands that look like this.

```
freq(anes20$V201320x, plot = FALSE)
```

The `plot = FALSE` part of the command instructs R not to create a bar chart for the variable. If it is dropped or if it is changed to `plot = TRUE`, R will produce a bar chart along with the frequency table. You still need to add commands to create labels and main titles and to make other adjustments, but you can do all of this within the frequency command. Go ahead, run the following code and see what you get! If you are working in RStudio, you should get the frequency table in the Console window and the bar chart in the Plots window.

```
freq(anes20$V201320x, plot = TRUE,
     names.arg = c("Increase/Lot", "Increase",
                   "Same", "Decrease", "Decrease/Lot"),
     xlab = "Increase or Decrease Spending on Welfare?",
     ylab = "Number of Cases",
     main = "Spending Preference")
```

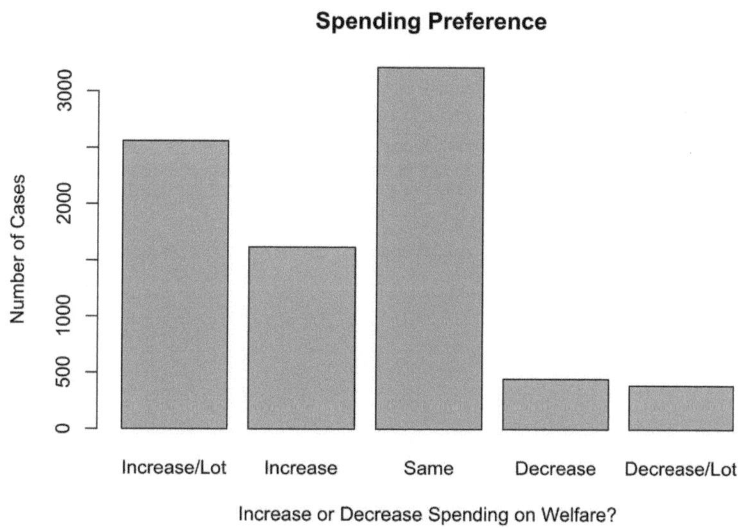


```
PRE: SUMMARY: Federal Budget Spending: aid to the poor
                          Frequency  Percent Valid  Percent
1. Increased a lot              2560  30.9179          31.125
2. Increased a little           1617  19.5290          19.660
3. Kept the same                3213  38.8043          39.064
```

4. Decreased a little	446	5.3865	5.422
5. Decreasaed a lot	389	4.6981	4.729
NA's	55	0.6643	
Total	8280	100.0000	100.000

Why not just always use the `freq` command to get bar charts? Why go through the extra steps? Two reasons, really. First, you don't need a frequency table every time you produce or modify a bar chart. More importantly, however, using the `barplot` command pushes you a bit more to understand "what's going on under the hood." For instance, telling R to use the results of the `table` command as input for a bar chart helps you understand a bit better what is happening when R creates the graph. If the graph just magically appears when you use the `freq` command, you are another step removed from the process and don't even need to think about it. You may recall from Chapter 2 that I discussed how some parts of the data analysis process seem a bit like a *black box* to students—something goes in, results come out, and we have no idea what's going on inside the box. Learning about bar charts via the `barplot` command gives you a little peek inside the black box.

Bar Chart Limitations

Bar charts work really well for most categorical variables because they do not assume any particular quantitative distance between categories on the *x*-axis and because categorical variables tend to have relatively few discrete categories. Bar charts generally do not work well with numeric data, for reasons to be explored in just a bit. That said, there are some instances where this general rule for numeric data doesn't hold up. Let's look at one such exception, using a state policy variable from the `states20` dataset. Following is a bar chart for `abortion_laws`, a variable measuring the number of legal restrictions on abortion in the states in 2020.[2]

```
barplot(table(states20$abortion_laws),
        xlab = "Number of laws Restricting Abortion Access",
        ylab = "Number of States",
        cex.axis = .8) #Reduces the axis labels by 80%.
```

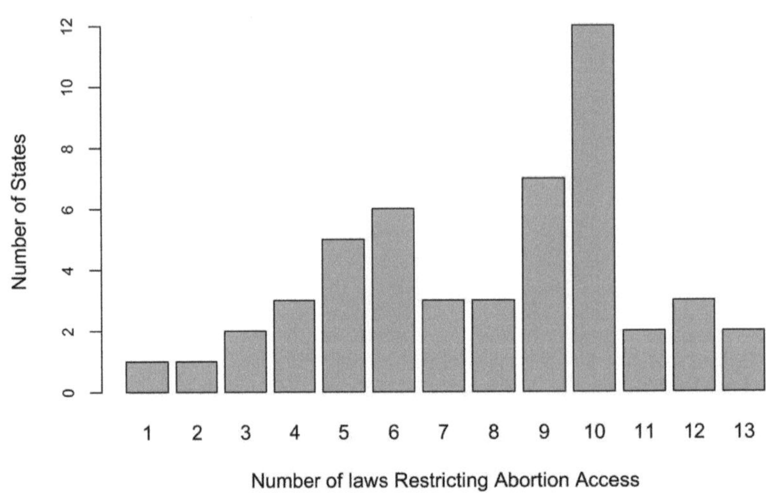

Okay, this is actually kind of a nice looking graph. It's easy to get a sense of how this variable is distributed: Most states have several restrictions, and the most common outcomes are states with 9 or 10 restrictions. It is also easier to comprehend than if we got the same information in a frequency table with thirteen categories (go ahead and get a frequency table to see if you agree).

The bar chart works in this instance because there are relatively few discrete categories, and the categories are consecutive, with no gaps between values. In most cases, numeric variables do not share these characteristics, and bar charts don't work well. This point is illustrated quite nicely in this graph of Joe Biden's percentage of the two-party vote in the states in the 2020 election.

```
par(las = 2) #This tells R to plot the x-axis labels vertically
barplot(table(states20$d2pty20),
        xlab = "Biden % of Two-party Vote",
        ylab = "Number of States",
        cex.names = .7) #Reduces the axis labels by 70%.
```

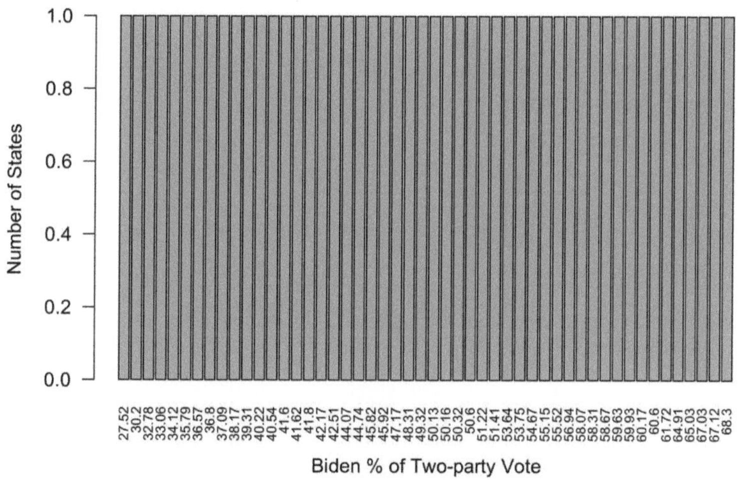


```
par(las = 0) #This tells R to use horizontal labels after running code
```

Not to put too fine a point on it, but this is a terrible graph, for many of the same reasons the initial frequency table for this variable was of little value. Other than telling us that the outcomes range from 27.52 to 68.3, there is nothing useful conveyed in this graph. What's worse, it gives the misleading impression that votes were uniformly distributed between the lowest and highest values. There are a couple of reasons for this. First, no two states had exactly the same outcome, so there are as many distinct outcomes and vertical bars as there are states, leading to a flat distribution. This is likely to be the case with many numeric variables, especially when the outcomes are continuous. Second, the proximity of the bars to each other reflects the rank order of outcomes, not the quantitative distance between categories. For instance, the two lowest values are 27.52 and 30.20, a difference of 2.68, and the third and fourth lowest values are 32.78 and 33.06, a difference of .28. Despite these quantitative differences in the distance between the first and second and third and fourth outcomes, the spacing between the bars in the bar chart makes it look like the distances are the same. Bottom line, bar charts are great for most categorical data but usually are not the preferred method for graphing numeric data.

Histograms

Histograms were introduced in Chapter 2 as a tool for visualizing how the outcomes of a variable are distributed. It is hard to overstate how useful histograms are for conveying information about the range of outcomes, whether outcomes tend to be concentrated or widely dispersed across that range and if there is something approaching a "typical" outcome. In short, histograms show us the shape of the data.[3]

Let's take another look at Joe Biden's percentage of the two-party vote in the states but this time using a histogram.

```
#Histogram for Biden's % of the two-party vote in the states
hist(states20$d2pty20,
     xlab = "Biden % of Two-party Vote",
     ylab = "Number of States",
     main = "Histogram of Biden Support in the States")
```

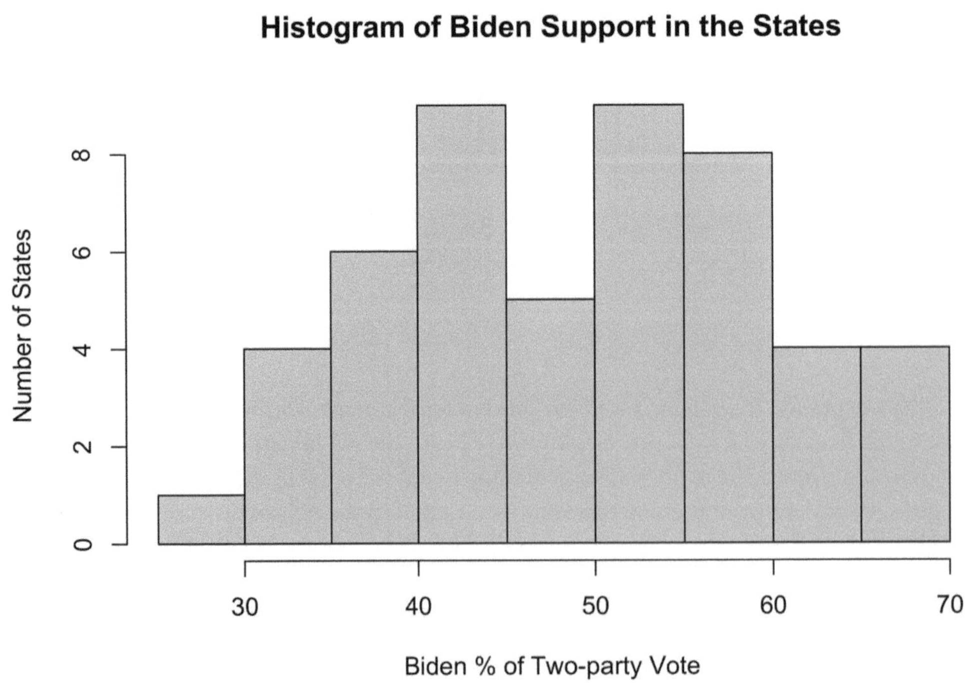

The width of the bars represents a range of values, and the height represents the number of outcomes that fall within that range. At the low end there is just one state in the 25 to 30 range, and at the high end, there are four states in the 65 to 70 range. More importantly, there is no clustering at one end or the other, and the distribution is somewhat bell shaped but with dip in the middle. It would be very hard to glean this information from the frequency table and bar chart for this variable presented earlier.

Just as bar charts are the graphic representation of the raw frequencies for each outcome, we can think of histograms as the graphic representation of the *binned frequencies*, similar to those produced by the `Freq` command. In fact, R uses the same rules for grouping observations for histograms as it does for the binned frequencies used in the `Freq` command used earlier.

You can also examine the *relative frequency* of histogram bins by adding `probability = T` to the r code. This produces "density" values on the vertical axis.

```
#Histogram for Biden's % of the two-party vote in the states
hist(states20$d2pty20,
     probability = T,
     xlab = "Biden % of Two-party Vote",
     ylab = "Density",
     main = "Histogram of Biden Support in the States")
```

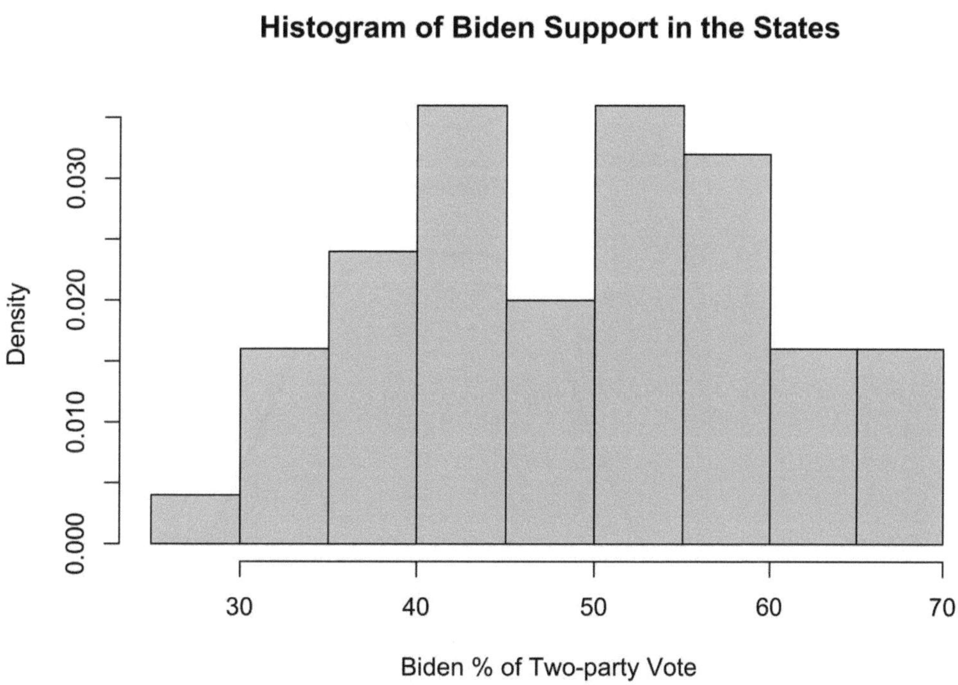

Density values are a particular expression of probability (or percentages), and their application to histograms is easy to understand if we walk through an example with a couple simple calculations. Let's focus on the fourth bin from the left in the histogram, which ranges from 40 to 45. The density on the vertical axis for this bin .036. If you multiply the value of this density times the width of the interval (5), you get .036 * 5 = .18. This value represents the proportion of states that are in the 40 to 45 bin. This is the same as the binned percentage from the frequency table shown earlier in this chapter, in which 18% of all states were in this interval. The takeaway is that higher density values reflect higher proportions (percentages), and lower density values reflect lower proportions (percentages).

Density Plots

The concept of density values can be extended to **density plots**, a close kin to histograms. One slight drawback to histograms is that the chunky nature of the bars can sometimes obscure the more general, continuous shape of the distribution. A density plot helps alleviate this problem by taking information from the variable's outcomes and generating a line that smooths out the bumpiness of the histogram and summarizes the shape of the distribution. The line should be thought of as an estimate of a theoretical distribution of a continuous variable, based on the underlying patterns in the data.

To get a density plot, you continue to use the `plot` function, as shown next. Here, the command tells R to create a plot of the density values of `states20$d2pty20`.

```
#Generate a density plot with no histogram
plot(density(states20$d2pty20),
     xlab = "Biden % of Two-Party Vote",
     main = "Biden Votes in the States",
     lwd = 3)
```

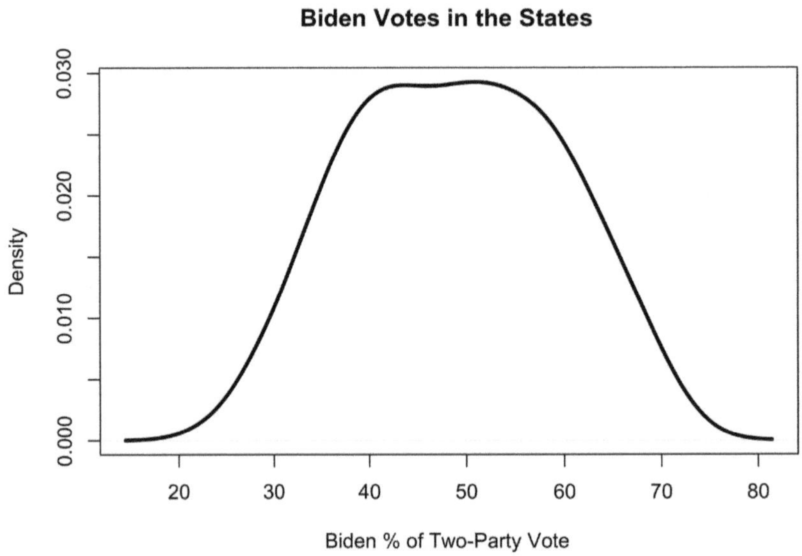

The smoothed density line reinforces the impression from the histogram that there are relatively few states with extremely low or high values, and the vast majority of states are clustered in the 40% to 60% range. It is not quite a bell-shaped curve—somewhat symmetric but a bit flatter than a bell-shaped curve. As previously discussed, the density values on the vertical axis reflect the relative frequency of the outcomes. That said, it is best to focus on the height on the *y*-axis for various points on the density line rather than on interpretation of specific density values. Also, note that the density plot is not limited to the same *x*-axis limits as in the histogram, and the solid line can extend beyond those limits *as if* there were data points out there.

Let's take a quick look at a density plot for the other numeric variable used in this chapter, abortion laws in the states.

```
#Density plot for Number of abortion laws in the states
plot(density(states20$abortion_laws),
     xlab = "Number of Laws Restricting Access to Abortions",
     main = "Abortion Laws in the States",
     lwd = 3) #Increase line width
```

This plot shows that the vast majority of the states have more than five abortion restrictions on the books, and the distribution is sort of bimodal (two primary groupings) at around five and ten restrictions.

A Few Add-Ons for Graphing

As you progress through the remaining chapters in this book, you will learn a lot more about how to use graphs to illuminate interesting things about your data. Before moving on to the next chapter, I want to show you a few things you can do to change the appearance of the simple bar charts and histograms we have been working with so far.

- `col = " "` is used to designate the color of the bars. Gray is the default color, but you can choose to use some other color if it makes sense to you. You can get a list of all colors available in R by typing `colors()` at the prompt in the console window.
- `horiz = T` is used if you want to flip a bar chart so the bars run horizontally from the vertical axis. This is not available for histograms.
- `breaks =` is used in a histogram to change the number of bars (bins) used to display the data. We used this command earlier in the discussion of setting specific bin ranges in frequency tables, but for right now, we will just specify a single number that determines how many bars will be used.

The following examples add some of this information to graphs we examined earlier in this chapter.

```
#Use five bins and color them white
hist(states20$d2pty20,
     xlab = "Biden % of Two-party Vote",
     ylab = "Number of States",
     main = "Histogram of Biden Support in the States",
     col = "white", #Use white to color the bars
     breaks = 5) #Use just five categories
```

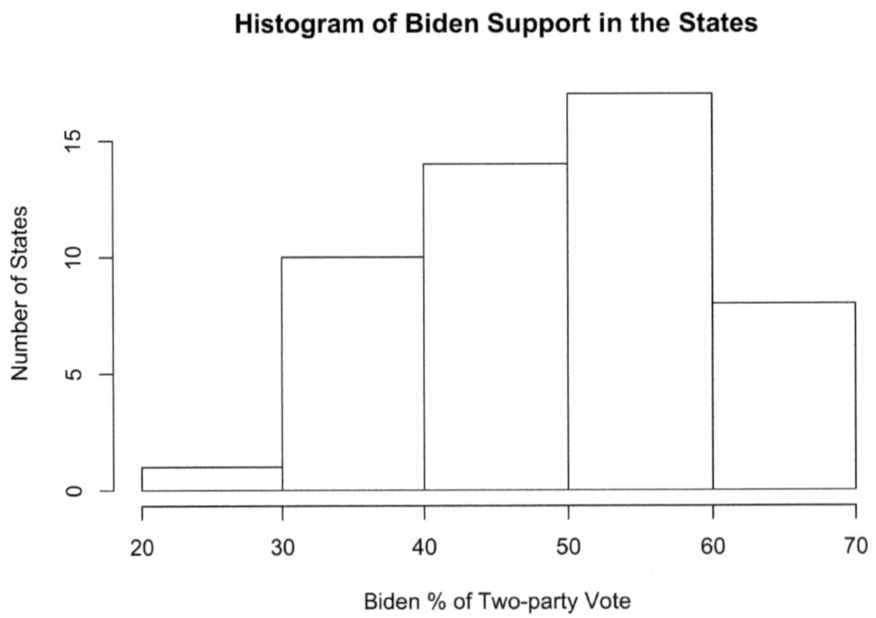

As you can see, things turned out well for the histogram with just five bins, though I think five is probably too few and obscures some of the important variation in the variable. If you decided you prefer a certain color but not the default gray, you just change `col = "white"` to something else. Go ahead, give it try!

The following graph is a first attempt at flipping the barplot for attitudes toward the spending on aid for the poor, using white horizontal bars.

```
#Horizontal bar chart with white bars
barplot(poorAid.tbl,
        names.arg = c("Increase/Lot", "Increase", "Same",
                     "Decrease", "Decrease/Lot"),
        xlab = "Number of Cases",
        ylab = "Increase or Decrease Spending on Aid to the Poor?",
        main = "Spending Preference",
        horiz = T, #Plot bars horizontally
        col = "white") #Use white to color the bars
```

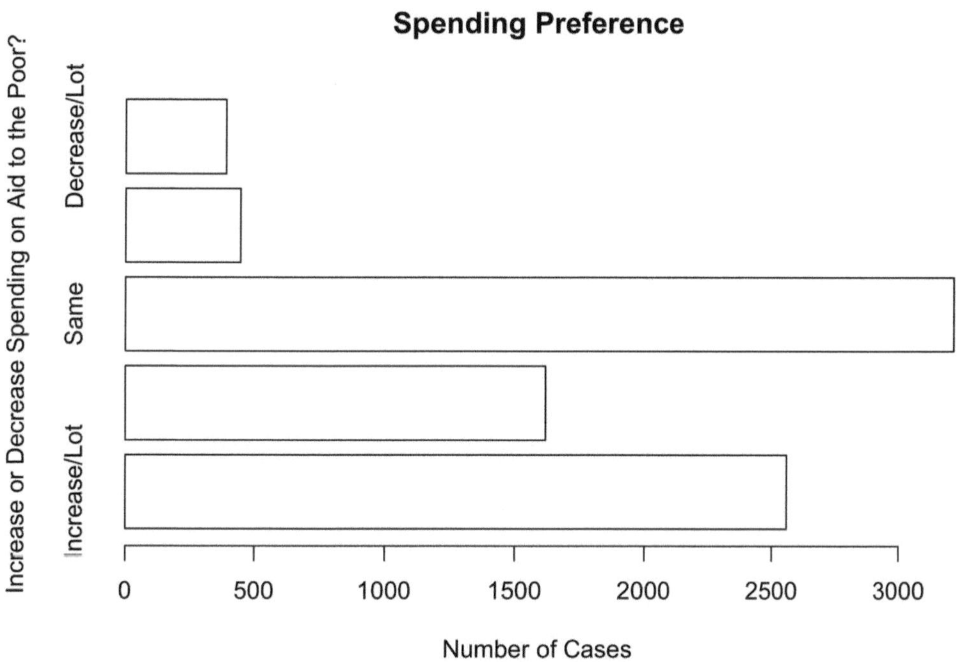

As you can see, the horizontal bar chart for spending preferences turns out to have a familiar problem: The value labels are too large and don't all print. This isn't horrible, but it could be improved by making room for all value labels, as I've done in the next section.[4]

First, `par(las = 2)` instructs R to print the value labels sideways. Anytime you see `par` followed by other terms in parentheses, it is a command to alter the graphic parameters. The labels were still a bit too long to fit, so I increased the margin on the left with `par(mar = c(5,8,4,2))`. This command sets the margin size for the graph, where the order of numbers is `c(bottom, left, top, right)`. Normally this is set to `mar = c(5,4,4,2)`, so increasing the second number to eight expanded the left margin area and provided enough room for value labels. However, the horizontal category labels overlapped with the *y*-axis title, so I dropped the axis title and modified the main title to help clarify what the labels represent.

```
#Change direction of the value labels
par(las = 2)
#Change the left border to make room for the labels
par(mar = c(5,8,4,2))
barplot(poorAid.tbl,
        names.arg = c("Increase/Lot", "Increase", "Same",
                      "Decrease", "Decrease/Lot"),
        xlab = "Number of Cases",
        main = "Spending Preference: Aid to the Poor",
        horiz = T,
        col = "white")
```

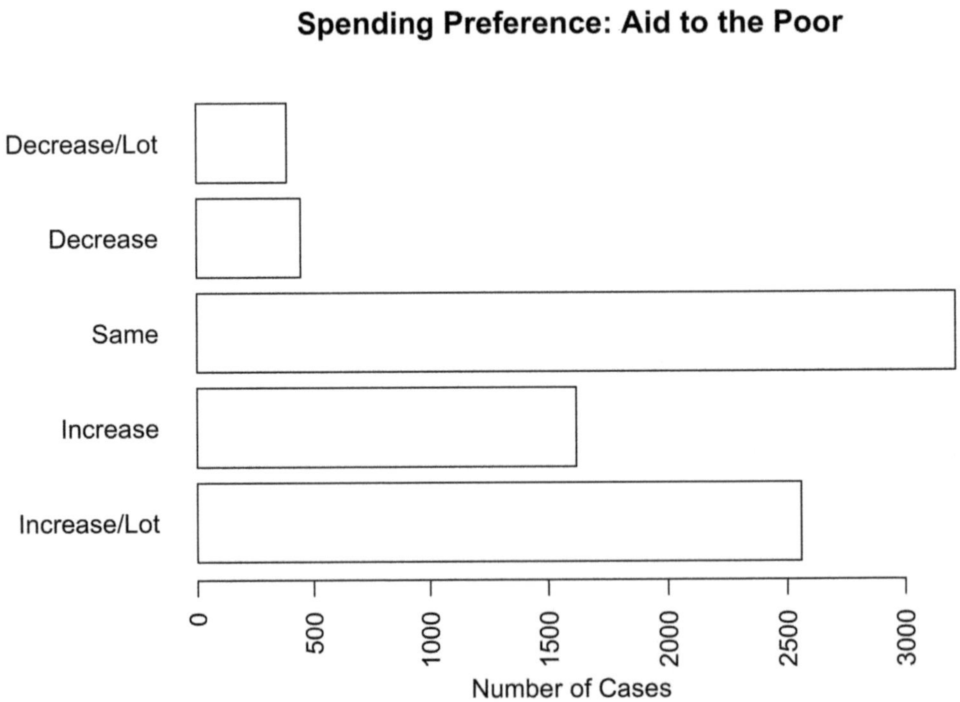

As you can see, changing the orientation of your bar chart could entail changing any number of other characteristics. If you think the horizontal orientation works best for a particular variable, then go for it. Just take a close look when you are done to make sure everything looks the way it should look. An important lesson to take away from this is that you sometimes have to work hard, trying many different things, to get a particular graph to look as nice as it should.

Whenever you change the graphing parameters as we have done here, you need to change them back to what they were originally. Otherwise, those changes will affect all of your subsequent work.

```
#Return graph settings to their original values
par(las = 0)
par(mar = c(5,4,4,2))
```

NEXT STEPS

Simple frequency tables, bar charts, and histograms can provide a lot of interesting information about how variables are distributed. Sometimes this information can tip you off to potential errors in coding or collecting data, but mostly it is useful for "getting to know" the data. Starting with this type of analysis provides researchers with a level of familiarity and connection to the data that can pay dividends down the road when working with more complex statistics and graphs.

As alluded to earlier, you will learn about a lot of other graphing techniques in subsequent chapters, once you've become familiar with the statistics that are used in conjunction with those graphs. Prior to that, though, it is important to spend a bit of time learning more about how to use R to transform variables so that you have the data you need to create the most useful graphs and statistics for your research. This task is taken up in the next chapter.

EXERCISES

Concepts and Calculations

1. You might recognize this list of variables from the exercises at the end of Chapter 1. Identify whether a histogram or bar chart would be most appropriate for summarizing the distribution of each variable. Explain your choice.

   - Course letter grade
   - Voter turnout rate (%)
   - Marital status (married, divorced, single, etc.)
   - Occupation (professor, cook, mechanic, etc.)
   - Body weight

- Total number of votes cast in an election
- #Years of education
- Subjective social class (poor, working class, middle class, etc.)
- % of people living below poverty level income
- Racial or ethnic group identification

2. This histogram shows the distribution of medical doctors per 100,000 population across the states.

    - Assume that you want to describe this distribution to someone who does not have access to the histogram. What do you tell them?
    - Given that the intervals in the histogram are right closed, what range of values are included in the 250 to 300 interval? How would this be different if the intervals were left closed.

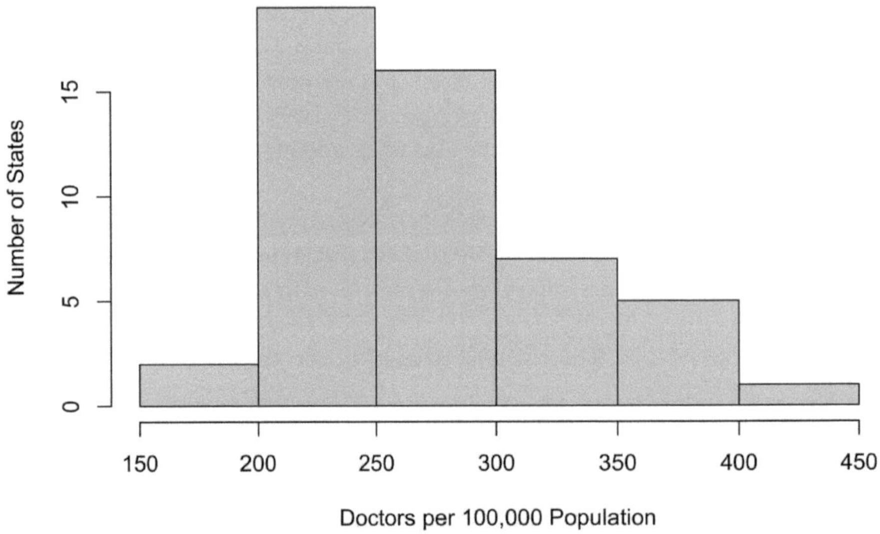

3. A group of students on a college campus interviewed 300 students to get their evaluations of different aspects of campus life. One question they asked was, "Overall, how satisfied are you with the quality of instruction in your courses?" The response categories and frequencies are presented in the following table. Fill in the missing information in the Percent and Cumulative Percent columns and describe the overall level of satisfaction with instruction at this university.

	Frequency	Percent	Cumulative %
Very Dissatisfied	45		
Somewhat Dissatisfied	69		
Somewhat Satisfied	120		
Very Satisfied	66		

R Problems

1. I've tried to get a bar chart for `anes20$V201119`, a variable that measures how happy people are with the way things are going in the United States, but I keep getting an error message. Can you figure out what I've done wrong? Diagnose the problem and present the correct barplot.

```
barplot(anes20$V201119,
        xlab = "How Happy with the Way Things Are Going?",
        ylab = "Number of Respondents")
```

2. Choose the most appropriate type of frequency table (`freq` or `Freq`) to summarize the distribution of values for the three variables listed. Make sure to look at the codebooks so you know what these variables represent. Present the tables and provide a brief summary of their contents. Justify your choice of tables.
   - Variables: `anes20$V202178`, `anes20$V202384`, and `states20$union`
3. For the same set of variables listed in Question 2, choose the appropriate type of graph (bar chart or histogram) for visualizing the distributions of the variables and discuss what you see. As in Question 2, make sure to justify your choice of graphing technique.
4. Create a histogram and a density plot for `states20$beer`, a variable that measures the per capita gallons of beer consumed in the states, making sure to create appropriate axis titles. In your opinion, is the density plot or the histogram easier to read and understand? Explain your response.

4 DATA PREPARATION

GET READY

To work through the examples as you read this chapter, make sure to load the `descr`, `dplyr`, and `Hmisc` libraries and the `states20.rda` and `anes20.rda` data files. You may have to install the packages if you have not done so already. When loading the datasets, you do not need to include the file path if the files are in your working directory.

INTRODUCTION

Before moving on to other graphing and statistical techniques, we need to take a bit of time to explore ways in which we can use R to transform the data to make it more appropriate for the tasks at hand. Some of the things we will do, such as assigning simpler, more intuitive names to variables and datasets, are very basic, while other things are a bit more complex, such as reordering the categories of a variable and combining several variables into a single variable. Everything presented in this chapter is likely to be of use to you in your course assignments and, if you continue on with data analysis, at some point in the future.

DATA TRANSFORMATIONS

When working with a dataset, even one you may have put together yourself, it is almost always the case that you will need to change something about the data in order to use it more effectively. Maybe there is just one variable you need to modify in some way, or there could be several variables that need some attention, or perhaps you need to modify the entire dataset. Not to fear, though, as these modifications are usually straightforward, especially with practice, and they should make for better research.

Don't Forget Script Files

As suggested before, one very important part of the data transformation process is record keeping. It is essential that you keep track of the transformations you make. This is where script files come in handy. Save a copy of all of the commands you use by creating and saving a script file in the Source window of RStudio (see Chapter 2 if this is unfamiliar to you). At some point, you are going to have to remember and possibly report the transformations you make. It could be when you turn in your assignment, when you want to use something you created previously, or perhaps when you are writing a final paper. You cannot and should not just work from memory. *Keep track of all changes you make to the data.*

Changing Variable Names

One of the simplest forms of data transformation is renaming objects. Sometimes, the datasets we use have what seem like cumbersome variable naming conventions, and it might be easier to work with the data if we renamed the variables. For instance, in the previous chapter, we worked with two variables from the 2020 ANES measuring spending preferences, V201314x (welfare programs) and V201320x (aid to the poor). At some level, these variable names are useful because they are relatively short, but at the same time, they can be hard to use because they have no inherent meaning as variable labels. Let's face it, "V201314x" might as well be "Vgibberish" for most new users of the ANES surveys. For experienced users, that may not be the case, as "V20" identifies the variable as coming from the 2020 election study, variables in the 1300-range are generally of a type (issue variables), and variables ending in "x" are summary variables that combine responses from previous variables. This is well and good, but for most users, it is helpful to use variable names that reflect the content of the variables. This is an easy thing to do and a good place to start learning about transformations. All you have to do is assign the content of the old variables to new objects and use descriptive names for the new objects, as shown next.

```
#Copy spending variables into new variables with meaningful names
anes20$welfare_spnd <- anes20$V201314x
anes20$poor_spnd <- anes20$V201320x
```

In these commands, we are telling R to take the contents of the two spending preference variables and put them into two new objects with different names (note the use of <-). One important thing to notice here is that the new variables were added to the `anes20` dataset. If we had created stand-alone objects (e.g., `welfare_spnd <- anes20$V201314x`), they would exist outside the `anes20` dataset and we would not be able to use any of the other `anes20` variables to analyze outcomes on the new variables. For instance, we would not be able to look at the relationship between religiosity and support for welfare spending if the new object was saved outside of the `anes20` dataset because there would be no way to connect the responses of the two variables.

It is always important to check your work. I've lost track of the number of times I thought I had transformed a variable correctly, only to discover later that I had made a silly mistake. A quick look at frequencies of the old and new variables confirms that we have created the new variables correctly.

```
#Original "Aid to poor"
table(anes20$V201320x)

      1. Increased a lot  2. Increased a little        3. Kept the same
                    2560                   1617                    3213
   4. Decreased a little     5. Decreasaed a lot
                     446                    389
```

```
#New "Aid to poor"
table(anes20$poor_spnd)

     1. Increased a lot  2. Increased a little        3. Kept the same
                   2560                   1617                    3213
   4. Decreased a little     5. Decreasaed a lot
                    446                    389
#Original "Welfare"
table(anes20$V201314x)

     1. Increased a lot  2. Increased a little        3. Kept the same
                   1289                   1089                    3522
   4. Decreased a little     5. Decreasaed a lot
                   1008                   1305
#New "Welfare"
table(anes20$welfare_spnd)

     1. Increased a lot  2. Increased a little        3. Kept the same
                   1289                   1089                    3522
   4. Decreased a little     5. Decreasaed a lot
                   1008                   1305
```

It looks like everything worked the way it was supposed to work; the outcomes for the new variables look exactly like the outcomes for their original counterparts.

Let's do this with a couple of other variables from anes20, the political party feeling thermometers. These variables are based on questions that asked respondents to rate various individuals and groups on a 0-to-100 scale, where a rating of 0 means you have very cold, negative feelings toward the individual or group and 100 means you have very warm, positive feelings. The variable names for the Democratic and Republican feeling thermometer ratings are anes20$V201156 and anes20$V201157, respectively. Again, these variable names do not exactly trip off the tongue, and you are probably going to have to look them up in the codebook every time you want to use one of them. So, we can copy them into new objects and give them more substantively meaningful names.

```
#Copy feeling thermometers into new variables
anes20$dempty_ft <- anes20$V201156
anes20$reppty_ft <- anes20$V201157
```

To demonstrate a simple way in which this can be very useful, let's take a quick look at the distributions of the original and renamed variables.

```
#This first bit tells R to use two rows and two columns
par(mfrow = c(2, 2))
hist(anes20$V201156,
     xlab = "Rating",
     ylab = "Frequency",
     cex.main = .8)
hist(anes20$V201157,
     xlab = "Rating",
     ylab = "Frequency",
     cex.main = .8)

hist(anes20$dempty_ft,
     xlab = "Rating",
     ylab = "Frequency",
     cex.main = .8)
hist(anes20$reppty_ft,
     xlab = "Rating",
     ylab = "Frequency",
     cex.main = .8)
par(mfrow = c(1, 1)) #reset par
```

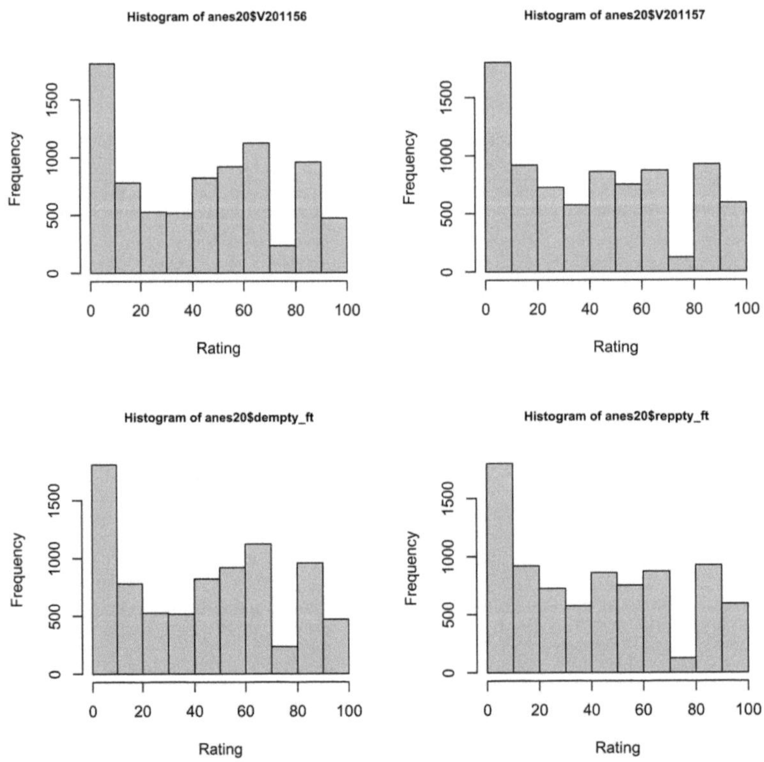

Besides the remarkable similarities between these Democratic and Republican distributions, what's also of note is that the original variable names used as titles in the first row of graphs give you absolutely no information about what the variables are, as "V201165" and "V201157" have no inherent meaning. The new variable names used in the second row of graphs make it easier to figure out which graph represents which party without having to refer to the codebook to check the variable names.

Changing Attributes

Variable Class

When importing the `anes20` dataset, R automatically classified non-numeric variables as *factor* variables. However, when doing this, no distinction was made between ordered and unordered factor variables. For instance, both federal spending variables are recognized as factor variables. Of course, they are factor variables, but they are also *ordered* variables.

```
#Check the class of spending preference variables
class(anes20$poor_spnd)
[1] "factor"
class(anes20$welfare_spnd)
[1] "factor"
```

This may not make much difference, if the alphabetical order of categories matches the substantive ordering of categories, since the default is to sort the categories alphabetically. For instance, the alphabetical ordering of the categories for the spending preference variables is in sync with their substantive ordering because the categories begin with numbers. You can check this using frequencies or by using the `levels` function.

```
#Check the category labels
levels(anes20$poor_spnd)
[1] "1. Increased a lot"     "2. Increased a little"  "3. Kept the same"
[4] "4. Decreased a little"  "5. Decreasaed a lot"

levels(anes20$welfare_spnd)
[1] "1. Increased a lot"     "2. Increased a little"  "3. Kept the same"
[4] "4. Decreased a little"  "5. Decreasaed a lot"
```

However, this is not always the case, and there could be some circumstances in which R needs to formally recognize that a variable is "ordered" for the purpose of performing some function. It's easy enough to change the variable class, as shown next, where the class of the two spending preference variables is changed to ordered.

```
#Change class of spending variables to "ordered"
anes20$poor_spnd <- ordered(anes20$poor_spnd)
anes20$welfare_spnd <- ordered(anes20$welfare_spnd)
```

Now, let's double-check to make sure the change in class worked.

```
class(anes20$poor_spnd)
[1] "ordered" "factor"
class(anes20$welfare_spnd)
[1] "ordered" "factor"
```

These are now treated as **ordered factors**.

We should also do our due diligence and verify that the feeling thermometers are recognized as numeric.

```
#Check class of feeling thermometers
class(anes20$dempty_ft)
[1] "numeric"
class(anes20$reppty_ft)
[1] "numeric"
```

All good with the feeling thermometers, but I knew this already because R would not have produced histograms for variables that were classified as factor variables. Occasionally, you may get an error message because the function you are using requires a certain class of data. For instance, if I try to get a histogram of `anes20$poor_spnd`, I get the following error: `Error in hist.default(anes20$poor_spnd): 'x' must be numeric`. When this happens, check to make sure the variable is properly classified, and change the classification if it is not. If it is properly classified, then use a more appropriate function (barplot in this case).

Value Labels

Occasionally, it makes sense to modify the names of *category* labels. For instance, in `anes20`, many of the category labels are very long and make graphs and tables look a bit clunky. The developers of the ANES survey created long, descriptive labels as a way of being faithful to the response categories that were used in the survey itself. This is an important goal, but sometimes this doesn't work well when displaying the data. You may recall that we already had to modify value labels when producing bar charts for the spending preference variables in Chapter 3 because the original labels were too long.

```
#names.arg is used to change category labels
barplot(table(anes20$V201320x),
        names.arg = c("Increase/Lot", "Increase", "Same", "Decrease",
                      "Decrease/Lot"),
        xlab = "Increase or Decrease Spending on Aid to the Poor?",
        ylab = "Number of Cases",
        main = "Spending Preference")
```

In this case, we used `names.arg` to temporarily replace the existing value labels with shorter labels that fit on the graph. We can use the `levels` function to change these labels permanently, so we don't have to add `names.arg` every time we want to create a bar chart. We used the `levels` function earlier to view value labels, and we can also use it to change those labels. We'll do this for both spending variables, applying the labels used in the `names.arg` function in the `barplot` command. The table that follows shows the original value labels and the replacement labels for the spending preference variables.

TABLE 4.1 ■ Original and Replacement Value Labels for Spending Preference Variables

Original	Replacement
1. Increased a lot	Increase/Lot
2. Increased a little	Increase
3. Kept the same	Same
4. Decreased a little	Decrease
5. Decreased a lot	Decrease/Lot

In the command that follows, we tell R to replace the value labels for `anes20$poor_spnd` with a defined set of alternative labels taken from Table 4.1.

```
#Assign levels to 'poor_spnd' categories
levels(anes20$poor_spnd) <- c("Increase/Lot", "Increase", "Same",
                              "Decrease", "Decrease/Lot")
```

It is important that each of the labels is enclosed in quotation marks and that they are separated by commas. We can check our handiwork to make sure that the transformation worked.

```
#Check levels
levels(anes20$poor_spnd)
[1] "Increase/Lot" "Increase"    "Same"     "Decrease" "Decrease/Lot"
```

Now, we can get a bar chart with value labels that fit on the horizontal axis without having to use `names.arg`. Note that the graph uses `anes20$poor_spnd`, which has the new labels, rather than the original variable, `anes20$V201320x`.

```
#Note that this uses new levels, not "names.arg"
barplot(table(anes20$poor_spnd),
        xlab = "Increase or Decrease Spending on Aid to the Poor?",
        ylab = "Number of Cases",
        main = "Spending Preference")
```

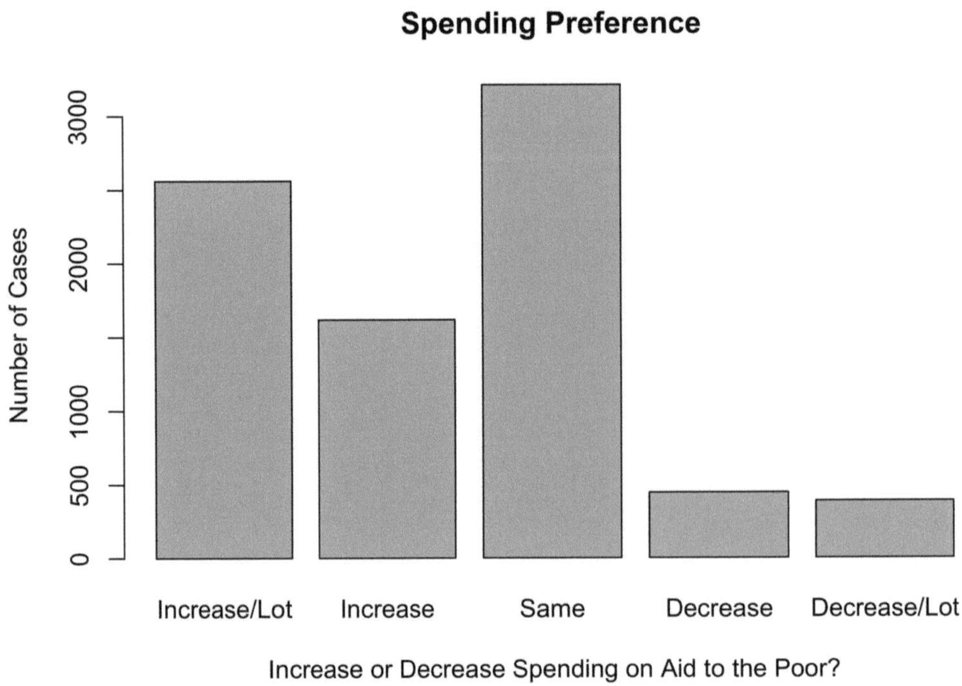

I like using the `levels` function to change value labels. There are other alternatives to this method, including using the `recode` command, as illustrated next with `anes20$welfare_spnd`. In this example, you must provide the variable name, list the existing labels, set them equal to the revised labels, and then assign the new labels to `anes20$welfare_spnd`.

```
anes20$welfare_spnd <- recode(anes20$welfare_spnd,
        "1. Increased a lot" = "Increase/Lot",
        "2. Increased a little" = "Increase",
        "3. Kept the same" = "Same",
        "4. Decreased a little" = "Decrease",
        "5. Decreasaed a lot" = "Decrease/Lot")

#Check levels
levels(anes20$welfare_spnd)
[1] "Increase/Lot" "Increase"    "Same"     "Decrease" "Decrease/Lot"
```

One very important thing that you should notice about how we changed these value labels is that we did not alter the value labels for the original variables (V201314x and V201320x). Those variables still exist in their original form and have the original category labels. Instead, we created new variables and replaced their value labels. One of the *first rules of data transformation* is to **never write over (replace) original data**. If you make a mistake in creating the new variables, you can always create them again. But if you make a mistake with the original data and save your changes, you can't go back and undo those changes.[1]

COLLAPSING AND REORDERING CATEGORIES

Besides changing value labels, you might also need to alter the number of categories or reorder the existing categories of a variable.

Collapsing Categories

Let's start with the two spending preference variables, anes20$poor_spnd and anes20$welfare_spnd, both of which have five categories, ranging from spending should be "increased a lot" to preferring that it be "decreased a lot." Now, suppose we are only interested in three discrete outcomes, whether people want to see spending increased, decreased, or kept the same. We can collapse the "Increase/Lot" and "Increase" into a single category and do the same for "Decrease/Lot" and "Decrease," resulting in a variable with three categories, "Increase," "Keep Same," and "Decrease."

Since we know from the preceding work how the categories are ordered, we can tell R to label the first two categories "Increase," the third category "Same," and the last two categories "Decrease." Let's start by copying the contents of anes20$poor_spnd to a three-category variable, anes20$poor_spnd.3.[2]

```
#create new ordered variable with appropriate name
anes20$poor_spnd.3 <- (anes20$poor_spnd)
#Then, write over existing five labels with three labels
levels(anes20$poor_spnd.3) <- c("Increase", "Increase", "Same",
                                "Decrease", "Decrease")
#Check to see if labels are correct
freq(anes20$poor_spnd.3, plot = F)

anes20$poor_spnd.3
          Frequency   Percent  Valid Percent  Cum Percent
Increase       4177   50.4469          50.78        50.78
Same           3213   38.8043          39.06        89.85
Decrease        835   10.0845          10.15       100.00
NA's             55    0.6643
Total          8280  100.0000         100.00
```

As shown in the following example, using anes20$welfare_spnd, the recode function can be used to accomplish the same task.

```
anes20$welfare_spnd.3 <- recode(anes20$welfare_spnd,
        "Increase/Lot" = "Increase",
        "Increase" = "Increase",
        "Same" = "Same",
        "Decrease" = "Decrease",
        "Decrease/Lot" = "Decrease")

#Check to see if labels are correct
freq(anes20$welfare_spnd.3, plot = F)

anes20$welfare_spnd.3
            Frequency   Percent  Valid Percent  Cum Percent
Increase         2378   28.7198          28.95        28.95
Same             3522   42.5362          42.88        71.84
Decrease         2313   27.9348          28.16       100.00
NA's               67    0.8092
Total            8280  100.0000         100.00
```

Note that in both of these transformations, I used the ".3" extension on the new variable name to remind me that this is the three-category version of the variable. Based on the frequencies, everything appears to be in order, and the comparison drives home that the level of support for social welfare programs depends on how you frame those programs, with many more people supporting "programs for the poor" than supporting "welfare" programs.

Reordering Categories

Using three categories clarifies things a bit, but you may have noticed that while the categories are ordered, there is a bit of disconnect between the magnitude of the labels and their placement on the horizontal axis. The ordered meaning of the label of the first category (Increase) is greater than the meaning of the label of the third category (Decrease). In this case, moving from the lowest to highest listed categories corresponds with moving from the highest to lowest labels, in terms of ranked meaning. Technically, this is okay, as long as you can keep it straight in your head, but it can get a bit confusing, especially if you are talking about how outcomes on this variable are related to outcomes on other variables. Fortunately, we can **reorder the categories** to run from the intuitively "low" value (Decrease) to the intuitively "high" value (Increase).

Let's do this with `poor_spnd.3` and `welfare_spnd.3` and then check the levels afterwords. Here, we use the `ordered` function to tell R to use the order of labels specified in the `levels` command. Note that we must use the already assigned labels, but we can reorder them.

```
#Use'ordered' and levels' function to reorder the categories
anes20$poor_spnd.3 <- ordered(anes20$poor_spnd.3,
                              levels = c("Decrease", "Keep Same",
                                         "Increase"))
```

```
#Use 'ordered' and 'levels' to reorder the categories
anes20$welfare_spnd.3 <- ordered(anes20$welfare_spnd.3,
                                 levels = c("Decrease", "Keep Same",
                                            "Increase"))
#Check the order of categories
levels(anes20$poor_spnd.3)
[1] "Decrease"    "Keep Same" "Increase"
levels(anes20$welfare_spnd.3)
[1] "Decrease"    "Keep Same" "Increase"
```

The transformations worked according to plan, and there is now a more meaningful match between the label names and their order on graphs and in tables when using these variables.

Sometimes, it is also necessary to reorder categories because the original order makes no sense, given the nature of the variable. Let's use `anes20$V201354`, a variable measuring whether respondents favor or oppose voting by mail, as an example.

```
#Frequency of "vote by mail" variable
freq(anes20$V201354, plot = F)

PRE: Favor or oppose vote by mail
                              Frequency   Percent  Valid Percent
1. Favor                           2142   25.8696          25.93
2. Oppose                          3083   37.2343          37.32
3. Neither favor nor oppose        3036   36.6667          36.75
NA's                                 19    0.2295
Total                              8280  100.0000         100.00
```

As currently constructed, the order of the three categories does not consistently increase or decrease in the level of some underlying scale. In particular, "Neither favor nor oppose" should be treated as a middle category rather than as the highest category. It would make more sense for this to be scaled as either "favor-neither-oppose" or "oppose-neither-favor." Given that "oppose" is a negative term and "favor" a positive one, it probably makes the most sense for this variable to be reordered to "oppose-neither-favor."

First, let's create a new variable name, `anes20$mail`, and replace the original labels to get rid of the numbers and shorten things a bit before we reorder the categories.

```
#Create new variable
anes20$mail <- (anes20$V201354)

#Create new labels
levels(anes20$mail) <- c("Favor", "Oppose", "Neither")

#Check levels
levels(anes20$mail)
[1] "Favor"     "Oppose"     "Neither"
```

Now, let's reorder these categories to create an ordered version of `anes20$mail` and then check our work.

```
#Reorder categories
anes20$mail <- ordered(anes20$mail,
                       levels = c("Oppose", "Neither", "Favor"))
#Check Levels
levels(anes20$mail)
[1] "Oppose"    "Neither" "Favor"

#Check Class
class(anes20$mail)
[1] "ordered" "factor"
```

Now we have an ordered version of the same variable that will be easier to use and interpret in the future.

COMBINING VARIABLES

It is sometimes very useful to combine information from separate variables to create a new variable. For instance, consider the two feeling thermometers, `anes20$dempty_ft` and `anes20$reppty_ft`. Rather than examining each of these variables separately, we might be more interested in how people rate the parties *relative* to each other, based on how they answered both questions. One way to do this is to measure the **net party feeling thermometer rating** by subtracting respondents' Republican Party rating from their Democratic Party rating. These sorts of mathematical transformations are possible when working with numeric data. In fact, you may recall doing something along these lines when you created `Approve21$net_approve` in one of the R problems back in Chapter 2.

```
#Subtract the Republican rating from the Democratic rating
anes20$netpty_ft <- anes20$dempty_ft-anes20$reppty_ft
```

Positive values on the net feeling thermometer rating, `anes20$netpty_ft`, indicate a higher rating for Democrats than for Republicans, negative values indicate a higher Republican rating, and values of 0 indicate that the respondent rated both parties the same. We can look at a histogram of this variable to get a sense of its distribution.

```
hist(anes20$netpty_ft,
     xlab = "Democratic Rating MINUS Republican Rating",
     ylab = "Number of Respondents",
     main = "Net Feeling Thermometer Rating")
```

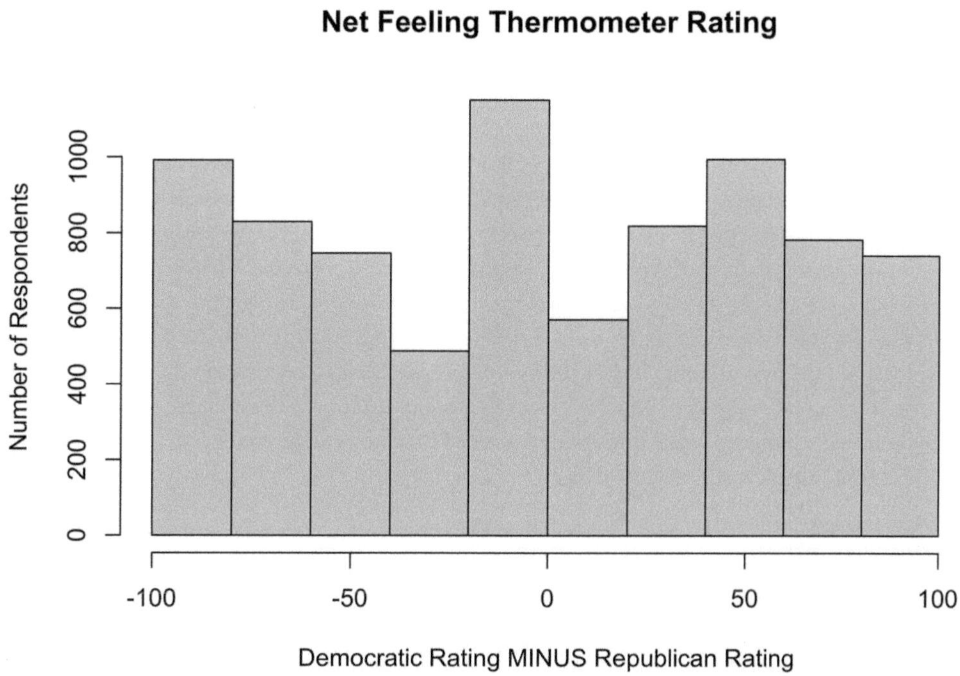

This is an interesting distribution. As you might expect, during times of political polarization, there are a lot of respondents at both ends of the distribution, indicating they really liked one party and really disliked the other. But at the same time, there are also a lot of respondents at or near 0, indicating they gave the parties very similar ratings.

Collapsing Numeric Variables

Suppose what you really want to know is which party the respondents preferred, and you don't care by how much. We could get this information by collapsing `netpty_ft` into a three-category ordinal variable that indicates whether people preferred the Republican Party, rated both parties the same, or preferred the Democratic Party.

We will do this using the `cut2` function (from the `Hmisc` package), which allows us to specify *cut points* for putting the data into different bins, similar to how we binned data using the `Freq` command in Chapter 3. In this case, we want one bin for outcomes less than zero, one for outcomes equal to zero, and one for outcomes greater than zero. Let's look at how we do this, then I'll explain how the cut points are determined.

```
#Collapse "netpty_ft" to three groups:-100 to 0, 0, 1 to 100
anes20$netpty_ft.3 <- ordered(cut2(anes20$netpty_ft, c(0, 1)))

#Check the class of the variable
class(anes20$netpty_ft.3)
[1] "ordered" "factor"
```

This command tells R to use the cut2 function to create three categories from anes20$netpty_ft and assign them to a new ordered object, anes20$netpty_ft.3. This is similar to creating bins with the Freq function in Chapter 3, except that the default in cut2 is to create left-closed bins.

Because I wanted three categories (< 0, 0, and > 0), I needed to specify two different cut points, 0 and 1. Using this information, R put all responses with values less than zero (the lowest cut point) into one group, all values between zero and one into another group, and all values equal to or greater than one (the highest cut point) into another group. When setting cut points in this way, you should specify *one fewer cut than the number of categories you want to create.* In this case, I wanted three categories, so I specified two cut points. Again, in the preceding example, I wanted the first group to be less than zero, so I specified zero as the cut point. Similarly, I wanted the second category to be zero, so I set the cut point at one. Always check the labels of the newly created variable to make sure the cuts worked they way you intended.

Let's take a look at this new variable.

```
freq(anes20$netpty_ft.3, plot = F)

anes20$netpty_ft.3
              Frequency  Percent  Valid Percent  Cum Percent
[-100,   0)        3308   39.952          40.76        40.76
    0               900   10.870          11.09        51.85
[   1, 100]        3908   47.198          48.15       100.00
NA's                164    1.981
Total              8280  100.000         100.00
```

First, it is reassuring to note that the result shows three categories with limits that reflect the desired groupings. Substantively, this shows that the Democratic Party holds a slight edge in feeling thermometer ratings, preferred by 48% compared to 41% for the Republican Party, and that about 11% rated both parties exactly the same. Looking at this table, however, it is clear that we could do with some better value labels. We know that the [-100, 0) category represents respondents who preferred the Republican Party, and respondents in the [1, 100] category preferred the Democratic Party, but this is not self-evident by the labels. So let's replace the numeric ranges with some more descriptive labels.

```
#Assign meaningful level names
levels(anes20$netpty_ft.3) <- c("Favor Reps", "Same", "Favor Dems")
freq(anes20$netpty_ft.3, plot = F)

anes20$netpty_ft.3
             Frequency  Percent  Valid Percent  Cum Percent
Favor Reps        3308   39.952          40.76        40.76
Same               900   10.870          11.09        51.85
Favor Dems        3908   47.198          48.15       100.00
NA's               164    1.981
Total             8280  100.000         100.00
```

In this latest example, we had a clear idea of exactly which cut points to use. But sometimes, you don't care so much about the specific values but are more interested in grouping data into a certain number of groups. This is easy to do using the `cut2` function. For instance, suppose we are using the `states20` dataset and want to classify the states according to state policy liberalism.

The values on the policy liberalism, `states20$pol_lib` are a bit hard to interpret on their own. Scores range from –2.53 to 2.51, and the mean is .04. The best way to think about these values is that the higher values indicate relatively liberal policies, lower values indicate relatively conservative policies, and values near zero indicate a mixed bag of policies. It might be better in some situations to collapse the values into low (conservative), middle (mixed), and high (liberal) categories. To do this, you still use the `cut2` command, except that instead of specifying the precise cut points, you just tell R to create three groups (*g* = 3). Let's do this and also assign new labels to the three categories.

```
#Collapse "policy_lib" into three roughly equal sized categories
states20$policy_lib.3 <- ordered(cut2(states20$policy_lib, g = 3))
#Check levels
levels(states20$policy_lib.3)
[1] "[-2.525, -0.927)" "[-0.927, 0.916)" "[0.916, 2.515]"

#Create meaningful level names
levels(states20$policy_lib.3) <- c("Conservative", "Mixed", "Liberal")

#Check levels
freq(states20$policy_lib.3, plot = F)

states20$policy_lib.3
             Frequency  Percent  Cum Percent
Conservative        17       34           34
Mixed               17       34           68
Liberal             16       32          100
Total               50      100
```

You should notice a couple of things about this transformation. First, although not exactly equal in size, each category has roughly one-third of the observations in it. It's not always possible to make the groups exactly the same size, especially if values are rounded to whole numbers, but using this method will get you close. Second, as in the first example, the original category labels were virtually meaningless, and the addition of the new labels was important to understanding the collapsed measure of state policy liberalism.

Creating an Index

Suppose you are interested in exploring how attitudes toward LGBTQ rights were related to vote choice in the 2020 presidential election. The 2020 ANES includes several questions related to LGTBQ rights. Respondents were asked if they think businesses should be required to provide services to same-sex couples (V201406), whether transgender people should use the bathroom of their birth or identified gender (V201409), whether they favor laws protecting gays and lesbians from job discrimination (V201412), whether gay and lesbian couples should be allowed to adopt children (V201415), and whether gays and lesbians should be allowed to marry (V201416).

Any one of these five variables might be a good indicator of support and opposition to LGBTQ rights, but it's hard to say that one of them is necessarily better than the others. One strategy researchers frequently use in situations such as this is to combine information from all of the variables into a single index. For instance, in this case, we could identify the liberal response category for each variable (using the `levels` function to see the categories) and then count the number of times respondents gave liberal responses across all five variables. A respondent who gives 0 liberal responses would be considered conservative on LGBTQ issues, and a respondent who gives 5 liberal responses would be considered liberal on LGBTQ issues.

Table 4.2 summarizes the information we need to begin the construction of the LGBTQ rights index.

TABLE 4.2 ■ Liberal Outcomes on LGBTQ Rights Variables in the 2020 ANES

Variable	Topic	Liberal Outcome
V201406	Services to same-sex couples	2. Should be required to provide services
V201409	Transgender bathroom policy	2. Bathrooms of their identified gender
V201412	Job discrimination Protections	1. Favor
V201415	Adoption by gays and lesbians	1. Yes
V201416	Gay marriage	1. Allowed to legally marry

The process for combining these five variables into a single index of support for LGBTQ rights involves two steps. First, for each question, we need to create an indicator variable with a value of 1 for all respondents who gave the liberal response and a value of 0 for all respondents who gave some other response. Let's do this first for `anes20$V201406` (services to same-sex couples) for demonstration purposes.

```
#Create indicator (0, 1) for liberal category of anes20$V201406
anes20$lgbtq1 <- as.numeric(anes20$V201406 ==
                    "2. Should be required to provide services")
```

In this command, we tell R to take the information it has for `anes20$V201406` and create a numeric variable, `anes20$lgbtq1`, in which respondents who gave a response of "2. Should be required to provide services" are given a score of 1, and all other valid responses are scored 0.

There are a couple of things to which you need to pay close attention. First, we used double equal signs (==). This is pretty standard in programming languages and is almost always used in R (one exception is if you use "=" instead of "<-" when creating new objects). Second, when identifying the category of interest, everything needs to be written *exactly* the way it appears in the original variable and must be enclosed in quotation marks. This includes spelling, punctuation, spacing, and letter case. I would have gotten an error message if I had written "2 Should be required to provide services" instead of "2. Should be required to provide services." Can you spot the difference?

So, now we have an indicator variable identifying respondents who thought businesses should be required to provide services to gays and lesbians, which scored 0 and 1. Before creating the other four indicator variables, let's compare the new variable to the original variable, just to make sure everything is the way it should be.

```
#Compare indicator variable to original variable
table(anes20$V201406)

          1. Should be allowed to refuse
                                    4083
2. Should be required to provide services
                                    4085
table(anes20$lgbtq1)

   0    1
4083 4085
```

Here, we see an even split in public opinion on this topic and that the raw frequencies in categories 0 and 1 on `anes20$lgbtq1` are exactly what we should expect based on the frequencies in `anes20$V201406`.

Just a couple of quick side notes before creating the other indicator variables for the index. First, these types of variables are also commonly referred to as *indicator*, *dichotomous*, or *dummy* variables. I tend to use all three terms interchangeably. Second, we have taken information from a factor variable and created a numeric variable. This is important to understand because it can open up the number and type of statistics we can use with data that are originally coded as factor variables.

The code that follows is used to create the other indicators.

```
#Create indicator for "bathroom" variable
anes20$lgbtq2 <- as.numeric(anes20$V201409 ==
                   "2. Bathrooms of their identified gender")

#Create indicator for "job discrimination" variable
anes20$lgbtq3 <- as.numeric(anes20$V201412 == "1. Favor")

#Create indicator for "adoption" variable
anes20$lgbtq4 <- as.numeric(anes20$V201415 == "1. Yes")

#Create indicator for "marriage" variable
anes20$lgbtq5 <- as.numeric(anes20$V201416 ==
                        "1. Allowed to legally marry")
```

It might be a good exercise for you to copy all of this code and check on your own to see that the transformations have been done correctly.

Now, the second step of the process is to combine all of these variables into a single index. Since these are numeric variables and they are all coded the same way, we can simply add them together. Just to make sure you understand how this works, let's suppose a respondent gave liberal responses (1) to the first, third, and fifth of the five variables and conservative responses (0) to the second and fourth ones. The index score for that person would be $1 + 0 + 1 + 0 + 1 = 3$.

Following, the dichotomous variables are added together to create the new index, `anes20$lgbtq_rights`, and a frequency table is included to check the outcomes.

```
#Combine five indicator variables into one index
anes20$lgbtq_rights <- (anes20$lgbtq1
                        + anes20$lgbtq2
                        + anes20$lgbtq3
                        + anes20$lgbtq4
                        + anes20$lgbtq5)

#Check the distribution of lgbtq_rights index
freq(anes20$lgbtq_rights, plot = F)

anes20$lgbtq_rights
       Frequency  Percent  Valid Percent
0            458    5.531           5.86
1            766    9.251           9.80
2            935   11.292          11.96
3           1272   15.362          16.27
4           1768   21.353          22.62
5           2617   31.606          33.48
NA's         464    5.604
Total       8280  100.000         100.00
```

The picture that emerges in the frequency table is that there is fairly widespread support for LGBTQ rights, with about 56% of respondents supporting liberal outcomes in at least four of the five survey questions and very few respondents at the low end of the scale. This point is, I think, made even more apparent by looking at a bar chart for this variable.[3]

```
barplot(table(anes20$lgbtq_rights),
        xlab = "Number of Rights Supported",
        ylab = "Number of Respondents",
        main = "Support for LGBTQ Rights")
```

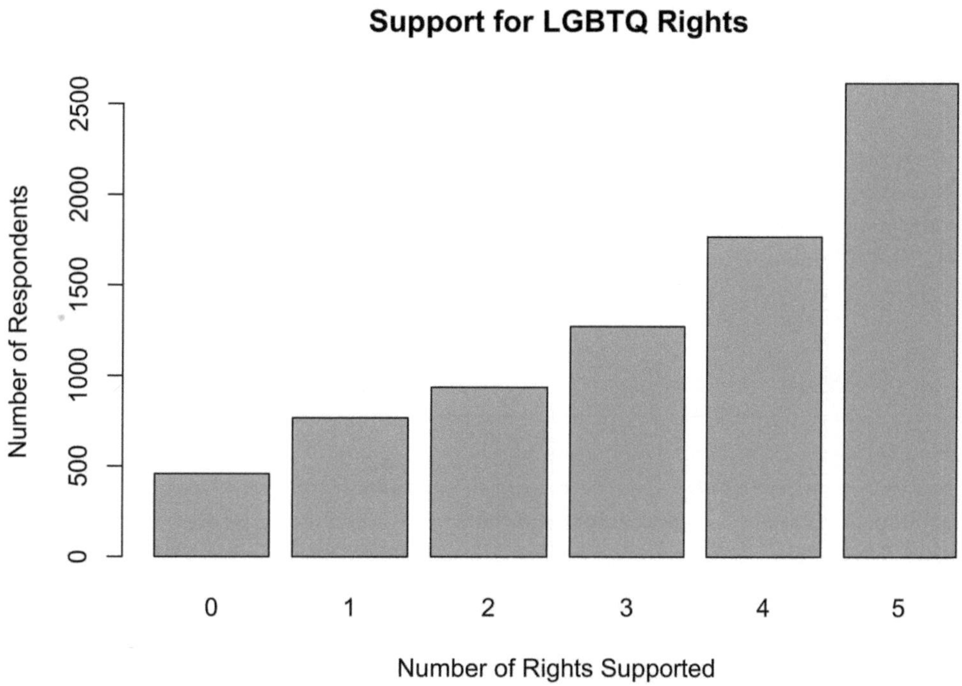

What's most important here is that we now have a more useful measure of support for LGBTQ rights than if we had used just a single variable out of the list of five previously reported. This variable is based on responses to five separate questions and provides a more comprehensive estimate of where respondents stand on LGBTQ rights.

SAVE YOUR CHANGES

If you want to keep all of the changes you've made, you can save them and write over `anes20`, or you can save them as a new file, presumably with a slightly altered file name, perhaps something like `anes20a`. You can also opt not to save the changes and, instead, rely on your script file if you need to recreate any of the changes you've made.

Recall from Chapter 2 that the general format for saving R data files is to specify the object you want to save and then the file name and the directory (if other than your working directory) where you want to store the file.

```
#Format of 'save' function
save(object, file = "<FilePath>/filename")
```

Remember to use `getwd` to see what your current working directory is (this is the place where the file will be saved if you don't specify a directory in the command) and `setwd` to change the working directory to a place where you want to save the results if you need to.

To save the anes20 dataset over itself, do the following.

```
save(anes20, file = "anes20.rda")
```

Saving the original file over itself is okay if you are only adding newly created variables. Remember, though, that if you transformed variables and replaced the contents of original variables (using the same variable names), you cannot retrieve the original data if you save now using the same file name. None of the transformations done in this chapter are replacing original variables, so you can go ahead and save this file as anes20.

NEXT STEPS

Hopefully, you are becoming more comfortable working with R and thinking about political and social data. These first four chapters form an important foundation of support for learning about political and social data analysis and the ways in which R can be deployed to facilitate that analysis. As you work your way through the next several chapters, you should see that the portions of the text focusing on R are easier to understand, becoming a bit more second nature. The next several chapters will emphasize the statistical elements of data analysis a bit more, but learning more about R still plays an important role.

What's up next is a look at some descriptive statistics that may be familiar to you, measures of central tendency and dispersion. As in the preceding chapters, we will focus on intuitive, practical interpretations of the relevant statistics and graphs. Along the way, there are a few formulas and important concepts to grasp in order to develop a solid understanding of what the data can tell you.

EXERCISES

Concepts and Calculations

1. The ANES survey includes a variable measuring marital status with six categories, as shown here.

```
[1] "1. Married: spouse present"
[2] "2. Married: spouse absent {VOL - video/phone only}"
[3] "3. Widowed"
[4] "4. Divorced"
[5] "5. Separated"
[6] "6. Never married"
```

   Create a table similar to Table 4.1 in which you show how you would map these six categories on to a new variable with three categories, "Married," "Never Married," and "Other." Explain your decision rule for the way you mapped categories from the old to the new variable.

2. The ANES survey also includes a seven-category variable that measures party identification. Create another table in which you map these seven categories on to a new measure of party identification with three categories, "Democrat," "Independent," and "Republican." Explain your decision rule for the way you mapped categories from the old to the new variable.

```
[1] "1. Strong Democrat"                "2. Not very strong Democrat"
[3] "3. Independent-Democrat"           "4. Independent"
[5] "5. Independent-Republican"         "6. Not very strong Republican"
[7] "7. Strong Republican"
```

3. One of the variables in the anes20 dataset (`V201382x`) measures people's perceptions of whether corruption decreased or increased under President Trump.

```
PRE: SUMMARY: Corruption increased or decreased since Trump
                                 Frequency  Percent  Valid Percent
1. Increased a great deal             3394   40.990         41.415
2. Increased a moderate amount        1068   12.899         13.032
3. Increased a little                  203    2.452          2.477
4. Stayed the same                    2371   28.635         28.932
5. Decreased a little                  284    3.430          3.466
6. Decreased a moderate amount         592    7.150          7.224
7. Decreased a great deal              283    3.418          3.453
NA's                                    85    1.027
Total                                 8280  100.000        100.000
```

I'd like to convert this 7-point scale into a 3-point scale coded in the following order: "Decreased," "Same," "Increased." Which of the two coding schemes presented next accomplishes this the best? Explain why one of the options would give incorrect information.

```
#Option 1
anes20$crrpt <- ordered(anes20$V201382x)
levels(anes20$crrpt) <- c("Decreased", "Decreased", "Decreased", "Same",
                          "Increased", "Increased", "Increased")
#Option 2
anes20$crrpt <- ordered(anes20$V201382x)
levels(anes20$crrpt) <- c("Increased", "Increased", "Increased",
                          "Same", "Decreased", "Decreased",
                          "Decreased")
anes20$crrpt <- ordered(anes20$crrpt, levels = c("Decreased", "Same",
                                                 "Increased"))
```

4. A member of the student research team that conducted the survey 300 students to get their evaluations of different aspects of campus life (see problems at the end of Chapter 3) has decided to collapse, relabel, and reorder the categories from instructional satisfaction question. The first of the following tables shows the original response categories, and the second table shows the revised variable. What do you think of the revised variable? Do you spot any potential problems?

	Frequency
Very Dissatisfied	45
Somewhat Dissatisfied	69
Somewhat Satisfied	120
Very Satisfied	66

	Frequency
Approve	186
Disapprove	114

5. You've had a chance to use both the `levels` and `recode` functions to revise and collapse categories for qualitative variables in this chapter. Which method makes the most sense to you or seems easiest to use? Why?

R Problems

1. Rats! I've done it again. I was trying to combine responses to two gun control questions into a single, three-category variable (`anes20$gun_cntrl`) measuring support for restrictive gun control measures. The original variables are `anes20$V202337` (should the federal government make it more difficult or easier to buy a gun?) and `anes20$V202342` (favor or oppose banning "assault-style" rifles). Make sure you take a close look at these variables before proceeding.

   I used the following code to create the new variable, but, as you can see when you try to run it, something went wrong (the resulting variable should range from 0 to 2). What happened? Once you fix the code, produce a frequency table for the new index and report how you fixed it.

```
anes20$buy_gun <- as.numeric(anes20$V202337 == "1. Favor")
anes20$ARguns <- as.numeric(anes20$V202342 == "1. More difficult")

anes20$gun_cntrl <- anes20$buy_gun + anes20$Arguns

freq(anes20$gun_cntrl, plot = F)

anes20$gun_cntrl
```

	Frequency	Percent	Valid Percent
0	7372	89.03	100
NA's	908	10.97	
Total	8280	100.00	100

2. Use the mapping plan you produced for Question 1 of the Concepts and Calculations problems to collapse the current six-category variable measuring marital status (`anes20$V201508`) into a new variable, `anes20$marital`, with three categories, "Married," "Never Married," and "Other":
   - Create a frequency table for both `anes20$V201508` and the new variable, `anes20$marital`. Do the frequencies for the categories of the new variable match your expectations, given the category frequencies for the original variable?
   - Create barplots for both `anes20$V201508` and `anes20$marital`.
3. For this problem, use `anes20$V201231x`, the 7-point ordinal scale measuring party identification:
   - Using the mapping plan you created for Question 2 of the Concepts and Calculations problems, create a new variable named `anes20$ptyID.3` that includes three categories, "Democrat," "Independent," and "Republican."
   - Create a frequency table for both `anes20$V201231x` and `anes20$ptyID.3`. Do you notice any difference in the impressions created by these two tables?
   - Now create bar charts for the two variables and comment on any differences in the impressions they make on you.
   - Do you prefer the frequency table or bar chart as a method for looking at these variables? Why?
4. The following table summarizes information about four variables from the `anes20` dataset that measure attitudes toward different immigration policies. Take a closer look at each of these variables so you are comfortable with them:
   - Use the information in the table to create four numeric indicator (dichotomous) variables (one for each item) and combine those variables into a new index of immigration attitudes named `anes20$immig_pol` (show all steps along the way).
   - Create a frequency table OR bar chart for `anes20immig_pol` and describe its distribution.

Variable	Topic	Liberal Response
V202234	Accept refugees	1. Favor
V202240	Path to citizenship	1. Favor
V202243	Send back undocumented	2. Oppose
V202246	Separate undocumented parents/kids	2. Oppose

5. Use the `cut2` function to collapse the feeling thermometer for police (anes20$V202171) into five groups, –100 to –51, –50 to –1, 0, 1 to 50, and 51 to 100, and assign the result to a new variable, `anes20$policeFT.5`. Then assign the following level labels, "Very Negative," "Negative," "Neutral," "Positive," and "Very Positive," in the same order, and generate a frequency table for the new variable. On balance, do people feel positive or negative about the police?

5 MEASURES OF CENTRAL TENDENCY

GET READY

In this chapter, we move on to examining measures of central tendency. The techniques learned here complement and expand upon the material from Chapter 3, providing a few more options for describing how the outcomes of variables are distributed. In order to follow along in this chapter, you should load the `anes20`, `countries2`, and `states20` datasets and attach the libraries for the following packages: `DescTools` and `descr`.

CENTRAL TENDENCY

While frequencies tell us a lot about the distribution of variables, we are also interested in more precise measures of the general tendency of the data. Do the data tend toward a specific outcome? Is there a "typical" value? What is the "expected" value? *Measures of central tendency* provide this information. There are a number of different measures of central tendency, and their role in data analysis depends in part on the level measurement for the variables you are studying.

MODE

The **mode** is the category or outcome that occurs most often, and it is most appropriate for nominal data because it does not require that the underlying variable be quantitative in nature. That said, the mode can sometimes provide useful information for ordinal and numeric data, especially if those variables have a limited number of categories.

Let's look at an example of the mode by using religious affiliation, a nominal-level variable that is often tied to important social and political outcomes. The frequency that follows shows the outcomes from `anes20$denom`, an eight-category recoded version of the original twelve-category variable measuring religious affiliation (`anes20$V201435`).

```
#Create new variable for religious denomination
anes20$denom <- anes20$V201435
#Reduce number of categories by combine some of them
levels(anes20$denom) <- c("Protestant", "Catholic", "OtherChristian",
                          "OtherChristian", "Jewish", "OtherRel",
                          "OtherRel", "OtherRel", "Ath/Agn", "Ath/Agn",
                          "SomethingElse", "Nothing")
```

```
#Checkout the new variable
freq(anes20$denom, plot = F)

PRE: What is present religion of R
                    Frequency   Percent   Valid Percent
Protestant               2113    25.519          25.778
Catholic                 1640    19.807          20.007
OtherChristian            267     3.225           3.257
Jewish                    188     2.271           2.294
OtherRel                  163     1.969           1.989
Ath/Agn                   796     9.614           9.711
SomethingElse            1555    18.780          18.970
Nothing                  1475    17.814          17.994
NA's                       83     1.002
Total                    8280   100.000         100.000
```

From this table, you can see in both the raw frequency and percent columns that "Protestant" is the modal category (recognizing that this category includes a number of different religions). Let's think about how this represents the central tendency or expected outcome of this variable. With nominal variables such as this, the concept of a "center" doesn't hold much meaning, strictly speaking. However, if we are a bit less literal and take "central tendency" to mean something more like the typical or expected outcome, it makes more sense.

Think about this in terms of guessing the outcome, and you need information that will minimize your error in guessing (this idea will also be important later in the book). Suppose you have all 8197 valid responses on separate strips of paper in a great big hat, and you need to guess the religious affiliation of each respondent, using the coding scheme from this variable. What's your best guess as you pull each piece of paper out of the hat? It turns out that your best strategy is to guess the modal category, "Protestant," for all 8197 valid respondents. You will be correct 2113 times and wrong 6084 times. That's a lot of error, but no other guess will give you less error, because "Protestant" is the most likely outcome. In this sense, the mode is a good measure of the "typical" outcome for nominal data.

Besides using the frequency table, you can also get the mode for this variable using the `Mode` command in R.

```
#Get the modal outcome, Note the upper-case M
Mode(anes20$denom)
[1] NA
attr(, "freq")
[1] NA
```

Oops, this result doesn't look quite right. That's because many R functions don't know what to do with missing data and will report `NA` instead of the information of interest. We get this error message because there are 83 missing cases for this variable (see the frequency). This is fixed in most cases by adding `na.rm = T` to the command line, telling R to remove the NAs from the analysis.

```
#Add "na.rm = T" to account for missing data
Mode(anes20$denom, na.rm = T)
[1] Protestant
attr(, "freq")
[1] 2113
8 Levels: Protestant Catholic OtherChristian Jewish OtherRel ... Nothing
```

This confirms that "Protestant" is the modal category, with 2113 respondents, and also lists all of the levels. In many cases, I prefer to get the mode from a frequency table because it provides a more complete picture of the variable, showing, for instance, that while "Protestant" is the modal category, "Catholic" is a very close second.

While the mode is the most suitable measure of central tendency for nominal-level data, it can be used with ordinal and interval-level data. Let's look at two variables we've used in earlier chapters, spending preferences on programs for the poor (`anes20$V201320x`)[1] and state abortion restrictions (`states20$abortion_laws`).

```
#Mode for spending on aid to the poor
Mode(anes20$V201320x, na.rm = T)
[1] 3. Kept the same
attr(, "freq")
[1] 3213
5 Levels: 1. Increased a lot 2. Increased a little ... 5. Decreasaed a lot

#Mode for # Abortion restrictions
Mode(states20$abortion_laws)
[1] 10
attr(, "freq")
[1] 12
```

Here, we see that the modal outcome for spending preferences is "Kept the same", with 3213 respondents, and the mode for abortion regulations is 10, which occurred twelve times. While the mode does provide another piece of information for these variables, there are better measures of central tendency for working with ordinal or interval/ratio data, especially continuous numeric variables, for which the mode is unlikely to provide much useful information. However, the mode is the preferred measure of central tendency for nominal data.

MEDIAN

The median cuts the sample in half: It is *the value of the outcome associated with the observation at the middle of the distribution when cases are listed in order of magnitude.* Because the median is found by ordering observations from the lowest to the highest value, the median is **not** an appropriate measure of central tendency for nominal variables.

If the cases for an ordinal or numeric variable are listed in order of magnitude, we can look for the point that cuts the sample exactly in half. The value associated with that observation is the median. For instance, if we had five observations ranked from lowest to highest, the middle observation would be the third one (two above it and two below it), and the median would be the value of the outcome associated with that observation. Just to be clear, the median in this example would not be 3, but the value of the outcome associated with the third observation.

Here is a useful formula for finding the middle observation:

$$\text{Middle Observation} = \frac{n+1}{2}$$

where n = number of cases.

If n is an odd number, then the middle is a single data point. If n is an even number, then the middle is between two data points and you should use the midpoint between those two values as the median. Figure 5.1 illustrates how to find the median, using hypothetical data for a small sample of cases.

FIGURE 5.1 ■ Finding the Median With Odd and Even Numbers of Cases

Mid-point is the sixth observation	6 7 9 12 14 15 16 18 21 23 27	Median = 15
Mid-point is between the fifth and sixth observations	6 7 9 12 14 15 16 18 21 23	Median = 14.5

In the first row of data, the sixth observation perfectly splits the sample, with five observations above it and five below. The value of the sixth observation, 15, is the median. In the second row of data, there is an even number of cases (10), so the middle of the distribution is between the fifth and sixth observations, with values of 14 and 15, respectively. The median is the midpoint between these values, 14.5.

Now, let's look at this with some real-world data, using the abortion laws variable from the `states20` dataset. There are 50 observations, so the midpoint is between the 25th and 26th observations ((50 + 1)/2). Obviously, we can get R to just show us the median, but it is instructive at this point to find it by "eyeballing" the data. All fifty observations for `states20$abortion_laws` are listed next in order of magnitude. The value associated with the 25th observation is 9, as is the value associated with the 26th observation. Since these are both the same, the median outcome is 9.

```
#Use "sort" to view outcomes in order of magnitude
sort(states20$abortion_laws)
[1]  1 2 3 3 4 4  4   5   5   5   5   5   6   6   6   6   6   6  7  7  7  8  8  8  9
[26]  9  9  9  9  9  9 10 10 10 10 10 10 10 10 10 10 10 10 11 11 12 12 12 13 13
```

We could also get this information more easily.

```
#Get the median value
median(states20$abortion_laws)
[1] 9
```

To illustrate the importance of listing the outcomes in order of magnitude, check the list of outcomes for `states20$abortion_law` when the data are listed in alphabetical order, by state name.

```
#list abortion_laws without sorting from lowest to highest
states20$abortion_laws
 [1]  8  8  1110  5  2  4  6  10  9  5  11  5  12  9  13  10  10  4  6  510  9  9  12
[26]  7  10  7  3  7  6  4  9  9  10  13  3  9  6  10  10  10  12  10  1  8  5  6  10  6
```

If you took the midpoint between the 25th (12) and 26th (7) observations, you would report a median of 9.5. While this is close to 9, by coincidence, it is incorrect.

Let's turn now to finding the median value for spending preferences on programs for the poor (`anes20$V201320x`). Since there are over 8000 observations for this variable, it is not practical to list them all in order, but we can do the same sort of thing using a frequency table. Here, I use the `Freq` command because I am particularly interested in the cumulative percentages.

```
# Use 'Freq' to get cumulative %
Freq(anes20$V201320x)

                    Level   freq   perc  cumfreq  cumperc
1       1. Increased a lot  2'560  31.1%    2'560    31.1%
2   2. Increased a little  1'617  19.7%    4'177    50.8%
3        3. Kept the same  3'213  39.1%    7'390    89.8%
4   4. Decreased a little    446   5.4%    7'836    95.3%
5      5. Decreasaed a lot    389   4.7%    8'225   100.0%
```

What we want to do now is use the cumulative percentage to identify the category associated with the 50th percentile, the middle of the distribution. In this case, you can see that the cumulative frequency for the category "Increased a little" is 50.8%, meaning the middle observation (50th percentile) is in this category, so this is the median outcome.

We can check this with the `median` command in R. One little quirk here is that R requires numeric data to calculate the median. That's fine. All we have to do is tell R to treat the values *as if* they were numeric, replacing the five levels with values 1, 2, 3, 4, and 5.

```
#Get the median, treating the variable as numeric
median(as.numeric(anes20$V201320x), na.rm = T)
[1] 2
```

This confirms that the median outcome is the second category, "Increased a little".

THE MEAN

The **mean** is usually represented as $\overline{x}$ and is also referred to as the average or the expected value of the variable. It is a good measure of the typical outcome for numeric data. The term "typical" outcome is interesting in this context because the mean value of a variable may not exist as an actual outcome.

The best way to think about the mean as a measure of typicality is somewhat similar to the discussion of the mode as your best guess if you want to minimize error in predicting the outcomes in a nominal variable. The difference here is that since the mean is used with numeric data, we don't judge accuracy in a dichotomous right/wrong fashion; instead, we can judge accuracy in terms of how close the mean is to each outcome. For numeric data, the mean is closer overall to the actual values than any other guess you could make. In this sense, the mean represents the typical outcome better than any other statistic.

The formula for the mean is

$$\overline{x} = \frac{\sum_{i=1}^{n} x_i}{n}.$$

This reads: Take the sum of the values of all observations (the numerical outcomes) of x, and divide that by the total number of valid observations (observations with real values). This formula highlights an important way in which the mean is different from both the median and the mode: It is based on information from all of the values of x, not just the middle value (the median) or the value that occurs most often (the mode). This makes the mean a more encompassing statistic than either the median or the mode.

Using this formula, we could calculate the mean number of state abortion restrictions in 2020 (`states20$abortion_laws`) something like this:

$$\frac{1+2+3+3+4+4+\cdots+11+12+12+12+13+13}{50}$$

We don't actually have to add up all fifty outcomes manually to get the numerator. Instead, we can tell R to sum up all of the values of x.

```
#Sum all of the values of 'abortion_laws'
sum(states20$abortion_laws)
[1] 394
```

So the numerator is 394. We divide through by the number of cases (50) to get the mean.

```
#Divide the sum or all outcomes by the number of cases
394/50
[1] 7.88
```

$$\overline{x} = \frac{394}{50} = 7.88$$

Of course, it is simpler, though not quite as instructive, to have R tell us the mean.

```
#Tell R to get the mean value
mean(states20$abortion_laws)
[1] 7.88
```

A very important characteristic of the mean is that it is the point at which the weight of the values is perfectly balanced on each side. It is helpful to think of the outcomes as numeric variables distributed according to their weight (value) on a plank resting on a fulcrum, with the fulcrum placed at the mean of the distribution, the point at which both sides are perfectly balanced (see Figure 5.2).[2] If the fulcrum is placed at the mean, the plank will not tip to either side, but if it is placed at some other point (the median, for instance), then the weight will not be evenly distributed and the plank will tip to one side or the other.

A really important concept here is the deviation of the values of x from the mean of x, represented as $x_i - \overline{x}$. For instance, the deviation from the mean number of abortion restrictions (7.88) for a state like Alabama, with eight abortion restrictions on the books is .12 units (8 – 7.88), and for a state like Colorado, with two restrictions on the books, the deviation is –5.88 (2 – 7.88). For any numeric variable, the distance between the mean and all values greater than the mean is perfectly offset by the distances between the mean and all values lower than the mean.

This means that summing up all of the deviations from the mean ($\sum_{i=1}^{n} x_i - \overline{x}$) is always equal to zero. This point is important for understanding what the mean represents, but it is also important for many other statistics that are based on the mean.

We can check this balance property in R. First, we express each observation as a deviation from the means.

FIGURE 5.2 ■ The Mean Perfectly Balances a Distribution

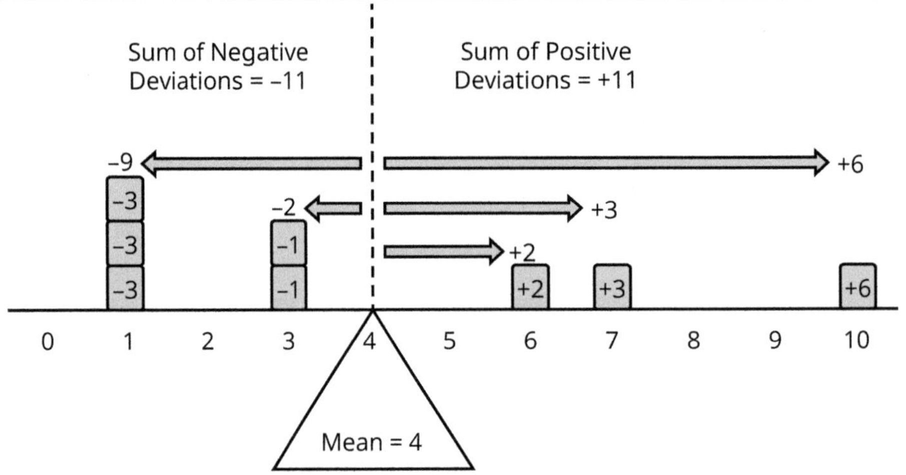

Note: From *Cross Validated: Explaining Mean, Median, Mode in Layman's Terms* by Vishal, 2016, Stack Exchange (https://stats.stackexchange.com/questions/200282/explaining-mean-median-mode-in-laymans-terms). CC BY-SA 3.0 (https://creativecommons.org/licenses/by-sa/3.0/).

```
#Subtract the mean of x from each value of x
dev_mean = (states20$abortion_laws- mean(states20$abortion_laws))

#Print the mean deviations
dev_mean
[1]    0.12  0.12  3.12  2.12 -2.88 -5.88 -3.88 -1.88  2.12  1.12 -2.88  3.12
[13]  -2.88  4.12  1.12  5.12  2.12  2.12 -3.88 -1.88 -2.88  2.12  1.12  1.12
[25]   4.12 -0.88  2.12 -0.88 -4.88 -0.88 -1.88 -3.88  1.12  1.12  2.12  5.12
[37]  -4.88  1.12 -1.88  2.12  2.12  2.12  4.12  2.12 -6.88  0.12 -2.88 -1.88
[49]   2.12 -1.88
```

As expected, some deviations from the mean are positive, and some are negative. Now, if we take the sum of all deviations, we get the following.

```
#Sum the deviations of x from the mean of x
sum(dev_mean)
[1] 5.329071e-15
```

That's a funny looking number. Because of the length of the resulting number, the sum of the deviations from the mean is reported using **scientific notation**. In this case, the scientific notation is telling us we need to move the decimal point 15 places to the left, resulting in .00000000000000532907. Not quite exactly 0, but essentially 0.

Note that the median does not share this "balance" property; if we placed a fulcrum at the median, the distribution would tip over because the positive deviations do not balance perfectly by the negative deviations, as shown here.

```
#Sum the deviations of x from the median of x
sum(states20$abortion_laws- median(states20$abortion_laws))
[1] -56
```

The negative deviations from the median outweigh the positive by a value of 56.

Dichotomous Variables

Here's an interesting bit of information that will be useful in the future: For dichotomous variables with all values scored either 0 or 1, the mean of the variable is the proportion of cases in Category 1. Let's take an example from one of the indicator variables we created in Chapter 4. There, we used the following code to convert `anes20$V201415`, a qualitative variable based on a survey question about whether gays and lesbians should be allowed to adopt children, into a numeric question scored 0 and 1.

```
#Code from Chapter 4. Create indicator for "adoption" variable
anes20$lgbtq4 <- as.numeric(anes20$V201415 == "1. Yes")
```

If we look at the outcomes for both variables, we can see that the 6554 respondents (80.4%) who said "Yes" on `anes20$V201415` are assigned a value of 1 on `anes20$lgbtq4`, and the 1598 (19.6%) respondents who said "No" are assigned a value of 0.

```
freq(anes20$V201415, plot = F)

PRE: Should gay and lesbian couples be allowed to adopt
          Frequency   Percent  Valid Percent
1. Yes         6554    79.155           80.4
2. No          1598    19.300           19.6
NA's            128     1.546
Total          8280   100.000          100.0

freq(anes20$lgbtq4, plot = F)

anes20$lgbtq4
          Frequency   Percent  Valid Percent
0              1598    19.300           19.6
1              6554    79.155           80.4
NA's            128     1.546
Total          8280   100.000          100.0
```

We now have a dichotomous numeric variable that distinguishes between those who think gays and lesbians should be allowed to adopt and those who think they should not. We can use the formula presented earlier to calculate the mean of this variable. Since the value 0 occurred 1598 times and the value 1 occurred 6554 times, the mean is equal to

$$\overline{x} = \frac{(0*1598) + (1*6554)}{8152} = \frac{6554}{8152} = .80397.$$

We can verify this with R.

```
#Get the mean of the indicator variable
mean(anes20$lgbtq4, na.rm = T)
[1] 0.8039745
```

The mean of this dichotomous variable is the proportion in Category 1. This should always be the case with similar dichotomous variables. This may not seem very intuitive to you at this point, but it will be very useful to understand it later on.

You might be questioning at this point how we can treat what seems like a nominal variable—whether people favor or oppose allowing gays and lesbians to adopt children—as a numeric variable. The way I like to think about this is that this variable measures the presence (1) or absence (0) of a characteristic, supporting adoption by gays and lesbians. In cases like this, you can think of the 0 value as a genuine zero point, indicating that the respondent does not have the characteristic you are measuring. Since the value 1 indicates having a unit of the characteristic, we can treat this variable as numeric. However, it is important to always bear in mind what the variable represents and that there are only two outcomes, 0 and 1. This becomes especially important in later chapters when we discuss using these types of variables in regression analysis.

MEAN, MEDIAN, AND THE DISTRIBUTION OF VARIABLES

As a general rule, the mean is most appropriate for interval-/ratio-level variables, and the median is most appropriate for ordinal variables. However, there are instances when the median might be a better measure of central tendency for numeric data, almost always because of *skewness* in the data. Skewness occurs in part as a consequence of one of the virtues of the mean. Because the mean takes into account the values of all observations, whereas the median does not actually emphasize "values" but uses the ranking of observations, the mean can be influenced by extreme values that "pull" it away from the middle of the distribution. Consider the following simple example, using five observations from a hypothetical variable:

0, 1, 2, 2, 5 → median = 2, mean = 2.0.

In this case, the mean and the median are the same. Now, watch what happens is we change the value 5 to 18, a value that is substantially higher than the rest of the values:

0, 1, 2, 2, 18 → median = 2, mean = 4.6.

The median is completely unaffected by the extreme value, but the mean is. The median and some other statistics are what we call **robust** statistics because their value is not affected by the weight of extreme outcomes. In some cases, the impact of extreme values is so great that it is better to use the median as a measure of central tendency when discussing numeric variables. The data are still balanced around the mean, and the mean is still your "best guess" (smallest error in prediction), but in terms of looking for an outcome that represents the general tendency of the data, it may not be the best option in some situations.

In general, when the *mean is significantly higher than the median*, this is a sign that the distribution of values is right (or positively) skewed, due to the influence of extreme values at the high end of the scale. This means that the extreme values are pulling the mean away from the middle of the observations. When the *mean is significantly lower than the median*, this indicates a left (or negatively) skewed distribution, due to some extreme values at the low end of the scale. When the mean and median are the same or nearly the same, this could indicate a bell-shaped distribution, but there are other possibilities as well. When the mean and median are the same, there is no skew.

Figure 5.3 provides caricatures of what these patterns might look like. The first graph shows that most data are concentrated at the low (left) end of the *x*-axis, with a few very extreme observations at the high (right) end of the axis pulling the mean out from the middle of the distribution. This is a positive skew. The second graph shows just the opposite pattern: Most data at the high end of the *x*-axis, with a few extreme values at the low (left) end dragging the mean to the left. This is a negatively skewed distribution.

FIGURE 5.3 ■ Illustrations of Different Levels of Skewness

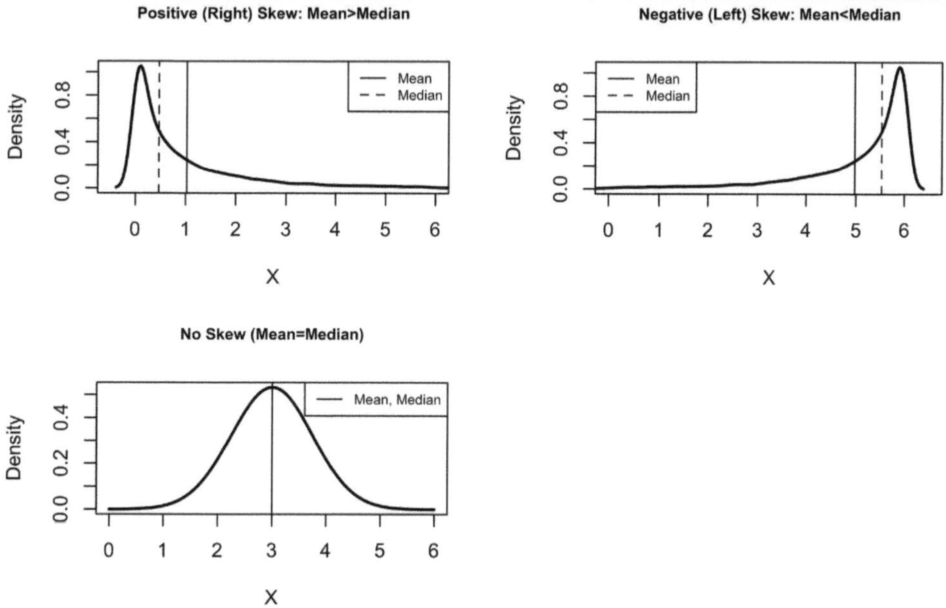

Finally, the third graph shows a perfectly balanced, bell-shaped distribution, with the mean and the median equal to each other and situated right in the middle of the distribution. There is no skewness in this graph. This bell-shaped distribution is not the only type of distribution with no skewness,[3] but it is an important type of distribution that we will take up later.

It has been my experience that once students hear that the median *might be preferred* over the mean when there is a significant difference between them, they tend to default to using the median whenever it is at all different from the mean. There will almost always be some difference between the mean and the median, so automatically defaulting to the median is not good practice. My best advice is that the mean should be your first choice. If, however, there is evidence that the mean is heavily influenced by extreme observations (see below), then you should also use the median. Of course, it doesn't hurt to present both statistics, as more information is usually good information.

Let's see what this looks like when using real-world data, starting with abortion laws in the states (`states20$abortion_laws`).

```
#get mean and median for 'abortion_laws'
mean(states20$abortion_laws)
[1] 7.88
median(states20$abortion_laws)
[1] 9
```

The mean is a bit less than the median, so we might expect to see signs of negative (left) skewness in the density plot.

```
plot(density(states20$abortion_laws),
     xlab = "# Abortion Restrictions in State Law",
     main = "")
#Insert vertical lines for mean and median
abline(v = mean(states20$abortion_laws))
abline(v = median(states20$abortion_laws), lty = 2) #use dashed line
```

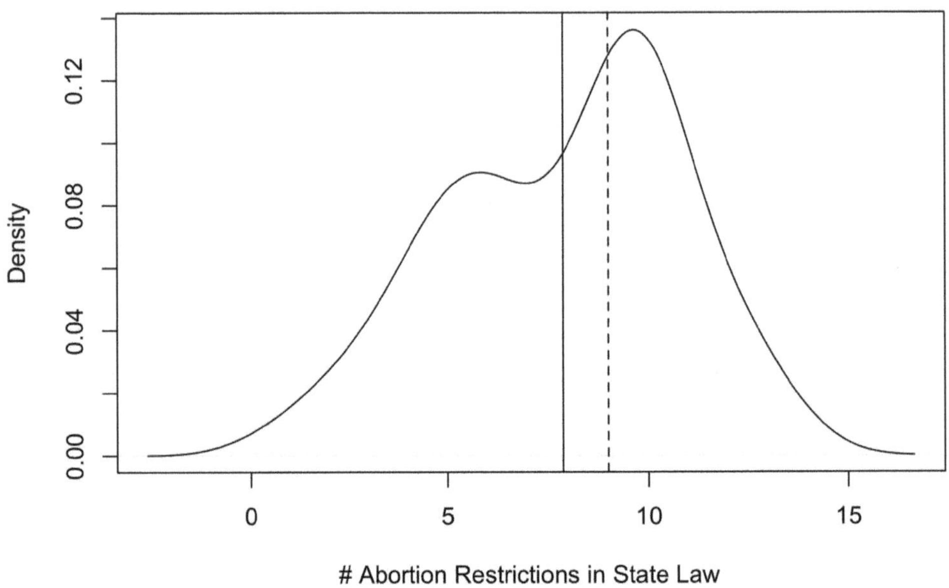

Here, we see the difference between the mean (solid line) and the median (dashed line) in the context of the full distribution of the variable, and the graph shows a distribution with a bit of negative skew. The skewness does not appear to be severe, but it is visible to the naked eye. Note that you get the same impression if you use a histogram.

```
hist(states20$abortion_laws,
     xlab = "# Abortion Restrictions in State Law",
     main = "")
#Insert vertical lines for mean and median
abline(v = mean(states20$abortion_laws))
abline(v = median(states20$abortion_laws), lty = 2) #use dashed line
```

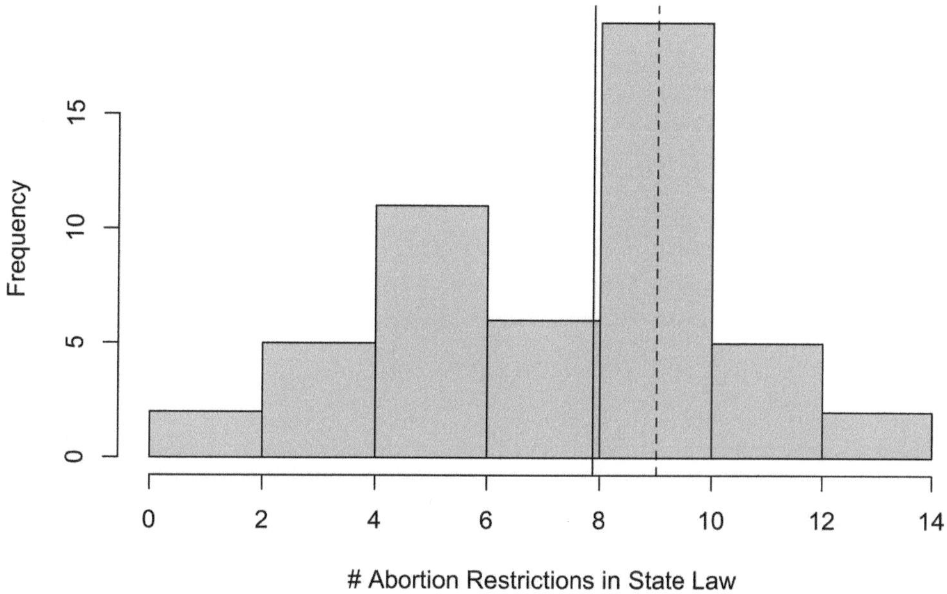

An R Code Digression

The preceding density plot includes the addition of two vertical lines, one for the mean and one for the median. To add these lines, I used the `abline` command. This command allows you to add lines to existing graphs. In this case, I want to add two vertical lines, so I use `v =` to designate where to put the lines. I could have put in the numeric values of the mean and median (`v = 7.88` and `v = 9`), but I chose to have R calculate the mean and median and insert the results in the graph (`v = mean(states20$abortion_laws)` and `v = median(states20$abortion_laws)`). Either way would get the same result. Also, note that for the median, I added `lty = 2` to get R to use "line type 2", which is a dashed line. The default line type (1) is a solid line, which is used for the mean.

Now, let's take a look at another distribution, this time for a variable we have not looked at before, the percentage of the state population who are foreign-born (`states20$fb`). This is an increasingly important population characteristic, with implications for a number of political and social outcomes.

```
mean(states20$fb)
[1] 7.042
median(states20$fb)
[1] 4.95
```

Here, we see a bit more evidence of a skewed distribution. Since the mean is greater than the median, it is reasonable to expect to see some positive skewness. The density plot supports (next) this expectation and looks somewhat more like a skewed distribution than in the example using state abortion laws.

```
#Density plot for % foreign-born
plot(density(states20$fb),
     xlab = "Percent foreign-born",
     main = "")

#Add lines for mean and median
abline(v = mean(states20$fb))
abline(v = median(states20$fb), lty = 2) #Use dashed line
```

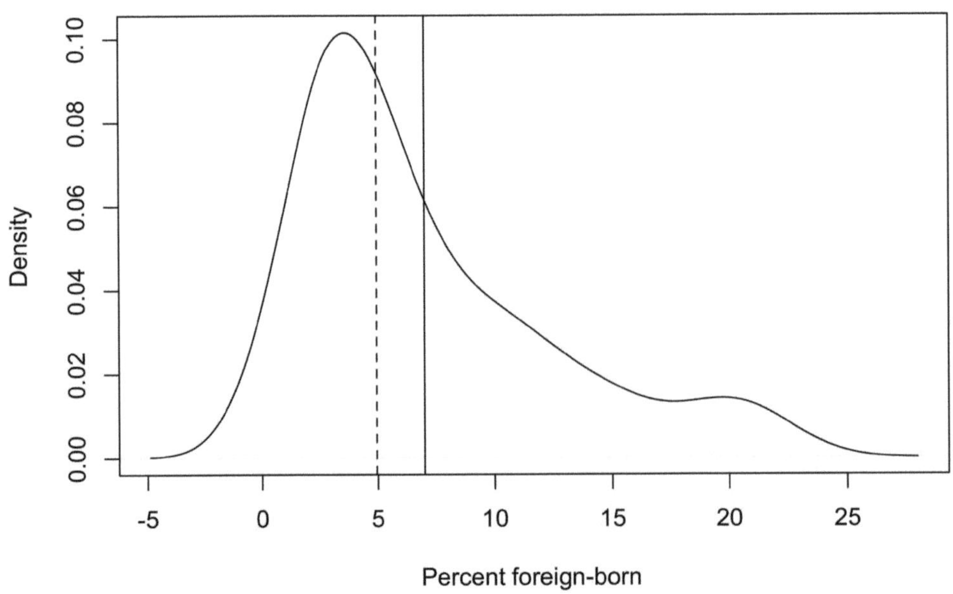

Finally, as a counterpoint to these skewed distributions, let's take a look at the distribution of percentage of the two-party vote for Joe Biden in the 2020 election.

```
mean(states20$d2pty20)
[1] 48.8044
median(states20$d2pty20)
[1] 49.725
```

Here, there is very little difference between the mean and the median, and there appears to be almost no skewness in the shape of the density plot (next).

```
plot(density(states20$d2pty20),
     xlab = "% of Two-Party Vote for Biden",
     main = "")
abline(v = mean(states20$d2pty20))
abline(v = median(states20$d2pty20), lty = 2)
```

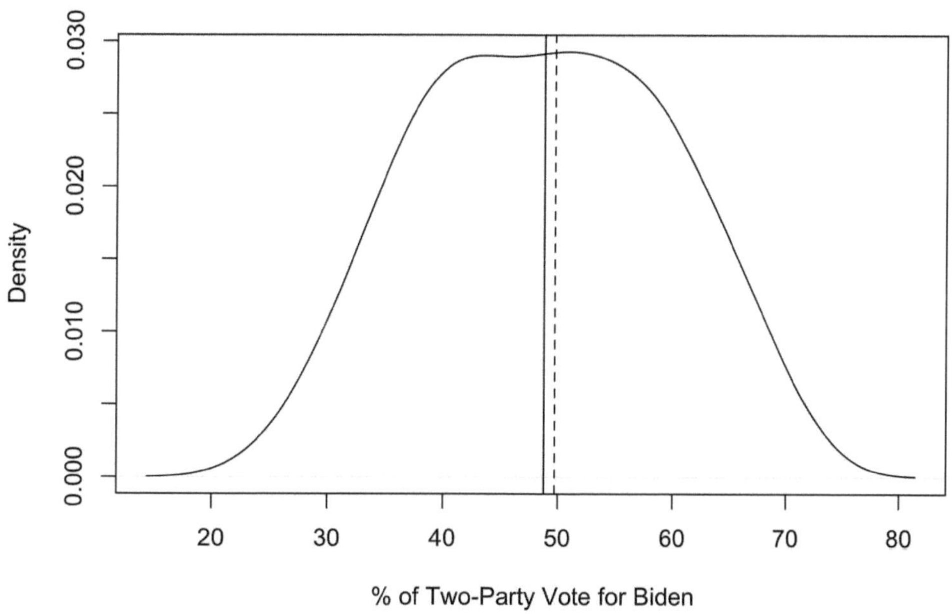

What's interesting here is that the distance between the mean and median for Biden's vote share (.92) is not terribly different than the distance between the mean and the median for the number of abortion laws (1.12). Yet there is not a hint of skewness in the distribution of votes, while there is clearly some negative skew to abortion laws. This, of course, is due to the difference in scale between the two variables. Abortion laws range from 1 to 13, while Biden's vote share ranges from 27.5 to 68.3, so *relative to the scale of the variable*, the distance between the mean and median is much greater for abortion laws in the states than it for Biden's share of the two-party vote in the states. This illustrates the real value of examining a numeric variable's distribution with a histogram or density plot alongside statistics like the mean and median. Graphs like these provide context for those statistics.

SKEWNESS STATISTIC

Fortunately, there is a skewness statistic that summarizes the extent of positive or negative skew in the data. There are several different methods for calculating skewness. The one used in R is based on the direction and magnitude of deviations from the mean, relative to the amount of variation in the data.[4] The nice thing about the skewness statistic is that there are some general guidelines for evaluating the direction and seriousness of skewness:

- A value of 0 indicates no skew to the data.
- Any negative value of skewness indicates a negative skew, and positive values indicate positive skew.
- Values lower than –2 and higher than 2 indicate extreme skewness that could pose problems for operations that assume a relatively normal distribution.

Let's look at the skewness statistics for the three variables just discussed.

```
#Get skew for three variables (upper-case S in "Skew" command)
Skew(states20$abortion_laws)
[1] -0.3401058

Skew(states20$fb)
[1] 1.202639

Skew(states20$d2pty20)
[1] 0.006792623
```

These skewness statistics make a lot of sense, given the earlier discussion of these three variables. There is a little bit of negative skew (–.34) to the distribution of abortion restrictions in the states, a more pronounced level of (positive) skewness to the distribution of the foreign-born population (1.2), and no real skewness in the distribution of Biden support in the states (.007). None of these results are anywhere near the –2, +2 cut-off points discussed earlier, so these distributions should not pose any problems for any analysis that uses these variables.

Log Transformations

Quite often, you will be using variables whose distributions look a lot like the three previously shown, maybe a bit skewed but nothing too outrageous. Occasionally, though, you will come across variables with severe skewness that can be detected visually and by using the skewness statistic. As a point of reference, consider the three distributions shown earlier in Figure 5.3, one with what appears to be severe positive skewness (top left), one with severe negative skewness (top right), and one with no apparent skewness (bottom left). The values of the skewness statistics for these three distributions are 4.25, –4.25, and 0, respectively.

In some instances, the skewness is so excessive, often due to just a few extreme outliers, that it can create serious problems for calculating some statistics. In many of these cases, it makes sense to transform the data so the extreme outliers do not have an out-sized effect on statistical results. One variable like this that is used in later chapters measures country-level population size in 2019, in millions of people (`countries2$pop19_M`). As you can see in the following histogram this variable's distribution is heavily influenced by a few extreme outliers.

```
hist(countries2$pop19_M, breaks = 13,
     xlab = "Population (Millions)",
     ylab = "Number of Countries",
     main = "")
```

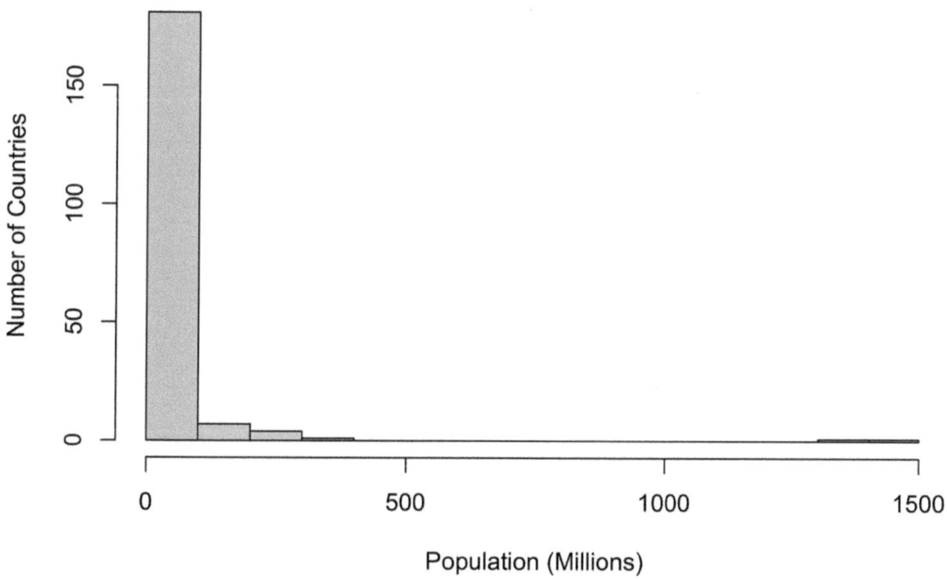


```
Skew(countries2$pop19_M)
[1] 8.320391
```

The vast majority of countries fall in the first bin (0 to 100,000, anchored by Nauru at 10,800 people), and a handful more are in the next three bins (the U.S.; the highest in this group 329 million), and then China and India are at the far right end of the scale, coming in at about 1.4 billion population each. The skewness statistic reflects this distribution with a value of 8.3.

One useful way to minimize the impact of the extreme values is to take logged values of each observation. Log transformations like this have many uses. In this instance, we want to minimize the impact of extreme values. Later, in Chapter 17, you will see how to use logged values to model certain types of relationships in regression analysis. Here, I will use a log(10) transformation. This means that all of the original values are transformed into their logged values, using a base of 10. A logged value is nothing more than the power to which you have to raise your log base (in this case, 10) in order to get the original raw score. For instance, for the U.S., the log10 of population is 2.517, since $10^{2.52} = 329$, the log10 of population for China is 3.1565 because $10^{3.1565} = 1434$, and so on. The following code shows how to get log10 values and how to translate logged values back to the original values.

```
log10(329)
[1] 2.517196

log10(1434)
[1] 3.156549

10^2.517196
[1] 329.0001

10^3.156549
[1] 1434
```

The result of the transformation is that relatively extreme values are brought closer to the middle, and the resulting distribution is usually easier to deal with. Check out the following histogram, which shows the log-transformed distribution.

```
hist(log10(countries2$pop19_M),
     xlab = "Log10 of Population (Millions)",
     ylab = "Number of Countries",
     main = "")
```

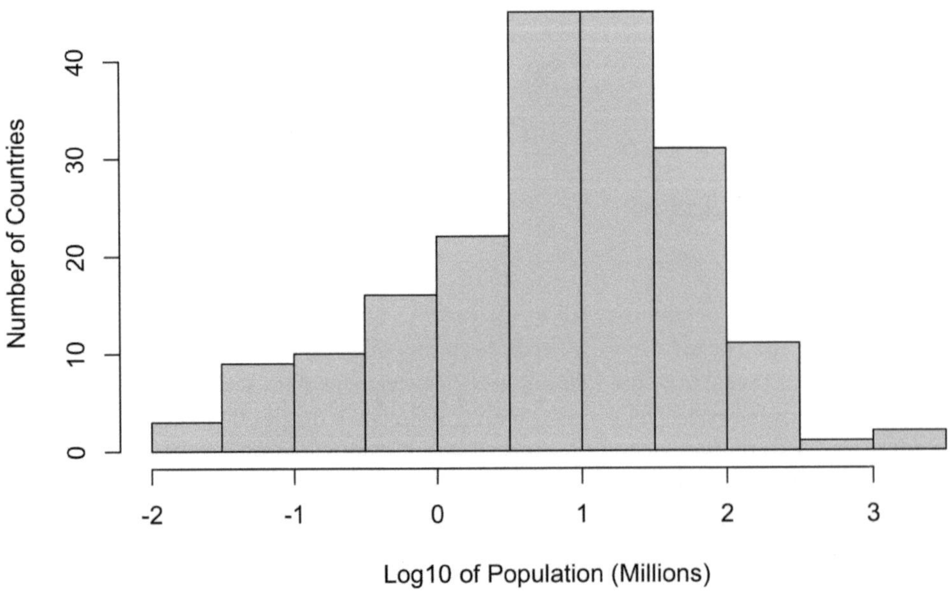

There will be more discussion of this transformation later, but for now, here are a couple of other things to know:

- This only works for variables with positive values. Taking log transformations of variables with negative or zero values will produce missing data.
- In the histogram command, I used `log10(countries2$pop19_M)`, where I told R to transform the original variable and use it in the histogram. Alternatively, you could create a transformed version of the variable (`countries2$lg10_pop <- log10(countries2$pop19_M)`) and use that variable in the histogram or elsewhere. Both approaches work.

ADDING LEGENDS TO GRAPHS

The vertical lines for the mean and median used in the preceding graphs are useful visual aids for understanding how those variables are distributed. Importantly, these lines were explained in the text, but there was no identifying information provided in the graphs themselves. When presenting information like this in a formal setting (e.g., on the job, for an assignment, or for a term paper), it is usually expected that you provide a **legend** with your graphs that identifies what the lines represent (see Figure 5.3 as an example). This can be a bit complex, especially if there are multiple lines in your graph. In the current case, however, it is not terribly difficult to add a legend.

This code is added to the density plot command used to generate the density plot for `states20$fb`.

```
#Create a legend to be put in the top-right corner, identifying
#mean and median outcomes with solid and dashed lines, respectively
legend("topright", legend = c("Mean", "Median"), lty = 1:2)
```

The first bit (`"topright"`) is telling R where to place the legend in the graph. In this case, I specified `topright` because I know that is where there is empty space in the graph. You can specify top, bottom, left, right, topright, topleft, bottomright, or bottomleft, depending on where the legend fits the best. The second piece of information, `legend = c("Mean", "Median")`, provides names for the two objects being identified, and the last part, `lty = 1:2`, tells R which line types to use in the legend (the same as in the graph). Let's add this to the command lines for the density plot for `states20$fb` and see what we get.

```
plot(density(states20$fb),
     xlab = "Percent foreign-born",
     main = "")
abline(v = mean(states20$fb))
```

```
abline(v = median(states20$fb), lty = 2)
#Add the legend
legend("topright", legend = c("Mean", "Median"), lty = 1:2)
```

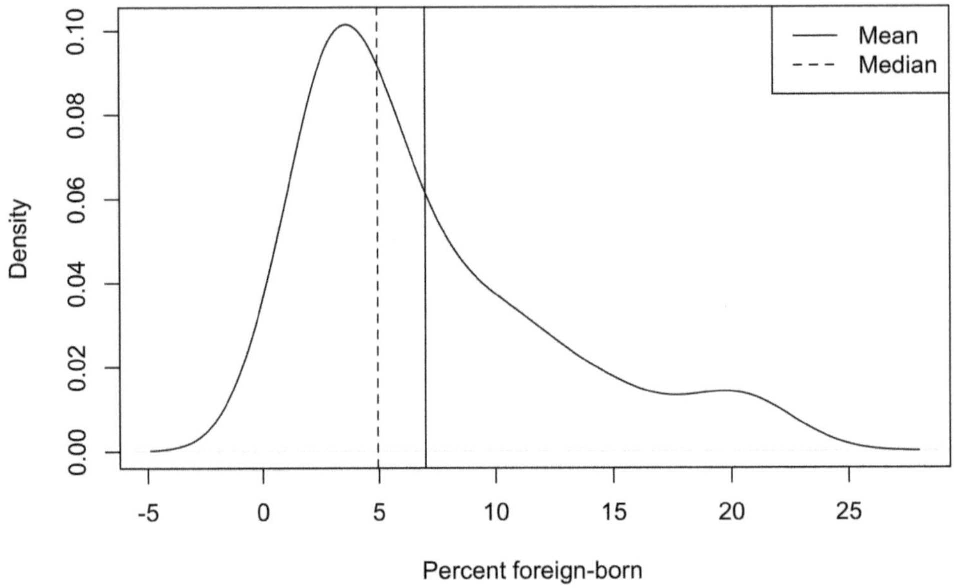

This extra bit of information fits nicely in the graph and aids in interpreting the pattern in the data.

Sometimes, it is hard to get the legend to fit well in a graph space. When this happens, you need to tinker a bit to get a better fit. There are some fairly complicated ways to achieve this, but I favor trying a couple of simple things first: Try different locations for the legend, reduce the number of words you use to name the lines, or add the `cex` command to reduce the overall size of the legend. When using `cex`, you might start with `cex = .8`, which will reduce the legend to 80% of its original size, and then change the value as needed to make the legend fit.

NEXT STEPS

The next chapter builds on the discussion of descriptive statistics by focusing on something we've dealt with tangentially in this chapter: measuring variability in the data. The measures of central tendency discussed in this chapter are important descriptive tools, but they also are essential for calculating and understanding many other statistics, including measures of dispersion, which we take up in Chapter 6. Following that, we spend some time on the more abstract topics of probability and statistical inference. These things might sound a bit intimidating, but you will be ready for them when you get there.

EXERCISES

Concepts and Calculations

1. Let's return to the list of variables used for exercises in Chapters 1 and 3. Identify what you think is the *most appropriate* measure of central tendency for each of these variables. Choose just one measure for each variable:
   - Course letter grade
   - Voter turnout rate (votes cast/eligible voters)
   - Marital status (married, divorced, single, etc.)
   - Occupation (professor, cook, mechanic, etc.)
   - Body weight
   - Total number of votes cast in an election
   - #Years of education
   - Subjective social class (poor, working class, middle class, etc.)
   - Poverty rate
   - Racial or ethnic group identification

2. Following is a list of voter turnout rates in twelve Wisconsin counties during the 2020 presidential election. Calculate the mean, median, and mode for the level of voter turnout in these counties. Which of these is the most appropriate measure of central tendency for this variable? Why? Based on the information you have here, is this variable skewed in either direction? Explain.

Wisconsin County	% Turnout
Clark	63
Dane	87
Forrest	71
Grant	63
Iowa	78
Iron	82
Jackson	65
Kenosha	71
Marinette	71
Milwaukee	68
Portage	74
Taylor	70

3. The following table provides the means and medians for four different variables from the `states20` dataset. Use this information to offer your best guess as to the level and direction of skewness in these variables. Explain your answer. Make sure to look at the `states20` codebook first, so you know what these variables are measuring.

Variable Name	Mean	Median
obesity	29.9	30.10
Cases10k	421.9	453.9
PerPupilExp	12,720	11,826
metro	64.2	68.5

4. The table that follows shows responses to a question from the `anes20` that asked if respondents agreed or disagreed that "immigrants are good for America's economy." What are the mode and median for this variable? Explain your answer.

```
POST: CSES5-Q05c: Out-group attitudes: immigrants good for America's economy
                              Frequency  Percent Valid  Percent
1. Agree strongly                  1870   22.585          25.418
2. Agree somewhat                  2770   33.454          37.651
3. Neither agree nor disagree      1876   22.657          25.500
4. Disagree somewhat                605    7.307           8.223
5. Disagree strongly                236    2.850           3.208
NA's                                923   11.147
Total                              8280  100.000         100.000
```

5. The table that follows shows the logged (base 10) values of the cumulative number of COVID-19 cases in five different states by the end of August 2021. Use this information to calculate the raw number of cases for each state. Comment on the difference in the impression you get about the differences in the prevalence of COVID-19 cases between the two columns of numbers.

State	Log10(Cases)	Number of Cases
Vermont	4.419	
Montana	5.102	
Arkansas	5.653	
Pennsylvania	6.113	
California	6.643	

R Problems

1. Use R to report all measures of central tendency that **are appropriate** for each of the following variables: The feeling thermometer rating for the National Rifle Association (`anes20$V202178`), Latinos as a percentage of state populations (`states20$latino`), party identification (`anes20$V201231x`), and region of the country where ANES survey respondents live (`anes20$V203003`). Where appropriate, also discuss potential skewness.

2. Using one of the numeric variables listed in the previous question, create a density plot that includes vertical lines showing the mean and median outcomes and add a legend. (Depending on which variable you use, you may have to add `na.rm = T` to the density command.) Describe what you see in the plot.

3. Get a histogram of `states20$pop2020` and describe its distribution. Then get a histogram of the logged (base 10) values of `states20$pop2020` and describe its distribution.

6 MEASURES OF DISPERSION

GET READY

This chapter examines another important consideration when describing how variables are distributed and the amount of variation or dispersion in the outcomes. We touched on this topic obliquely by considering skewness in the last chapter, but this chapter addresses the issue more directly. In order to follow along in this chapter, you should attach the libraries for the following packages: `descr`, `DescTools`, and `Hmisc`.

You should also load the `anes20`, `states20`, and `cces20` datasets. The Cooperative Congressional Election Study (`cces20`) is similar to the ANES in that it is a large-scale, regularly occurring survey of political attitudes and behaviors, with both a pre- and post-election wave.

INTRODUCTION

While measures of central tendency provide some sense of the typical outcome of a given variable, it is possible for variables with the same mean, median, and mode to have very different looking distributions. Take, for instance, the three graphs that follow, in Figure 6.1; all three are perfectly symmetric (no skew), with the same means and medians, but they vary considerably in the level of dispersion around the mean. On average, the data are most tightly clustered around the mean in the first graph, spread out the most in the third graph, and somewhere in between in the second graph. These graphs vary not in central tendency but in how concentrated the distributions are around the central tendency.

The concentration of observations around the central tendency is an important concept, and we are able to measure it in a number of different ways using *measures of dispersion*. There are two different, related types of measures of dispersion, those that summarize the overall spread of the outcomes and those that summarize how tightly clustered the observations are around the mean.

MEASURES OF SPREAD

Although we are ultimately interested in measuring dispersion around the mean, it can be useful sometimes to understand how spread out the observations are. Measures of spread focus on the upper and lower limits of the data, either over the full range of outcomes or some share of the outcomes.

FIGURE 6.1 ■ Distributions With Identical Central Tendencies but Different Levels of Dispersion

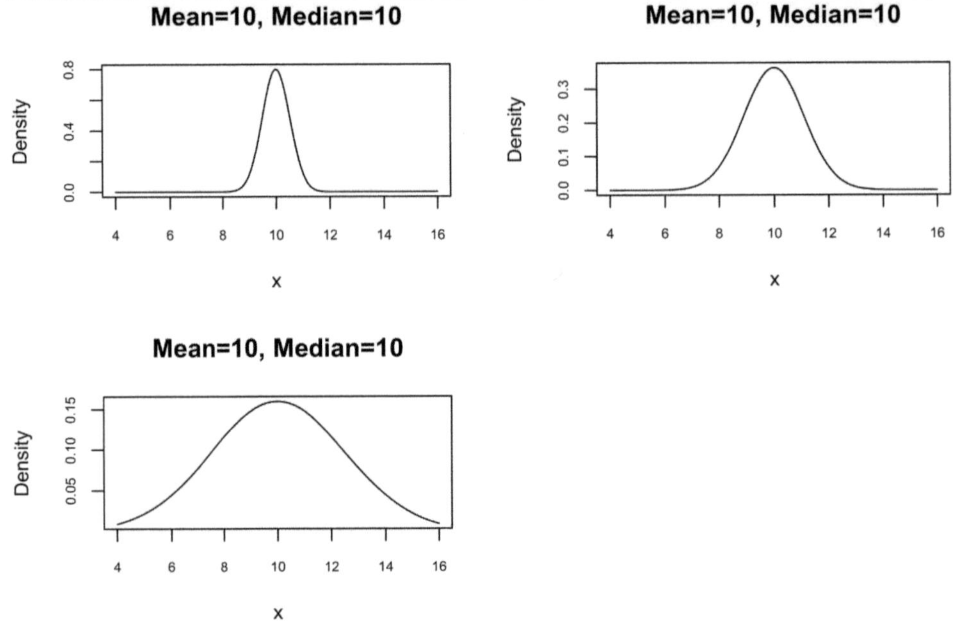

Range

The range is a measure of dispersion that does not use information about the central tendency of the data. It is simply the difference between the lowest and highest values of a variable. To be honest, this is not always a useful or interesting statistic. For instance, all three of the graphs shown in Figure 6.1 have the same range (4 to 16) despite the differences in the shape of the distributions. Still, the range does provide some information and could be helpful in alerting you to the presence of outliers or help you spot coding problems with the data. For example, if you know the realistic range for a variable measuring age in a survey of adults is roughly 18 to 100(ish), but the data show a range of 18 to 950, then you should look at the data to figure out what went wrong. In a case like this, it could be that a value of 950 was recorded rather than the intended value of 95.

Following, we examine the range for the age of respondents to the `cces20` survey. R provides the minimum and maximum values but does not show the width of the range.

```
#First, create cces20$age, using cces20$birthyr
cces20$age <- 2020-cces20$birthyr
#Then get the range for `age` from the cces20 data set.
range(cces20$age, na.rm = T)
[1] 18 95
```

Here, we see that the range in age in the cces20 sample is from 18 to 95, a perfectly plausible age range (only adults were interviewed), a span of 77 years. Other than this, there is not much to say about the range of this variable.

Interquartile Range (IQR)

One extension of the range is the interquartile range (IQR), which focuses on the middle 50% of the distribution. Technically, the IQR is the range between the value associated with the 25th percentile and the value associated with the 75th percentile. You may recall seeing this statistic supplied by colleges and universities as a piece of information when you were doing your college search: The middle 50% range for ACT or SAT scores. This is a great illustration of how the interquartile range can be a useful and intuitive statistic.

The IQR command in R estimates how wide the IQR is but does not provide the upper and lower limits. The width is important, but it is the values of the upper and lower limits that are more useful for understanding what the distribution looks like. The summary command provides the 25th ("1st Qu." in the output) and 75th ("3rd Qu.") percentiles, along with other useful statistics.

```
#Get inter-quartile range
IQR(cces20$age)
[1] 30

#Get quartiles for better understanding of IQR
summary(cces20$age)

   Min. 1st Qu.  Median    Mean 3rd Qu.    Max.
  18.00   33.00   49.00   48.39   63.00   95.00
```

The 25th percentile ("1st Qu") of cces20$age is associated with the value 33 and the 75th percentile ("3rd Qu") with the value 63, so the IQR is from 33 to 63 (a difference of 30).

One interesting aspect of the interquartile range is that you can think of it as both a measure of dispersion and a measure of central tendency. The upper and lower limits of the IQR define the middle of the distribution, the very essence of central tendency, while the difference between the upper and lower limits is an indicator of how spread out the middle of the distribution is, a central concept to measures of dispersion.

We can use tools learned earlier to visualize the interquartile range of cces20$age, using a histogram with vertical lines marking its upper and lower limits, as shown next.

```
#Age Histogram
hist(cces20$age, xlab = "Age",
     main = "Histogram of Age with Interquartile Range")
```

```
#Add lines for 25th and 75th percentiles
abline(v = 33, lty = 2, lwd = 2)
abline(v = 63, lwd = 2)
#Add legend
legend("topright", legend = c("1st Qu.", "3rd Qu."), lty = c (2, 1))
```

The histogram follows something similar to a bi-modal distribution, with a dip in the frequency of outcomes in the 40- to 55-years-old range. The thick vertical lines represent the 25th and 75th percentiles, the interquartile range. What's interesting to think about here is that if the distribution were more bell-shaped, the IQR would be narrower, as more of the sample would be in the 40- to 55-years-old range. Instead, because the center of the distribution is somewhat collapsed, the data are more spread out.

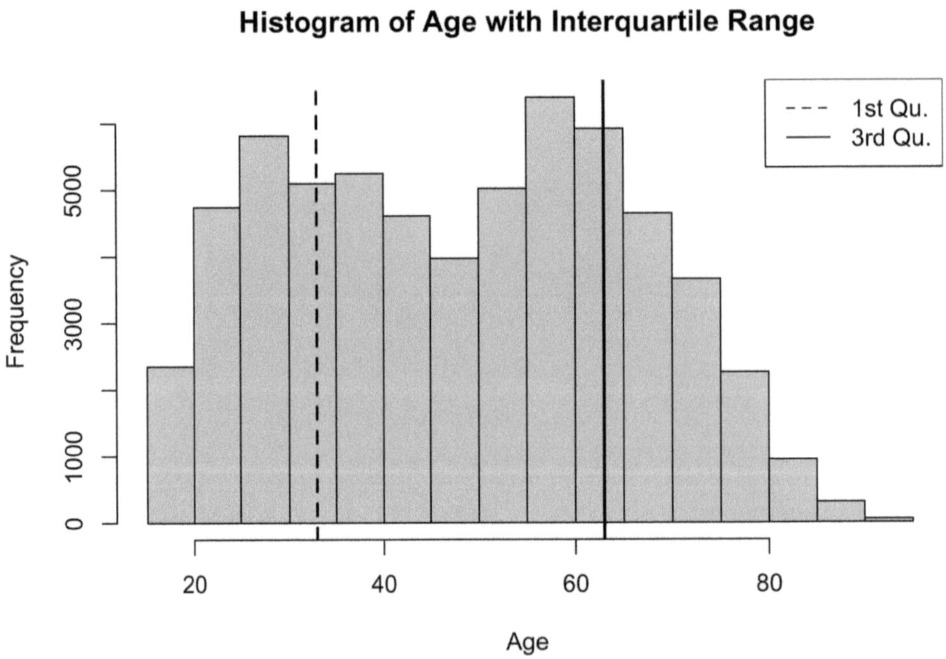

Boxplots

A nice graphing method that focuses explicitly on the range and IQR is the boxplot, shown next. The boxplot is a popular and useful tool, one that is used extensively in subsequent chapters.

```
#Boxplot Command
boxplot(cces20$age, main = "Boxplot for Respondent Age", ylab = "Age")
```

I've added annotation to the output in Figure 6.2 to make it easier for you to understand the contents of a boxplot. The dark horizontal line in the box is the median, the box itself represents the interquartile range (middle fifty percent,) and the two end-caps usually represent the lowest and highest values. In cases where there are extreme outliers, they will be represented with a small circle outside the upper and lower limits.[1] Similar to what we saw in the histogram, the boxplot shows that the middle 50% of outcomes is situated fairly close to the middle of the range, indicating a low level of skewness.

Now, let's look at boxplots and associated statistics using some of the variables from the `states20` dataset that we looked at in Chapter 5: percentage of the state population who are foreign-born and the number of abortion restrictions in state law. Recall from Chapter 5 the skewness statistics for percentage foreign born and state abortion restrictions were 1.20 and –.34, respectively. First, you can use the `summary` command to get the relevant statistics, starting with percentage of foreign born.

FIGURE 6.2 ■ Annotated Boxplot for Respondent Age

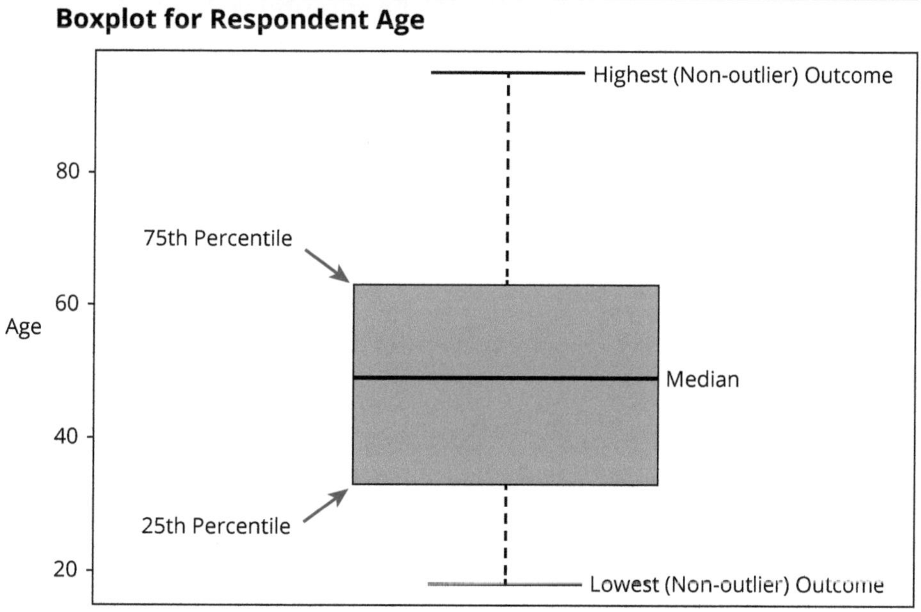


```
#Get summary statistics
summary(states20$fb)

   Min.  1st Qu.  Median    Mean  3rd Qu.    Max.
  1.400    3.025   4.950   7.042    9.850  21.700
```

The range for this variable is from 1.4% of the state population (West Virginia) to 21.7% (California), for a difference of 20.3, and the interquartile range is from 3.025% to 9.85%, for a difference of 6.825. The elements of the boxplot—the interquartile range, the median, and minimum and maximum values—can help illustrate the level of skewness in the data. From the summary statistics, we know that 50% of the outcomes have values between 3.025 and 9.85, that 75% of the observations have values of 9.85 (the high end of the IRQ) or lower, and that the full range of the variable runs from 1.4 to 21.7. This means that most of the observations are clustered at the low end of the scale, which you expect to find when data are positively skewed. The fact that the midpoint of the data (the median) is 4.95, less than a quarter of the highest value, also indicates a clustering at the low end of the scale.

The boxplot provides a visual representation of all of these things.

```
#Boxplot for % foreign born
boxplot(states20$fb, ylab = "% Foreign-born", main = "Boxplot of % Foreign-born")
```

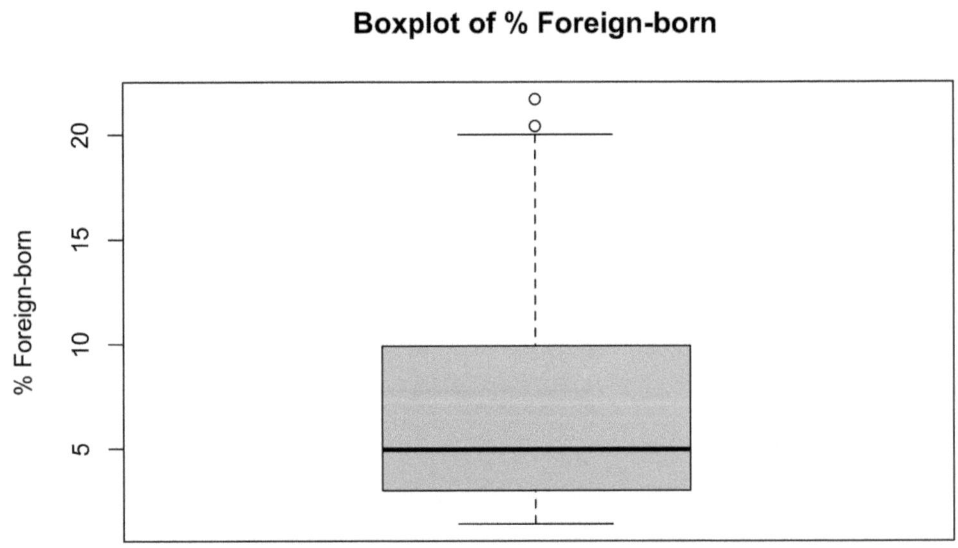

Here, you can see that the "box" is located at the low end of the range and that there are a couple of extreme outliers (the circles) at the high end. This is what a positively skewed variable looks like in a boxplot. Bear in mind that the level of skewness for this variable is not extreme (1.2), but you can still detect it fairly easily in the boxplot. Contrast this to the boxplot for `cces20$age` discussed earlier, which offers no strong hint of skewness.

Sometimes, it is easier to detect skewness in a boxplot by flipping the plot on its side, so the view is more similar to what you get from a histogram or a density plot. I think this is pretty clear in the horizontal boxplot for `states20$fb`.

```
#Horizontal boxplot
boxplot(states20$fb, xlab = "% Foreign-born",
        main = "Horizontal Boxplot of % Foreign-born",
        horizontal = T) #Horizontal orientation
```

Let's look at the same information for `states20$abortion_laws`. We already know from Chapter 5 that there is a modest level of negative skew in this variable, so it might be a bit harder to spot it in the boxplot.

```
#Boxplot for abortion restriction laws
boxplot(states20$abortion_laws,
        xlab = "Number of Abortion Restrictions",
        main = "Boxplot of State Abortion Restrictions",
        horizontal = T)
```

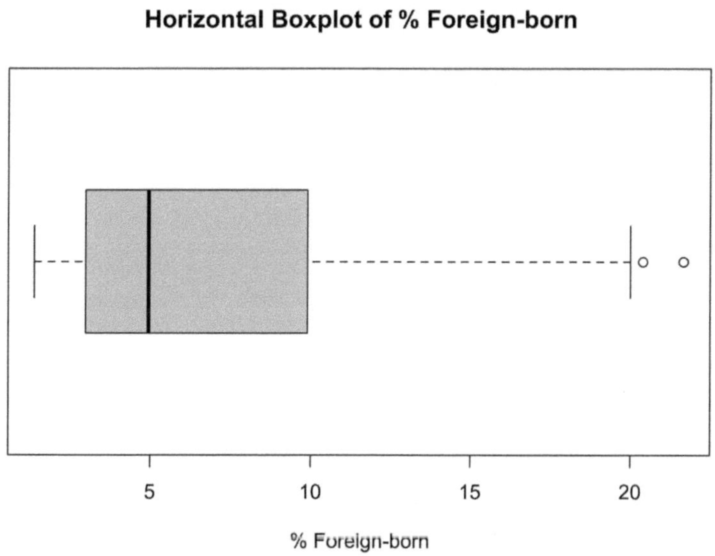

The pattern of skewness is not quite as clear as it was in the example using the foreign-born population, but you can pick up hints of a modest level of negative skewness based on the position of the "box" along with the location of the median, relative to the range of outcomes.

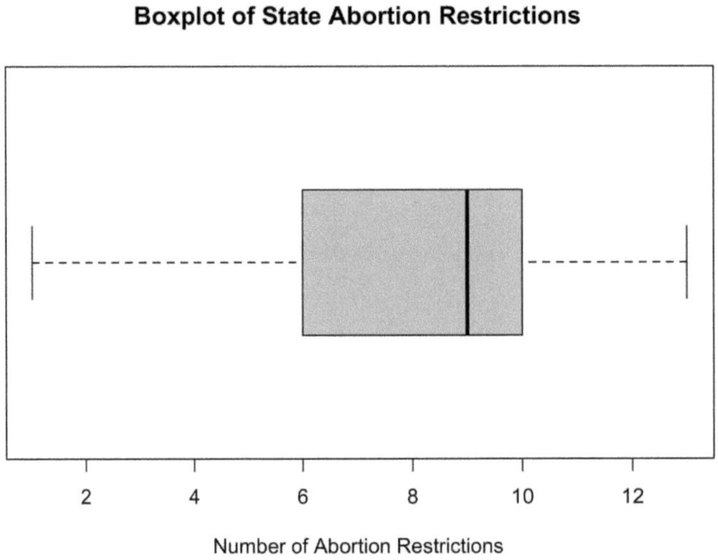

The IQR With Ordered Variables

The interquartile range can also be used for ordinal variables, with limits. For instance, for the ANES question on preferences for spending on the poor (`anes20$V201320x`), we can determine the IQR from the cumulative relative frequency.

```
Freq(anes20$V201320x)

                      Level    freq     perc   cumfreq   cumperc
1        1. Increased a lot   2'560    31.1%     2'560     31.1%
2   2. Increased a little   1'617    19.7%     4'177     50.8%
3          3. Kept the same   3'213    39.1%     7'390     89.8%
4   4. Decreased a little     446     5.4%     7'836     95.3%
5      5. Decreasaed a lot     389     4.7%     8'225    100.0%
```

For this variable, the 25th percentile is in the first category ("Increased a lot") and the 75 percentile is in the third category ("Kept the same"). The language for ordinal variables is a bit different and not quantitative as when using numeric data. For instance, in this case, it is not appropriate to say the interquartile range is 2 (from Categories 1 to 3), as the concept of numeric difference doesn't work well here. Instead, it is more appropriate to say the interquartile range is from "increased a lot" to "kept the same," or that the middle 50% hold opinions ranging from "increased a lot" to "kept the same."

DISPERSION AROUND THE MEAN

If we go back to the three distributions presented earlier in Figure 6.1, it should be clear that what differentiates the plots is not the range of outcomes but the amount of variation or dispersion *around the means*. The range and interquartile range have their uses, but they do not measure dispersion around the mean. Instead, we need a statistic that summarizes the typical deviation from the mean. This is not possible for nominal and ordinal variables, but it is an essential concept for numeric variables

Average Deviation From the Mean

Conceptually, what we are interested in is the average deviation from the mean. However, as you are no doubt saying to yourself right now, taking the average deviation from the mean is a terrible idea, as this statistic is always equal to zero.

$$\text{Average Deviation} = \frac{\sum_{i=1}^{n} x_i - \bar{x}}{n} = 0$$

The sum of the positive deviations from the mean will always be equal to the sum of the negative deviations from the mean, and the average deviation is zero. So, in practice, this is not a useful statistic, even though, conceptually, the typical deviation from the mean is what we want to measure. What we need to do, then, is somehow treat the negative values as if they were positive, so the sum of the deviations does not always equal 0.

Mean Absolute Deviation

One solution is to express deviations from the mean as absolute values. So, a deviation of –2 (meaning two units less than the mean) would be treated the same as a deviation of +2 (meaning two above the mean). Again, conceptually, this is what we need, the typical deviation from the mean but without offsetting positive and negative deviations.

$$\text{M.A.D} = \frac{\sum_{i=1}^{n} |x_i - \bar{x}|}{n}$$

Let's calculate this using the percentage foreign-born (`states20$fb`).

```
#Calculate absolute deviations
absdev = abs(states20$fb-mean(states20$fb))

#Sum deviations and divide by n (50)
M_A_D <- sum(absdev)/50
```

```
#Display M.A.D
M_A_D
[1] 4.26744
```

We can get the same result using the `MeanAD` function from the `DescTools` package.

```
#Mean absolute deviation from R
MeanAD(states20$fb, center = mean)
[1] 4.26744
```

This result shows that, on average, the observations for this variable are within 4.27 units (percentage points, in this case) of the mean.

This is a really nice statistic. It is relatively simple and easy to understand, and it is a direct measure of the typical deviation from the mean. However, there is one drawback that has sidelined the mean absolute deviation in favor of other measures of dispersion. One of the important functions of a mean-based measure of dispersion is to be able to use it as a basis for other statistics, and certain statistical properties associated with using absolute values make the mean absolute deviation unsuitable for those purposes. Still, this is a nice, intuitively clear statistic.

Variance

An alternative to using absolute deviations is to square the deviations from the mean to handle the negative values in a similar way. By squaring the deviations from the mean, all deviations are expressed as positive values. For instance, if you do this for deviations of –2 and +2, both are expressed as +4, the square of their values. These squared deviations are the basis for calculating the variance:

$$\text{Variance}\,(S^2) = \frac{\sum_{i=1}^{n}(x_i - \overline{x})^2}{n-1}$$

This is a very useful measure of dispersion and forms the foundation for many other statistics, including inferential statistics and measures of association. Let's review how to calculate it, using the percentage foreign-born (`states20$fb`).

Create squared deviations.

```
#Express each state outcome as a deviation from the mean outcome
fb_dev <- states20$fb-mean(states20$fb)

#Square those deviations
fb_dev.sq = fb_dev^2
```

Sum the squared deviations and divide by $n - 1$ (49), and you get the following.

```
#Calculate variance
sum(fb_dev.sq)/49
[1] 29.18004
```

Of course, you don't have to do the calculations on your own. You could just ask R to give you the variance of `states20$fb`.

```
#Use 'var' function to get variance
var(states20$fb)
[1] 29.18004
```

One difficulty with the variance, at least from the perspective of interpretation, is that because it is expressed in terms of squared deviations, it can sometimes be hard to connect the number back to the original scale. As important as the variance is as a statistic, it comes up a bit short as an intuitive descriptive device for conveying to people how much dispersion there is around the mean. For instance, the resulting number (29.18) is greater than the range of outcomes (20.3) for this variable. This makes it difficult to relate the variance to the actual data, especially if it is supposed to represent the "typical" deviation from the mean. This brings us to the most commonly used measure of dispersion for numeric variables.

Standard Deviation

The standard deviation is the square root of the variance. Its important contribution to consumers of descriptive statistical information is that it returns the variance to the original scale of the variable.

$$S = \sqrt{\frac{\sum_{i=1}^{n}(x_i - \overline{x})^2}{n - 1}}$$

In the current example,

$$S = \sqrt{29.18004} = 5.40.$$

Or get it from R.

```
# Use 'sd' function to get the standard deviation
sd(states20$fb)
[1] 5.401855
```

Note that the number 5.40 makes a lot more sense as a measure of typical deviation from the mean for this variable than 29.18 does. Although this is not exactly the same as the average deviation from the mean, it should be thought of as the "typical" deviation from the mean. You may have noticed that the standard deviation is a bit larger than the mean absolute deviation. This will generally be the case because squaring the deviations assigns a bit more weight to relatively high or low values and also because the denominator for the standard deviation is $n - 1$ instead of n, as it is for the mean absolute deviation.

Importance of Scale

Although taking the square root of the variance returns value to the scale of the variable, there is still a wee problem with interpretation. Is a standard deviation of 5.40 a lot of dispersion, or is it a relatively small amount of dispersion? In isolation, the number 5.40 does not have a lot of meaning. Instead, we need to have some standard of comparison. The question should be, "5.40, compared to what?" It is hard to interpret the magnitude of the standard deviation on its own, and it's always risky to make a comparison of standard deviations across different variables. This is because the standard deviation reflects two things: the amount of dispersion around the mean and the scale of the variable.

The standard deviation just reported comes from a variable that ranges from 1.4 to 21.7, and that scale profoundly affects its value. To illustrate the importance of scale, let's suppose that instead of measuring the foreign-born population as a *percentage* of the state population, we measure it as a *proportion* of the state population.

```
#Express foreign-born as a proportion
states20$fb.prop <- states20$fb/100
```

Now, if we get the standard deviation of `states20$fb.prop`, we see that it is 100 times smaller than the standard deviation for `states20$fb`.

```
#Get standard deviation of proportion foreign born
sd(states20$fb.prop)
[1] 0.05401855
```

Does that mean that there is less dispersion around the mean for `fb.prop` than for `fb`? No. Relative to their scales, both variables have the same amount of dispersion around their means. The difference is in the scales, nothing else.

So, if we are interested in getting a better intuitive sense of the magnitude of a single standard deviation value, we need to adjust the standard deviation so it is expressed *relative* to the scale of the variable. One simple but useful statistic for this is the **coefficient of variation**:

$$CV = \frac{S}{\bar{x}}$$

Here, we express the standard deviation relative to the mean of the variable. Values less than 1 indicate relatively low levels of variation; values greater than 1 indicate high levels of variation.

For states20$fb, the coefficient of variation is as follows.

```
CoefVar(states20$fb)
[1] 0.767091
```

This tells us that the outcomes for the percentage foreign-born in the states are relatively concentrated around the mean. One important caveat regarding the coefficient of variation is that it should only be used with variables that have all positive values.

Don't Make Bad Comparisons

The fact that the size of standard deviation, in part, reflects the scale of the variable also means that we should not make comparisons of it and other measures of dispersion across variables measured on different scales. To understand the issues related to comparing measures of dispersion a bit better, let's look at side-by-side distributions for three different variables: foreign-born as a percentage of the state population, Biden's share of the two-party vote in the states, and the number of abortion restrictions in the states.

```
#Three side-by-side boxplots, with labels for the variables
boxplot(states20$fb, states20$d2pty20, states20$abortion_laws,
        main = "A Truly Bad Comparison of Variables",
        names = c("Foreign-born", "Biden %", "Abortion Laws"))
```

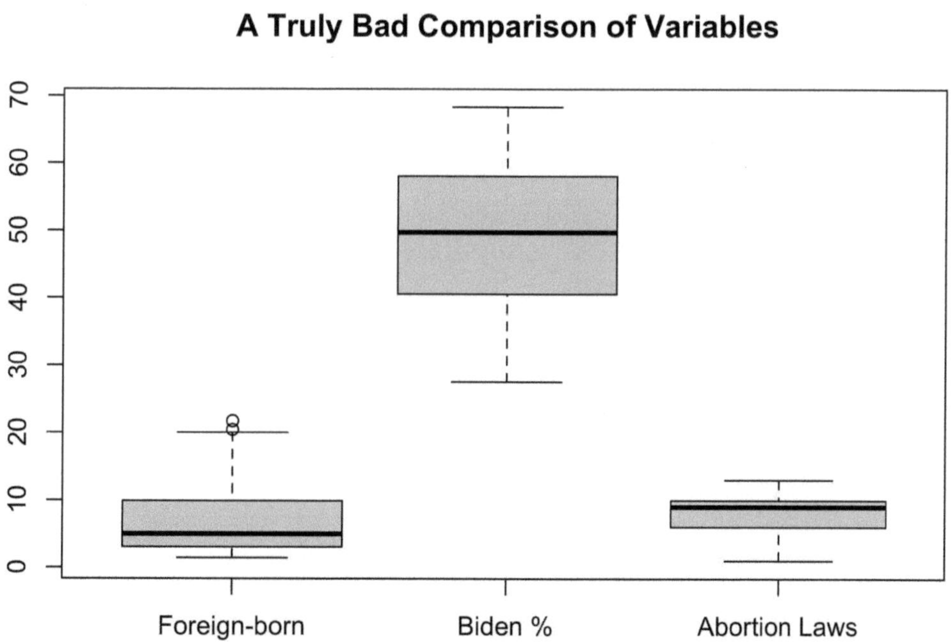

Which of these would you say has the least amount of variation, just based on "eyeballing" the plot? It looks like there is hardly any variation in abortion laws, a bit more variation in the foreign-born percentage of the population, and the greatest variation in Biden's percentage of the two-party vote.

The most important point to take away from this graph is that **it is a terrible graph**! You should never put three such disparate variables, measuring such different outcomes and using different scales, in the same distribution graph. It's just not a fair comparison, due primarily to differences in scale. Of course, that's the point of the graph.

You run into the same problem using the standard deviation as the basis of comparison.

```
#get standard deviations for all three variables
sd(states20$fb)
[1] 5.401855

sd(states20$d2pty20)
[1] 10.62935

sd(states20$abortion_laws)
[1] 2.946045
```

Here we get the same impression regarding relative levels of dispersion across these variables, but this is *still not a fair comparison* because one of the primary determinants of the standard deviation is the scale of the variable. The larger the scale, the larger the variance and standard deviation are likely to be.

The coefficient of variation is a useful statistic, precisely for this reason. Following, we see that when considering variation *relative to the scale* of the variables, using the coefficient of variation, the impression of these three variables changes.

```
#Get coefficient of variation for all three variables
CoefVar(states20$fb)
[1] 0.767091

CoefVar(states20$d2pty20)
[1] 0.2177949

CoefVar(states20$abortion_laws)
[1] 0.3738636
```

These results paint a different picture of the level of dispersion in these three variables: The percentage foreign-born exhibits the most variation, relative to its scale, followed by abortion laws and then by Biden's vote share. The differences in the boxplot were due largely to differences in the scale of the variables. Once that scale is taken into account by the coefficient of variation, the impression of dispersion levels changes a lot.

DICHOTOMOUS VARIABLES

It is also possible to think of variation in dichotomous variables. Consider, for example, the numeric indicator variable we created and used in Chapters 4 and 5 measuring whether respondents in the ANES 2020 survey supported (1) or opposed (0) adoption by gay and lesbian couples. Since all observations are scored 0 or 1, it might seem strange to think about variation around the mean in this variable. Still, because we are working with numeric data, it is possible to calculate the variance and standard deviation for these types of variables.

First, let's reconstruct and examine this variable.

```
anes20$lgbtq4
        Frequency  Percent  Valid Percent
0            1598   19.300           19.6
1            6554   79.155           80.4
NA's          128    1.546
Total        8280  100.000          100.0
```

About 19.6% in 0 (no) and 80.4% in 1 (yes). Does that seem like a lot of variation? It's really hard to tell without understanding what a dichotomous variable with a lot of variation would look like.

First, think about what this variable would look like if there were no variation in its outcomes. All of the observations (100%) would be in one category. Okay, so what would it look like if there was maximum variation? Half the observations would be in 0 and half in 1; it would be 50/50. The distribution for anes20$lgbtq4 seems closer to no variation than to maximum variation.

We use a different formula to calculate the variance in dichotomous variables:

$$S^2 = p(1 - p),$$

where p is the proportion in category 1.

In this case, S^2 = (.196*.804) = .1576.

The standard deviation, of course, is the square root of the variance:

$$S = \sqrt{p(1 - p)} = \sqrt{.1576} = .397$$

Let's check our work in R.

```
#Get the standard deviation for 'lgbtq4'
sd(anes20$lgbtq4, na.rm = T)
[1] 0.3970124
```

What does this mean? How should it be interpreted? With dichotomous variables like this, the interpretation is a bit harder to grasp than with a continuous numeric variable. One way to think about it is that if this variable exhibited maximum variation (50% in one category, 50%

in the other), the value of the standard deviation would be $\sqrt{(p(1-p)} = .50$. To be honest, though, the real value in calculating the standard deviation for a dichotomous variable is not in the interpretation of the value but in using it to calculate other statistics in later chapters.

DISPERSION IN CATEGORICAL VARIABLES?

Is it useful to think about variation in multi-category nominal-level variables for which the values have little or no quantitative meaning? Certainly, it does not make sense to think about calculating a mean for categorical variables, so it doesn't make sense to think in terms of dispersion around the mean. However, it does make sense to think of variables in terms of how *diverse* the outcomes are, whether they tend to cluster in one or two categories (concentrated) or spread out more evenly across categories (diverse or dispersed), similar to the logic used with evaluating dichotomous variables.

Let's take a look at a five-category nominal variable from the 2020 ANES survey measuring the race and ethnicity of the survey respondents (`anes20$raceth` is a recoded version of `anes20V201549x`). The response categories are white (non-Hispanic), Black (non-Hispanic), Hispanic, Asian and Pacific Islander (non-Hispanic), and a category that represents other identities.

```
#Create five-category race/ehtnicity variable
anes20$raceth.5 <- anes20$V201549x
levels(anes20$raceth.5) <- c("White(NH)", "Black(NH)", "Hispanic",
                             "API(NH)", "Other", "Other")

freq(anes20$raceth.5, plot = F)

PRE: SUMMARY: R self-identified race/ethnicity
          Frequency  Percent  Valid Percent
White(NH)      5963   72.017         72.915
Black(NH)       726    8.768          8.877
Hispanic        762    9.203          9.318
API(NH)         284    3.430          3.473
Other           443    5.350          5.417
NA's            102    1.232
Total          8280  100.000        100.000
```

What we need is a measure that tells how dispersed the responses are *relative to a standard that represents maximum dispersion*. Again, we could ask what this distribution would look like if there was no variation at all in responses. It would have 100% in a single category. What would maximum variation look like? Since there are five valid response categories, the most diverse set of responses would be to have 20% in each category. This would mean that responses were spread out evenly across categories, as diverse as possible. The responses for this variable are concentrated in "White(NH)" (73%), indicating not a lot of variation.

The **index of qualitative variation** (IQV) can be used to calculate how close to the maximum level of variation a particular distribution is.

The formula for the IQV is

$$IQV = \frac{K}{K-1} * \left(1 - \sum_{k=1}^{n} p_k^2\right).$$

Where

K = number of categories,

k = specific categories, and

p = proportion in category k.

This formula is saying to sum up all of the squared category proportions, subtract that sum from 1 and multiply the result times the number of categories divided by the number of categories minus 1. This last part adjusts the main part of the formula to take into account the fact that it is harder to get to maximum diversity with fewer categories. There is not an easy-to-use R function for calculating the IQV, so let's do it the old-fashioned way. (Note: The proportions are taken from the valid percentages in the frequency table.)

$$\text{IQV} = \frac{5}{4} * (1 - (.72915^2 + .08877^2 + .09318^2 + .03473^2 + .05417^2)) = .56$$

We should interpret this as meaning that this variable is about 56% as diverse as it could be, compared to maximum diversity. In other words, not terribly diverse.

THE STANDARD DEVIATION AND THE NORMAL CURVE

One of the interesting uses of the standard deviation lies in its application to the normal curve (also known as the normal distribution, sometimes as the bell curve). The normal curve can be thought of as a density plot that represents the distribution of a theoretical variable that has several important characteristics:

- It is single-peaked
- The mean, median, and mode have the same value
- It is perfectly symmetrical and "bell" shaped
- Most observations are clustered near the mean
- Its tails extend infinitely

Before describing how the standard deviation is related to the normal distribution, I want to be clear about the difference between a **theoretical** distribution and an **empirical** distribution. A theoretical distribution is one that *would* exist if certain mathematical conditions were satisfied. It is an idealized distribution. An empirical distribution is based on the measurement of concrete, tangible characteristics of a variable that actually exists. The normal distribution is derived from a mathematical formula and has all of the characteristics just listed. In addition, another important characteristic for our purposes lies in its relationship to the standard deviation:

- 68.26% of the area under the curve lies within one standard deviation of the mean (if we think of the area under the curve as representing the outcomes of a variable, about 68% of all outcomes are within one standard deviation of the mean)
- 95.24% of the area under the curve lies within two standard deviations of the mean
- 99.72% of the area under the curve lies within three standard deviations of the mean

This relationship is presented in Figure 6.3, using the standard normal distribution.[2]

FIGURE 6.3 ■ The Standard Deviation and the Standard Normal Distribution

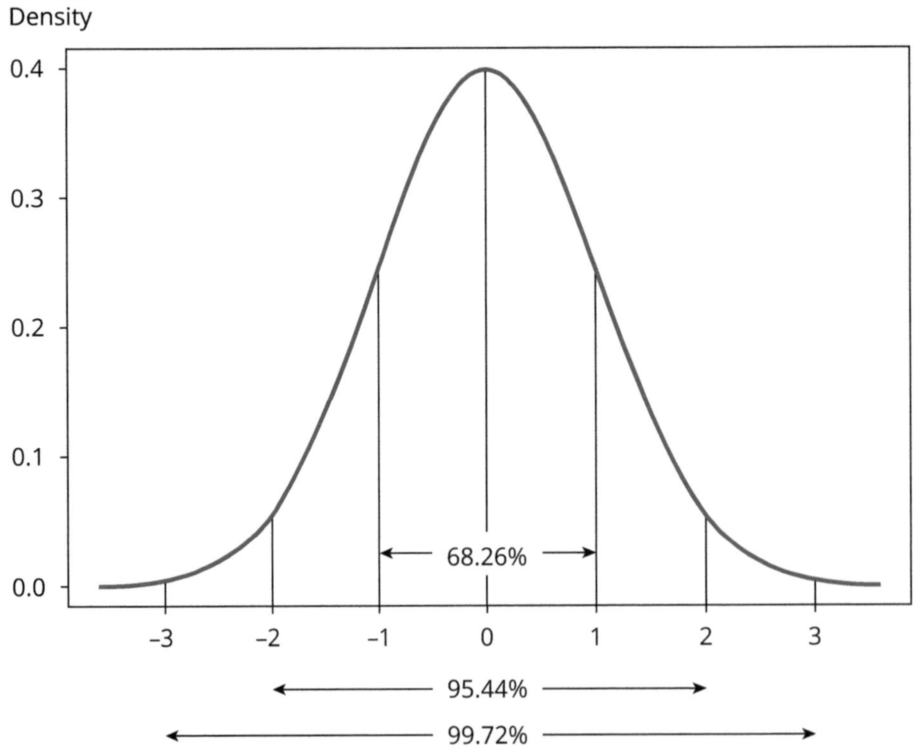

Note that these guidelines apply to areas above and below the mean together. So, for instance, when we say that 68.26% of the area under the curve falls within one standard deviation of the mean, we mean that 34.13% of the area is found between the mean and one standard deviation above the mean, and 34.13% of the area is found between the mean and one standard deviation below the mean.

Let's apply the guidelines in Figure 6.3 to some real data. Suppose that we know that one of the respondents from the `cces20` sample is 66 years old, and someone wants to know if that respondent is relatively old, given the sample. Of course, you know from earlier in the chapter that the sample does not include anyone under 18 years of age and that the mean is 48.39 years old, so you're not so sure that this respondent is really that old, relative to the rest of the sample.

One way to get a sense of how old they are, relative to the rest of the sample, would be to assume that age is normally distributed and then apply what we know about the standard deviation and the normal curve to this variable. To do this, we would need to know how many standard deviations above the mean the 66-year-old is. The result statistic is called a ***z*-score.** The *z*-scores transform the original (raw) values of a numeric variable into the number of standard deviations above or below the mean that those values are: *z*-scores are calculated as follows:

$$Z_i = \frac{x_i - \overline{x}}{S}$$

The values of any numeric variable can be transformed into *z*-scores. Knowing the *z*-score for a particular outcome on most variables gives you a sense of how typical it is. If your *z*-score on a midterm exam is .01, then you're very close to the average score. If your z-score is +2, then you did a great job, relative to everyone else. If your *z*-score is –1, then you struggled on the exam, relative to everyone else.

Let's transform the 66-year-old respondent's age into a *z*-score. First, we take the raw difference between the mean and the respondent's age, 66.

```
#Calculate deviation from the mean
(66-48.39)
[1] 17.61
```

So, the 66-year-old respondent is just about 18 years older than the average respondent. This seems like a lot, but thinking about this *relative* to the rest of the sample, it depends on how much variation there is in the age of respondents. For this particular sample, we can evaluate this with the standard deviation, which is 17.66.

```
#Get std dev of age
sd(cces20$age, na.rm = T)
[1] 17.65902
```

So, now we just need to transform the raw difference (17.61) into a *z*-score that expresses the difference in terms of how many standard deviations this 66-year-old respondent is above the mean age in this sample. Since the respondent is 17.61 years older than the average respondent and the standard deviation for this variable is 17.66, we know that the 66-year-old is close to one standard deviation above the mean.

```
#Express deviation from mean, relative to standard deviation
z <- (66-48.39)/17.66
z
[1] 0.9971687
```

Let's think back to the normal distribution in Figure 6.3 and assume that age is a normally distributed variable. We know that the 66-year-old respondent is about one standard deviation above the mean. What is the area under the curve to the left of +1 standard deviations? Since we know that 50% of the area under the curve (think of this as 50% of the observations of a normally distributed variable) lies to the left (below) of the mean and 34.13% lies between the mean and +1 standard deviation, we can also say that 84.13% (50 + 34.13) of the area under the curve lies to the left of +1 standard deviation above the mean. Therefore, we might expect that the 66-year-old respondent is somewhere in the neighborhood of the 84th percentile for age in this sample.

Of course, empirical variables such as age are unlikely to be normally distributed. Still, for most variables, if you know that an outcome is one standard deviation or more above the mean, you can be confident that, relative to that variable's distribution, the outcome is a fairly high value. Likewise, an outcome that is one standard deviation or more below the mean is relatively low. And, of course, an outcome that is two standard deviations above (below) the mean is very high (low), relative to the rest of the distribution.

I suggested that the relationship between the standard deviation and the normal curve could be applied to empirical variables to get a *general* sense of how typical different outcomes are. Consider how this works for the age variable we used above. While I estimated (with the *z*-score) that a 66-year-old in the sample would be in the 84th percentile, the actual distribution for age in the `cces20` sample shows that 82.1% of the sample is 66 years old or younger. I was able to find this number by creating a new variable that had two categories: one for those respondents less than or equal to 66 years old and one for those who are older than 66.

```
#Create a new variable for age (18-66 and 67 + )
cces20$age_66 <- factor(cut2(cces20$age, c(67)))
levels(cces20$age_66) <- c("18-66", "67 and older")

freq(cces20$age_66, plot = F)

cces20$age_66
```

	Frequency	Percent
18-66	50050	82.05
67 and older	10950	17.95
Total	61000	100.00

One reason the estimate from this empirical sample is so close to the expectations based on the theoretical normal distribution is that the distribution for age does not deviate drastically from normal (see following histogram, Figure 6.4). The solid curved line in Figure 6.4 is the normal distribution, and the vertical line identifies the cutoff point for 66-years-old.

Really Important Caveat

Just to be clear, the guidelines for estimating *precise areas* under the curve, or percentiles for specific outcomes, apply strictly **only for normal distributions**. That is how we will apply them in later chapters. Still, as a rule of thumb, the guidelines give a sense of how *relatively* high or low outcomes are for empirical variables.

FIGURE 6.4 ■ Comparing the Empirical Histogram for Age With the Normal Curve

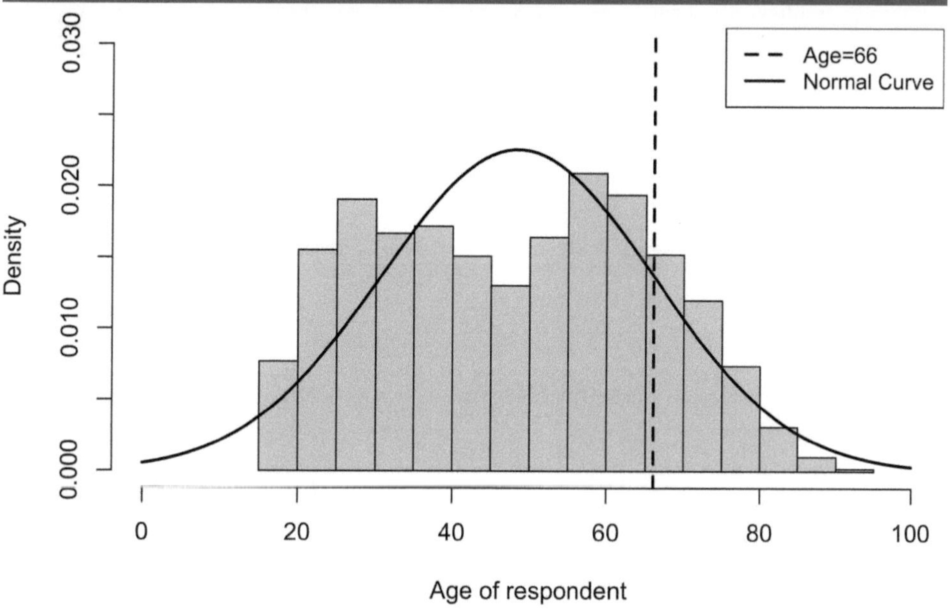

CALCULATING AREA UNDER A NORMAL CURVE

We can use a couple of different methods to calculate the area under a normal curve above or below any given z-score. We can go "old school" and look at a standard z-distribution table, like the one shown in Table 6.1.3 In this table, the left side column represents different *z*-score values, to one place to the right of the decimal point, and the column headers are the second *z*-score digit to the right of the decimal. The numbers inside the table represent the area under the curve to the left of any given *z*-score (given by the intersection of the row and column headers).

So, for instance, if you look at the intersection of 1.0 on the side and 0.00 at the top ($z = 1.0$), you see the value of .8413. This means, as we already know, that approximately 84% of the area under the curve is below a *z*-score of 1.0. Of course, this also means that approximately 16% of the area lies above $z = 1.0$ and approximately 34% lies between the mean and a *z*-score of 1.0. Likewise, we can look at the intersection of 1.9 (row) and .06 (column) to find the area under the curve for *z*-score of 1.96. The area to the left of $z = 1.96$ is 95.7%, and the area to the right of $z = 1.96$ is 2.5% of the total area under the curve. Let's walk through this once more, using $z = 1.44$. What is the area to the left, to the right, and between the mean and $z = 1.44$:

- To the left, .9251 (found at the intersection of 1.4 (row) and .04 (column))
- To the right, (1 – .9251) = .0749
- Between the mean and 1.44, (.9252 – .5) = .4251

If these results confuse you, make sure to check in with your professor.

So, that's the "old-School" way to find areas under the normal distribution. I think it is helpful to use a table like this when you first learn this material, but there is an easier and more precise way to do this using R. You can use the `pnorm` function to find areas under the curve for any given *z*-score. By default, this function displays the area under the curve *to the left* of specified *z*-scores. Let's check our work for $z = 1.44$.

```
#Get area under the curve to the left of z = 1.44
pnorm(1.44)
[1] 0.9250663
```

So far, so good. R can also give you the area to the right of a *z*-score. Simply add `lower.tail = F` to the pnorm command.

```
#Get area under the curve for to the right of z = 1.44
#Add "lower.tail = F"
pnorm(1.44, lower.tail = F)
[1] 0.0749337
```

And to get the area between the mean and $z = 1.44$

```
pnorm(1.44)-.5
[1] 0.4250663
```

TABLE 6.1 ■ Areas Under the Normal Curve to the Left of Positive z-scores

	Second Decimal for z-score									
z	0	0.01	0.02	0.03	0.04	0.05	0.06	0.07	0.08	0.09
0.0	0.5000	0.5040	0.5080	0.5120	0.5160	0.5119	0.5239	0.5279	0.5319	0.5359
0.1	0.5398	0.5438	0.5478	0.5517	0.5557	0.5596	0.05636	0.5675	0.5714	0.5753
0.2	0.5793	0.5832	0.5871	0.5910	0.5948	0.5987	0.6026	0.6064	0.6103	0.6141
0.3	0.6179	0.6217	0.6255	0.6293	0.6331	0.6368	0.6406	0.6443	0.6480	0.6517
0.4	0.6554	0.6591	0.6628	0.6664	0.6700	0.6736	0.6772	0.6808	0.6844	0.6879
0.5	0.6915	0.6950	0.6985	0.7019	0.7054	0.7088	0.7123	0.7157	0.7190	0.7224
0.6	0.7257	0.7291	0.7324	0.7357	0.7389	0.7422	0.7454	0.7486	0.7517	0.7549
0.7	0.7580	0.7611	0.7642	0.7673	0.7704	0.7734	0.7764	0.7794	0.7823	0.7852
0.8	0.7881	0.7910	0.7939	0.79867	0.7995	0.8023	0.8051	0.8078	0.8106	0.8133
0.9	0.8159	8186	0.8212	0.8238	0.8264	0.8289	0.8315	0.8340	0.8365	0.8389
1.0	**0.8413**	0.8438	0.8461	0.8485	0.8508	0.8531	0.8554	0.8577	0.8599	0.8621
1.1	0.8643	0.8665	0.8686	0.8708	0.8729	0.8749	0.8770	0.8790	0.8810	0.8830
1.2	0.8849	0.8869	0.8888	0.8907	0.8925	0.8944	0.8962	0.8980	0.8997	0.9015
1.3	0.9032	0.9049	0.9066	0.9082	0.9099	0.9115	0.9131	0.9147	0.9162	0.9177
1.4	0.9192	0.9207	0.9222	0.9236	**0.9251**	0.9265	0.9279	0.9292	0.9306	0.9319
1.5	0.9332	0.9345	0.9357	0.9370	0.9382	0.9394	0.9406	0.9418	0.9429	0.9441
1.6	0.9452	0.9463	0.9474	0.9484	0.9495	0.9505	0.9515	0.9525	0.9535	0.9545

(Continued)

TABLE 6.1 ■ Areas Under the Normal Curve to the Left of Positive *z*-scores (*Continued*)

	Second Decimal for *z*-score									
z	0	0.01	0.02	0.03	0.04	0.05	0.06	0.07	0.08	0.09
1.7	0.9554	0.9564	0.9573	0.9582	0.9591	0.9599	0.9608	0.9616	0.9625	0.9633
1.8	0.9641	0.9649	0.9656	0.9664	0.9671	0.9678	0.9686	0.9693	0.9699	0.9706
1.9	0.9713	0.9719	0.9726	0.9732	0.9738	0.9744	**0.9750**	0.9756	0.9761	0.9767
2.0	0.9772	0.9778	0.9783	0.9788	0.9793	0.9798	0.9803	0.9808	0.9812	0.9817
2.1	0.9821	0.9826	0.9830	0.9834	0.9838	0.9842	0.9846	0.9850	0.9854	0.9857
2.2	0.9861	0.9864	0.9868	0.9871	0.9875	0.9878	0.9881	0.9884	0.9887	0.9890
2.3	0.9893	0.9896	0.9898	0.9901	0.9904	0.9906	0.9909	0.9911	0.9913	0.9916
2.4	0.9918	0.9920	0.9922	0.9925	0.9927	0.9929	0.9931	0.9932	0.9934	0.9936
2.5	0.9938	0.9940	0.9941	0.9943	0.9945	0.9946	0.9948	0.9949	0.9951	0.9952
2.6	0.9953	0.9955	0.9956	0.9957	0.9959	0.9960	0.9961	0.9962	0.9963	0.9964
2.7	0.9965	0.9966	0.9967	0.9968	0.9969	0.9970	0.9971	0.9972	0.9973	0.9974
2.8	0.9974	0.9975	0.9976	0.9977	0.9977	0.9978	0.9979	0.9979	0.9980	0.9981
2.9	0.9981	0.9982	0.9982	0.9983	0.9984	0.9984	0.9985	0.9985	0.9986	0.9986
3.0	0.9987	0.9987	0.9987	0.9988	0.9988	0.9989	0.9989	0.9989	0.9990	0.9990

Note: Code for creating this table was adapted from *Generating Your Own Normal Distribution Table* by Arthur Charpentier, 2013, Freakonometrics, https://freakonometrics.hypotheses.org/9404

ONE LAST THING

Thus far, you have used specific functions, such as `sd()`, `var`, `IQR`, and `range`, to get individual statistics or `summary` to generate some of these statistics together. There is another important function that provides all of these statistics, plus many more, which can save you the bother of running several different commands to get the information you want. `Desc` is part of the `DescTools` package and provides summaries of the underlying statistical properties of variables. If you drop `plot = F` from the command (shown in the following section), you also get a histogram, boxplot, and density plot, though they are not quite "publication ready." Let's check this out using `states20$fb`.

```
#Use 'Desc' function to get several descriptive statistics
Desc(states20$fb, plot = F)

------------------------------------------------------------------

states20$fb (numeric)

  length       n     NAs  unique      0s    mean  meanCI'
      50      50       0      45       0   7.042    5.507
          100.0%    0.0%            0.0%            8.577

     .05     .10     .25  median     .75     .90      .95
   2.145   2.290   3.025   4.950   9.850  14.990   19.370

   range      sd   vcoef     mad     IQR    skew     kurt
  20.300   5.402   0.767   3.262   6.825   1.203    0.449

lowest : 1.4, 1.8, 2.1, 2.2 (2), 2.3 (2)
highest: 15.8, 18.6, 20.0, 20.4, 21.7

' 95%-CI (classic)
```

`Desc` provides you with almost all of the measures of central tendency and dispersion you've learned about so far, plus a few others. A couple of quick notes: The lower limit of the IQR is displayed under .25 (25th percentile), the upper limit under .75 (75th percentile), and the width is reported under IQR. Also, "mad" reported here is not the mean absolute deviation from the mean (it is the median absolute deviation).

NEXT STEPS

Measures of dispersion—in particular the variance and standard deviation—will make guest appearances in almost all of the remaining chapters of this book. As I just suggested, the importance of these statistics goes well beyond describing the distribution of variables.

In the next chapter, you will see how the standard deviation and normal distribution can be used to estimate probabilities. Following that, we examine the crucial role the standard deviation plays in statistical inference. Several chapters later, we will also see that the variances of independent and dependent variables are used to form the basis for correlation and regression analyses.

The next two chapters address issues related to probability, sampling, and statistical inference. You will find this material a bit more abstract than what we've done so far, and I think you will also find it very interesting. The content of these chapters is very important to understanding concepts related to something you may have heard of before, even if just informally, "statistical significance." For many of you, this will be strange, new material: Embrace it and enjoy the journey of discovery!

EXERCISES

Concepts and Calculations

As usual, when making calculations, show the process you used.

1. Use the information provided about three hypothetical variables to determine if they appear to be skewed in one direction or the other. Explain your conclusions.

Variable	Range	IQR	Median
Variable 1	1 to 1200	500 to 800	650
Variable 2	30 to 90	40 to 50	45
Variable 3	3 to 35	20 to 30	23

2. The list of voter turnout rates in twelve Wisconsin counties that you used for an exercise in Chapter 5 are reproduced next, with two empty columns added: one for the deviation of each observation from the mean and another for the square of that deviation. Fill in this information and complete the following:
   - Calculate the mean absolute deviation and the standard deviation. Interpret these statistics. Which one do you find easiest to understand? Why?
   - Next, calculate the coefficient of variation for this variable. How do you interpret this statistic?

Wisconsin County	% Turnout	$X_i - \bar{X}$	$(X_i - \bar{X})^2$
Clark	63		
Dane	87		
Forrest	71		
Grant	63		

Wisconsin County	% Turnout	$X_i - \bar{X}$	$(X_i - \bar{X})^2$
Iowa	78		
Iron	82		
Jackson	65		
Kenosha	71		
Marinette	71		
Milwaukee	68		
Portage	74		
Taylor	70		

3. Across the fifty states, the average cumulative number of COVID-19 cases per 10,000 population in August of 2021 was 1161, and the standard deviation was 274. The cases per 10,000 were 888 in Virginia and 1427 in South Carolina. Using what you know about the normal distribution and assuming this variable follows a normal distribution, what percentage of states do you estimate had values equal to or less than Virginia's, and what percentage do you expect to have had values equal to or greater than South Carolina's? Explain how you reached your conclusions.

4. The average voter turnout rate across the fifty states in the 2020 presidential election was 67.4% of eligible voters, and the standard deviation was 5.8. Calculate the *z*-scores for the following list of states and identify which state is most extreme and which state is least extreme.

State	Turnout (%)	*z*-score
OK	54.8	
WV	57.0	
SC	64.0	
GA	67.7	
PA	70.7	
NJ	74.0	
MN	79.6	

5. This horizontal boxplot illustrates the distribution of per capita income across the states. Discuss the distribution, paying attention to the range, the interquartile range, and the median. Are there any signs of skewness here? Explain.

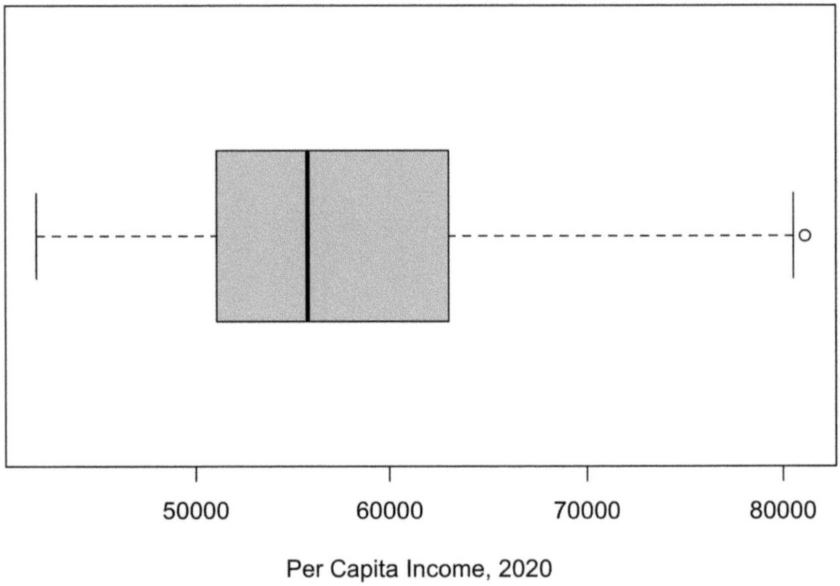

6. The survey of 300 students you used for one of the assignments in Chapter 3 included a question about the political leanings of students. Based on the responses provided, how would you describe the amount of variation in student political leanings. Calculate and use the index of qualitative variation to answer this question.

Political Leaning	Frequency	Percent
Far Left	35	11.67
Liberal	100	33.33
Moderate	75	25.0
Conservative	66	22.0
Far Right	24	8.0

R Problems

1. Using the `pnorm` function, estimate the area under the normal curve for each of the following:
   - Above $z = 1.8$
   - Below $z = -1.3$
   - Between $z = -1.3$ and $z = 1.8$

2. For the remaining problems, use `countries2` dataset. One important variable in the `countries2` dataset is `lifexp`, which measures life expectancy. Create a histogram, a boxplot, and a density plot, and describe the distribution of life expectancy across countries. Which of these graphing methods do you think is most useful for getting a

sense of how much variation there is in this variable? Why? What about skewness? What can you tell from these graphs? Be specific.

3. Use the results of the `Desc` command to describe the amount of variation in life expectancy, focusing on the range, interquartile range, and standard deviation. Make sure to provide interpretations of these statistics.

4. Now, suppose you want to compare the amount of variation in life expectancy to variation in gross domestic product (GDP) per capita (`countries2$gdp_pc`). What statistic should you use to make this comparison? Think about this before proceeding. Use `Desc` to get the appropriate statistic. Justify your choice and discuss the difference between the two variables.

5. Choose a different numeric variable from the `countries2` dataset that interests you, then create a histogram for that variable that includes vertical lines for the lower and upper limits of the interquartile range, add a legend for the vertical lines, and add a descriptive *x*-axis label. Describe what you see in the graph you created.

7 PROBABILITY

GET READY

This chapter provides a brief look at some of the important terms and definitions related to the concept of probability. To follow along, you should load the following datasets: `anes20.rda`, which you've worked with in previous chapters, and `grades.rda`, a dataset that summarizes final grades over several semesters from a course I teach on data analysis. You should also make sure to attach the libraries for the following R packages: `descr`, `DescTools`, and `Hmisc`.

PROBABILITY

You probably use the language of probability in your everyday life. Whether we are talking or thinking about the likelihood, odds, or chance that something will occur, you are using probabilistic language. At its core, *probability* is about whether something is likely or unlikely to happen.

Probabilities can be thought of as relative frequencies that express how often a given outcome (X) occurs, relative to the number of times it could occur:

$$P(X) = \frac{\text{Number } X \text{ outcomes}}{\text{Number possible } X \text{ outcomes}}$$

Probabilities range from 0 and 1, with zero meaning an event never happens and 1 meaning the event always happens. As you move from 0 to 1, the probability of an event occurring increases. When the probability value is equal to .50, there is an even chance that the event will occur. This connection between probabilities and everyday language is summarized below in Figure 7.1.[1] It is also common for people to use the language of percentages when discussing probabilities—for example, 0% to 100% range of possible outcomes. For instance, a .75 probability that something will occur might be referred to as a 75% probability that the event will occur.

Consider this example. Suppose we want to know the probability that basketball phenom and NBA Champion Giannis Antetokounmpo, of the Milwaukee Bucks, will make any given free throw attempt. We can use his performance from the 2020–2021 regular season as a guide. Giannis (if I may) had a total of 581 free throw attempts (# of possible outcomes) and made 398 (number of X outcomes). Using the preceding formula, we get the following:

$$P(success) = \frac{389}{581} = .685$$

FIGURE 7.1 ■ Substantive Meaning of Probability Values

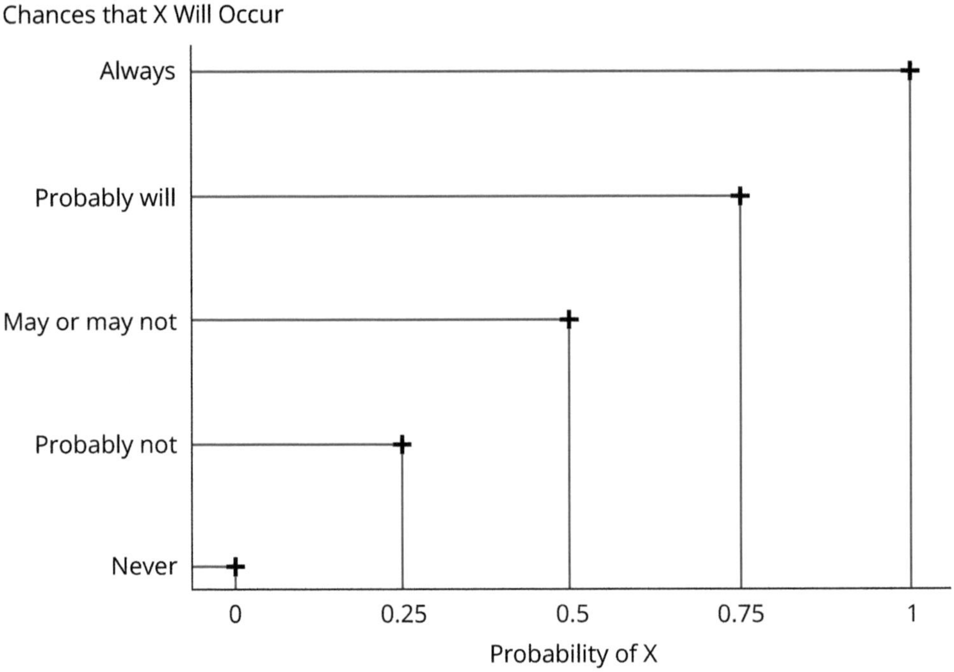

Note: Inspired by a figure on p. 74 of *Statistical Analysis* by S. K. Kachigan, 1986, Radius Press.

The probability of Giannis making any given free throw during the season was .685. This outcome is usually converted to a percentage and known as his free throw percentage (68.5%).

Is this a high probability or a low probability? Sometimes, the best way to evaluate probabilities is comparatively; in this case, that means comparing Giannis' outcome to others. If you compared Giannis to me, he would seem like a free throw superstar, but that is probably not the best standard for evaluating a professional athlete. Instead, you could also compare him to the league average for the 2020–2021 season (.778) or his own average in the previous year (.633). By making these comparisons, we are **contextualizing** the probability estimate, so we can better understand its meaning in the setting of professional basketball.

It is also important to realize that this probability estimate is based on hundreds of "trials" and does not rule out the possibility of "hot" or "cold" shooting streaks. For instance, Giannis made 4 of his 11 free throw attempts in Game 5 against the Phoenix Suns (.347) and was 17 of 19 (.897) in series-clinching Game 6. In those last two games, he was 21 of 30 (.70), just a wee bit higher than his regular season average.

THEORETICAL PROBABILITIES

It is useful to distinguish between *theoretical* and *empirical* probabilities. Theoretical probabilities can be determined, or solved, on logical grounds. There is no need to run experiments or gather data to estimate these probabilities. Sometimes, these are referred to as *a priori* (think of this as something like "based on assumptions") probabilities.

Consider the classic example of a coin toss. There are two sides to a coin, heads and tails, so if we assume the coin is fair (not designed to favor one side or the other), then the probability of getting heads on a coin flip is 1/2 or .5. The probability of coming up heads on the next toss is also .50 because the two tosses are independent, meaning that one toss does not affect subsequent tosses.

Likewise,

- We can estimate the probability of rolling a 4 on a fair six-sided die with sides numbered 1 through 6. Since 4 is one of six possibilities, the probability is 1/6, or .167.
- The same idea applies to estimating the probability of drawing an ace from a well-shuffled deck of cards. Since there are four aces out of the 52 cards, the probability is 4/52, or .077.
- The same for drawing a heart from a well-shuffled deck of cards. Since there are 13 cards for each suit (hearts, diamonds, clubs, and spades), the probability is 13/52, or .25.
- What about the probability of drawing an ace of hearts from a deck of cards? There is only one ace of hearts in the 52-card deck, so the probability is 1/52, or .019. Note that this is the probability of an ace (.077) multiplied times the probability of a heart (.25).

Sometimes, these probabilities may *appear* to be at odds with what we see in the real world, especially when we have a small sample size. Think about the flip of a coin. Theoretically, unless the coin is rigged, $P(H) = P(T) = .50$. So, if we flip the coin 10 times, should we expect 5 heads and 5 tails? If you've ever flipped a coin ten times in a row, you know that you don't always end up with five heads and five tails. But, in the *long run*, with many more than ten coin flips, the outcomes of coin tosses converge on $P(H) = .50$. I illustrate the idea of long run expected values below using small and large samples of coin tosses and rolls of a die.

Large and Small Sample Outcomes

The graph shown in Figure 7.2 illustrates the simulated[2] coin toss outcomes using different numbers of coin tosses, ranging from ten to two-thousand tosses. In the first four simulations (10, 20, 60, and 100 tosses), the results are not very close to expected outcome (50% heads), ranging from 30% to 58%. However, for the remaining ten simulations (150 to 2000 tosses), the results are either almost exactly on target or very slightly different from the expected outcome.

FIGURE 7.2 ■ Simulated Results From Large and Small Coin Toss Samples

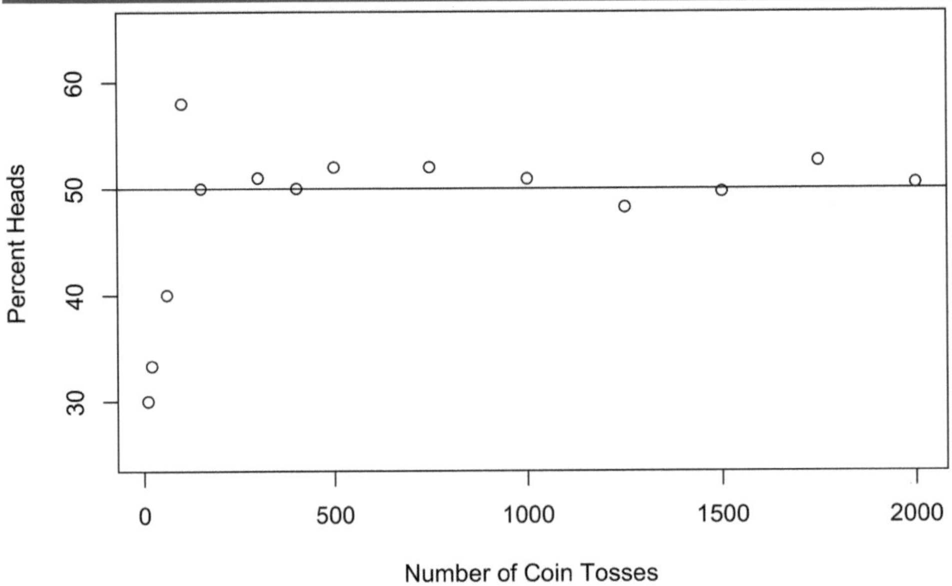

We can work through the same process for rolling a fair six-sided die, where the probability of each of the six outcomes is 1/6 = .167. The results in Figure 7.3 summarize the outcomes of rolling a six-sided die 100 times (on the left) and 2000 times (on the right), using a solid horizontal line to indicate the expected value (.167). Each of the six outcomes should occur approximately the same number of times. On the left side, based on just 100 rolls of the die, the outcomes deviate quite a bit from the expected outcomes, ranging from .08 of outcomes for the number 3, to .21 of outcomes for both numbers 2 and 4. This is not what we expect from a fair, six-sided die. But if we increase the number of rolls to 2000 (right side of Figure 7.3, we see much more consistency across each of the six numbers on the die, and all of the proportions are very close to .167, ranging from .161 for the number 1, to .175 for the number 4.

The coin toss and die rolling simulations are important demonstrations of the **law of large numbers**: *If you conduct multiple trials or experiments of a random event, the average outcome approaches the theoretical (expected) outcome as the number of trials increases and becomes large.*

This idea is very important and comes into play again in Chapter 8.

EMPIRICAL PROBABILITIES

Empirical probability estimates are based on *observations* of the relative occurrence of events in the past, as we discussed in the opening example. These probabilities are also sometimes referred to as *posterior* probabilities because they cannot be determined, or solved, on logical grounds; we need data from past experiences in order to estimate them.

FIGURE 7.3 ■ Simulated Results From Large and Small Samples of Die Rolls

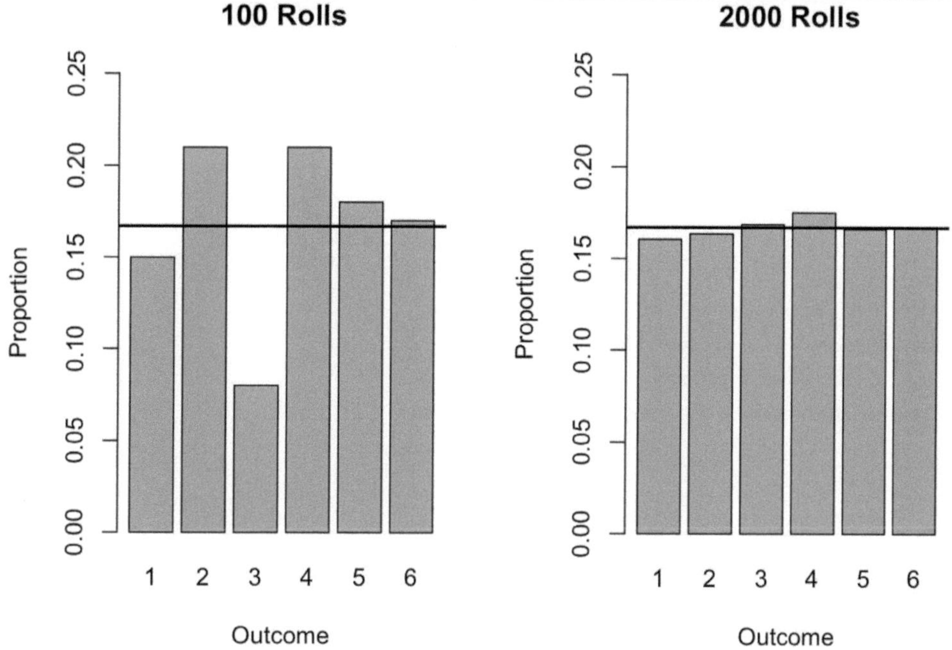

For instance,

- We know from earlier that the probability of Giannis making any given free throw during the 2020–2021 season was *P(success)* = (389/581) = .685, based on *observed* outcomes from the season.
- From the past several semesters of teaching data analysis, I know that 109 of 352 students earned a grade in the B range, so the *P(B)* = 109/352 = .31.
- Brittney Griner, the center for the Phoenix Mercury, made 248 of the 431 shots she took from the floor in the 2021 season, so the probability of her making any shot she took during that season was *P(Swish)* = 248/431 = .575.
- Joe Biden won the presidential election in 2020 with 51.26% of the popular vote, so the probability that a 2020 voter drawn at random voted for Biden is .5126.

To arrive at these empirical probabilities, I used data from the real world and observed the relative occurrence of different outcomes. These probabilities are the relative frequencies expressed as proportions.

Empirical Probabilities in Practice

We can also use the results of *joint frequency* distributions to estimate the probabilities of different outcomes occurring, either separately or at the same time. Let's look at a joint frequency table (cross tabulation) for vote choice and level of education in the 2020 U.S. presidential election, using data from the ANES.

First, we need to prepare the variables by modifying and collapsing some category labels.

```
#Create Vote variable
anes20$vote <- factor(anes20$V202073)
#Collapse levels to Biden, Trump, Other
levels(anes20$vote) <- c("Biden", "Trump", "Other", "Other",
                        "Other", "Other", "Other", "Other",
                        "Other")
#Create education variable
anes20$educ <- ordered(anes20$V201511x)
#Edit the levels so they fit in the table
levels(anes20$educ) <- c("LT HS", "HS", "Some Coll",
                        "4yr degr", "Grad degr")
```

Now, we can get a crosstabulation of both vote choice and education level.

```
#Get a crosstab of vote by education, using a weight, no plot
crosstab(anes20$vote, anes20$educ, weight = anes20$V200010b, plot = F)
```

You should note the addition of `weight = anes20$V200010b` in the command line for the crosstab. Often, despite best efforts at drawing a good sample, some sample characteristics differ significantly from known population characteristics. Using *sample weights* is a way to align certain sample characteristics with their known values in the population. The idea is to assign less weight to over-represented characteristics and more weight to under-represented characteristics. The ANES uses a complex weighting method that corrects for sampling bias related to sex, race, education, size of household, region, and other characteristics. The weighting information is activated by using `anes20$V200010b`. We won't use the weight function in most cases, but it does help improve the accuracy of the vote margins a bit (though not perfectly).

Each row in the table represents an outcome for vote choice, and each column represents a different level of educational attainment. Each of the interior cells in the table represents the intersection of a given row and column. At the bottom of each column and the end of each row, we find the row and column totals, also known as the *marginal frequencies*. These totals are the frequencies for the dependent and independent variables (note, however, that the sample size is now restricted to the 5183 people who gave valid responses to both survey questions).

```
   Cell Contents
|-------------------------|
|                   Count |
|-------------------------|
```

anes20$vote	anes20$educ LT HS	HS	Some Coll	4yr degr	Grad degr	Total
Biden	124	599	741	801	518	2783
Trump	117	644	770	493	232	2256
Other	10	20	47	47	20	144
Total	251	1263	1558	1341	770	5183

From this table we can calculate a number of different probabilities. To calculate the vote probabilities, we just need to divide the raw vote (row) totals by the sample size:

$$P(Biden) = 2783/5183 = .5369$$
$$P(Trump) = 2256/5183 = .4353$$
$$P(Other) = 144/5183 = .0278$$

Note that even with the sample weights applied, these estimates are not exactly equal to the population outcomes (.513 for Biden and .468 for Trump). Some part of this is due to sampling error, which you will learn more about in the next chapter.

We can use the column totals to calculate the probability that respondents have different levels of education.

$$P(\text{LT HS}) = 251/5183 = .0484$$
$$P(\text{HS}) = 1263/5183 = .2437$$
$$P(\text{Some Coll}) = 1558/5183 = .3006$$
$$P(\text{4yr degr}) = 1341/5183 = .2587$$
$$P(\text{grad degr}) = 770/5183 = .1486$$

Intersection of Two Probabilities

Sometimes, we are interested in the probability of two things occurring at the same time, the *intersection* of two probabilities represented as $P(A \cap B)$. If the events are independent, such as the *P*(*Ace*) and *P*(*Heart*), which we looked at earlier, then $P(A \cap B)$ is equal to $P(A)*P(B)$. By *independent*, we mean that the probability of one outcome occurring is not affected by another outcome. In the case of drawing an ace of hearts, the suit and value of cards are independent, because each suit has the same thirteen cards. Similarly, coin tosses are independent because the outcome of one toss does not affect the outcome of the other toss.

Suppose that we are interested in estimating the probability of a respondent being a Trump voter AND someone whose highest level of education is a high school diploma or equivalent, the intersection of Trump and HS (P(Trump ∩ HS). Vote choice and education levels in the following table are not independent of each other, so we can't use the simple multiplication rule to calculate the intersection of two probabilities.[3] Although we can't use the multiplication rule, we can use the information in the joint frequency distribution, focusing on that part of the table where the Trump row intersects with the HS column. We take the number of respondents in the Trump/HS cell (644) and divide it by the total sample size (5183).

$$P(Trump \cap HS) = 644/5183 = .1243$$

The Union of Two Probabilities

Sometimes, you might want to estimate the probability of one event OR another happening, the union of two probabilities, represented as $P(A \cup B)$. For instance, we can estimate the probability of Trump OR HS occurring (the union of Trump and HS) by adding the probability of each event occurring and then subtracting the intersection of the two events:

$$P(Trump \cup HS) = P(Trump) + P(HS) - P(Trump \cap HS)$$
$$P(Trump \cup HS) = .4353 + .2437 - .1243 = .5547$$

Why do we have to subtract $P(Trump \cap HS)$? If we didn't, we would be *double counting* the 644 people in the Trump/HS cell since they are in both the Trump row and HS column.

Of course, you can also get $P(Trump \cup HS)$ by focusing on the raw frequencies in the joint frequency table: Add the total number of Trump voters (2256) to the total number of people in the HS column (1263), subtract the total number of respondents in the cell where the Trump row intersects that HS column (644), and divide the resulting number (2875) by the total number of respondents in the table (5183):

$$P(Trump \cup HS) = (2256 + 1263 - 644)/5183 = .5547$$

Conditional Probabilities

These probabilities are well and good, but let's face it, sometimes they are not terribly interesting on their own, at least in the current example. They don't tell much of a story on their own. They tell us how likely it is that certain outcomes will occur, but not about how the outcomes of one variable might affect the outcomes of the other variable. From my perspective, it is far more interesting to learn how the probability of voting for Trump or Biden varies across levels of educational attainment. In other words, does educational attainment seem to have an impact on vote choice? To answer this, we need to ask something like, "What is the probability of voting for Trump, GIVEN that a respondent's level of educational attainment is a high school degree or equivalent?" and "How does that compare with the probability of voting for Trump GIVEN other levels of education?" These sorts of probabilities are referred to as *conditional probabilities* because they are saying the probability of an outcome on one variable is *conditioned* by or depends upon the outcome of another variable. This is what we mean when we say that two events are not independent—that the outcome on one depends upon the outcome on the other.

Let's start with the probability of voting for Trump given that someone is in the HS category on educational attainment: $P(Trump \mid HS)$. There are two ways to calculate this using the data we have. First, there is a formula based on other probability estimates:

$$P(Trump \mid HS) = \frac{P(Trump \cap HS)}{P(HS)}$$
$$(Trump \mid HS) = \frac{.12425}{.24368} = .5099$$

A more intuitive way to get this result, if you are working with a joint frequency distribution, is to divide the total number of respondents in the Trump/HS cell by the total number of people in the HS column: $\frac{644}{1263} = .5099$. By limiting the frequencies to the HS column, we are in effect calculating the probability of respondents voting for Trump *given* that high school equivalence is their highest level of education. Using the raw frequencies has the added benefit of skipping the step of calculating additional probabilities.

Okay, so the probability of someone voting for Trump in 2020, given their level of educational attainment was a high school degree, is .5099. So what? Well, first, we note that this is higher than the overall probability of voting for Trump ($P(Trump) = .4353$), so we know that having a high school level of educational attainment is associated with a higher than average probability of casting a Trump vote. We can get a broader sense of the impact of educational attainment on the probability of voting for Trump by calculating the other conditional probabilities:

$$P(Trump \mid LT\,HS = 117/251 = .4661$$
$$P(Trump \mid Some\ Coll) = 770/1558 = .4937$$
$$P(Trump \mid 4yr\ degr) = 493/1341 = .3676$$
$$P(Trump \mid Grad\ degr) = 232/770 = .3013$$

What this shows us is a very clear relationship between educational attainment and vote choice in the 2020 presidential election. Specifically, voters in the lowest three educational attainment categories had a somewhat higher than average probability of casting a Trump vote, while voters in the two highest educational attainment groups had a substantially lower probability of casting a Trump vote. People whose highest level of educational attainment is a high school degree had the highest probability of supporting President Trump's reelection (.5099), while those with graduate degrees had the lowest probability (.3013). Although the pattern is not perfect, generally, as education increases, the probability of voting for Trump decreases.

THE NORMAL CURVE AND PROBABILITY

The normal distribution can also be used to calculate probabilities, based on what we already know about the relationship between the standard deviation and the normal curve. In the last chapter, when we calculated areas to the right or left of some z-score or between two z-scores, we were essentially calculating the probability of outcomes occurring in those areas.

For instance, if we know that 68% of all values for a normally distributed variable lie within 1 standard deviation from the mean, then we know that the probability of a value, chosen at random, being within 1 standard deviation of the mean is about .68. What about the probability of a value being within 2 standard deviations of the mean? (About .95.) More than 2 standard deviations from the mean? (About .05.) These cutoff points are familiar to us from Chapter 6.

We can use the normal distribution to solve probability problems with real world data. For instance, suppose you are a student thinking about enrolling in my Political Data Analysis class and you want to know how likely it is that a student chosen at random would get a grade of A- or higher for the course (at least 89% of total points). Let's further suppose that you know the mean (75.12) and standard deviation (17.44) from the previous several semesters. What you want to know is, based on data from previous semesters, what is the probability of any given student getting an overall score of 89 or better?

In order to solve this problem, we need to make the assumption that course grades are normally distributed, *convert the target raw score (89) into a z-score*, and then calculate the probability of getting that score or higher, based on the area under the curve to the right of that z-score. Recall that the formula for a z-score is

$$Z_i = \frac{x_i - \bar{x}}{S}.$$

In this case x_i is 89, the score you need to earn in order to get at least an A-, so

$$Z_{89} = \frac{89 - 75.15}{17.44} = \frac{13.85}{17.44} = .7942.$$

We know that $Z_{89} = .7942$, so now we need to calculate the area to the right of .7942 standard deviations on a normally distributed variable.

```
#Get area to the right of z = .7942
pnorm(.7942, lower.tail = F)
[1] 0.2135395
```

The probability of a student chosen at random getting a grade of A- or higher is about .2135. You could also interpret this as meaning that the expectation is that about 21.35% of students will get a grade of A- or higher.

Again, it is important to recognize that this probability estimate is based on assumptions we make about a theoretical distribution, even though we are interested in estimating probabilities for an empirical variable. As discussed in Chapter 6, the estimates derived from a normal distribution are usually in the ballpark of what you find in empirical distributions, provided that the empirical distribution is not too oddly shaped. Since we have the historical data on grades, we can get a better sense of how well our theoretical estimate matches the overall observed probability of getting at least an A-, based on past experiences (empirical probability).

```
#Let's convert the numeric scores to letter grades (with rounding)
grades$A_Minus <- ordered(cut2(grades$grade_pct, c(88.5)))
#assign levels
levels(grades$A_Minus) <- c("LT A-", "A-/A")
#show the frequency distribution for the letter grade variable.
freq(grades$A_Minus, plot = F)

grades$A_Minus
      Frequency Percent Cum Percent
LT A-       276 78.41         78.41
A-/A         76 21.59        100.00
Total       352 100.00
```

Not Bad! Using real-world data and collapsing the point totals into two bins, we estimate that the proportion of students earning a grade of A- or higher is .216. This is very close to the probability estimate based on assuming a theoretical normal distribution (.214).

There is one important caveat: This bit of analysis disregards the fact that the probability of getting an A- is affected by a number of factors, such as effort and aptitude for this kind of work. If you have no other information about any given student, your best guess is that they have about a .22 probability of getting an A- or higher. Ultimately, it would be better to think about this problem in terms of conditional probabilities, whereby *P*(A-) is affected by other variables. Do you imagine that the probability of getting a grade of the A- or higher might be affected by how much time students are able to put into the course? Maybe those who do all of the reading and spend more time on homework have a higher probability of getting a grade in the A- range. In other words, to go back to the language of conditional probabilities, we might want to say that the probability of getting a grade in the A- range is **conditioned by** student characteristics.

NEXT STEPS

The material in this and the next two chapters is essential to understanding one of the important concepts from Chapter 1, the idea of level of confidence in statistical results. In the next chapter, we expand upon the idea of the law of large numbers by exploring related concepts, such as sampling error, the central limit theorem, and statistical inference. Building on this content, Chapter 9 kicks off a several-chapters-long treatment of hypothesis testing, beginning with testing hypotheses about the differences in mean outcomes between two groups. The upcoming material is a bit more abstract than what you have read so far, and I think you will find the shift in focus interesting. My sense from teaching these topics for several years is that this is going to be new material for most of you. Although this might make you nervous, I encourage you to look at it as an opportunity!

EXERCISES

Concepts and Calculations

1. Use the table that follows, showing the joint frequency distribution for attitudes toward the amount of attention given to sexual harassment (a recoded version of `anes20$V202384`) and vote choice for this problem.

```
   Cell Contents
|-------------------------|
|                   Count |
|-------------------------|

=====================================================================
                    anes20$harass
anes20$vote         Not Far Enough   About Right   Too Far      Total
---------------------------------------------------------------------
Biden                         1385          1108       293       2786
---------------------------------------------------------------------
Trump                          402           984       869       2255
---------------------------------------------------------------------
Other                           61            51        33        145
---------------------------------------------------------------------
Total                         1848          2143      1195       5186
=====================================================================
```

- Estimate the following probabilities:

$$P(\textit{Too Far})$$
$$P(\textit{About Right})$$
$$P(\textit{Not Far Enough})$$
$$P(\textit{Trump} \cap \textit{Not Far Enough})$$
$$P(\textit{Trump} \cup \textit{Not Far Enough})$$
$$P(\textit{Trump} \mid \textit{Too Far})$$
$$P(\textit{Trump} \mid \textit{About Right})$$
$$P(\textit{Trump} \mid \textit{Not Far Enough})$$

- Using your estimates of the conditional probabilities, summarize how the probability of voting for President Trump was related to how people felt about the amount of attention given to sexual harassment.

2. I flipped a coin 10 times and it came up heads only twice. I say to my friend that the coin seems biased toward tails. They say that I need to flip it a lot more times before I can be confident that there is something wrong with the coin. I flipped the coin 1000 times, and it came up heads 510 times. Was my friend right? What principle is involved here? In other words, how do you explain the difference between 2/10 on my first set of flips and 510/1000 on my second set?

3. In an analysis of joint frequency distribution for vote choice and whether people support or oppose banning assault-style rifles in the 2020 election, I find that $P(Biden) = .536$ and $P(Oppose) = .305$. However, when I apply the multiplication rule $((P(Biden) * P(Oppose))$ to find $P(Biden \cap Oppose)$ I get .1638, while the correct answer is .087. What did I do wrong? Why didn't the multiplication rule work?

4. Identify each of the following as a theoretical or empirical probability.
   - The probability of drawing a red card from a deck of 52 playing cards.
   - The probability of being a victim of violent crime.
   - The probability that 03 39 44 54 62 19 is the winning Powerball combination.
   - The probability of being hospitalized if you test positive for COVID-19.
   - The probability that Sophia Smith, of the Portland Thorns FC, will score a goal in any given match in which she plays.

5. The table that follows is a crosstabulation of two variables from the `anes20` dataset, respondent age and respondent opinion on whether business owners should be allowed to refuse services to same sex couples. After calculating the conditional probabilities, describe how age affects the probability that respondents chose the first option, should be `Allowed to Refuse`.

```
   Cell Contents
|-------------------------|
|                   Count |
|-------------------------|

==========================================================================
                                  PRE: SUMMARY: Respondent age
PRE: Services to same sex coupls  18-29  30-44  45-60  61-75  76+   Total
--------------------------------------------------------------------------
Allowed to Refuse                   388    932   1036   1135  401    3892
--------------------------------------------------------------------------
Required to Provide                 612   1068   1020    971  286    3957
--------------------------------------------------------------------------
Total                              1000   2000   2056   2106  687    7849
==========================================================================
```

R Problems

1. Use the code that follows to create a new object in R called `coin` and assign two different outcomes to the object, `Heads` and `Tails`. Double-check to make sure you've got this right.

```
coin <- c("Heads", "Tails")
coin
```

2. Next, you need to use R to simulate the results of tossing a coin ten times. Copy and run the code that follows to randomly choose between the two outcomes of `coin (Heads and Tails)` ten times, saving the outcomes in a new object, `coin10`. These outcomes represent the same thing as ten coin tosses. Use `table(coin10)` to see how many `Heads` and `Tails` outcomes you have from the ten-coin tosses. Is the outcome you got close to the 5 heads/5 tails you should expect to see?

```
#Take a random sample of outcomes from ten tosses
coin10 <- sample(coin, 10, rep = T)
table(coin10)
```

3. Now, repeat the R commands in Question 2 nine more times, recording the number of heads and tails you get from each simulation. Show your results and sum up the number of `Heads` outcomes from the ten simulations. If the probability of getting `Heads` on any given toss is .50, then you should have approximately 50 `Heads` outcomes. Discuss how well your results match the expected 50/50 outcome. Also, comment on the range of outcomes—some close to 50/50, some not at all close—you got across the ten simulations.

8 SAMPLING AND INFERENCE

GET READY

This chapter explores both theoretical and practical issues related to sampling and statistical inference. The material covered here is critical to understanding hypothesis testing, which is taken up in the next chapter, but is also interesting in its own right. To follow along with the graphics and statistics used here, you need to load the `county20.rda` dataset, as well as the `DescTools` library.

STATISTICS AND PARAMETERS

Almost always, social scientists are interested in making general statements about some social phenomenon. By general statements, I mean that we are interested in making statements that can be applied to a population of interest. For instance, I study voters and would like to make general statements that apply to all voters. Others might study funding patterns in local governments and would like to be able to generalize their findings to all local governments, or someone who studies aversion to vaccines among a certain group would like to be able to make statements about all members of that group.

Unfortunately, gathering information on entire populations is usually not possible. Therefore, social scientists, as well as natural scientists, rely on samples drawn from the population. Hopefully, the statistics generated from these samples provide an accurate reflection of the underlying population values. We refer to the calculations we make with sample data as *statistics*, and we assume that those statistics are good representations of population *parameters*. The connection between sample statistics and population parameters is summarized in Table 8.1, using the already familiar statistics: the mean, variance, and standard deviation.

TABLE 8.1 ■ Symbols and Formulas for Sample Statistics and Population Parameters

Measure	Sample Statistic	Formula	Population Parameter	Formula
Mean	$\bar{x}$	$\frac{\sum_{i=1}^{n} x_i}{n}$	μ	$\frac{\sum_{i=1}^{N} x_i}{N}$
Variance	S^2	$\frac{\sum_{i=1}^{n}(x_i-\bar{x})^2}{n-1}$	σ^2	$\frac{\sum_{i=1}^{N}(x_i-\mu)^2}{N}$
Standard Deviation	S	$\sqrt{\frac{\sum_{i=1}^{n}(x_i-\bar{x})^2}{n-1}}$	σ	$\sqrt{\frac{\sum_{i=1}^{N}(x_i-\mu)^2}{N}}$

When generalizing from a sample statistic to a population parameter, we are engaging in the process of *statistical inference*. We are inferring something about the population based on an estimate from a sample. For instance, we are interested in the population parameter, μ (mu), propronounced like a French cow saying "moo", but must settle for the sample statistic ($\bar{x}$), which we hope is a good approximation of μ. Or, since we know from Chapter 7 that it is also important to know how much variation there is around any given mean, we calculate the sample variance (S^2) and standard deviation (S), expecting that they are good approximations of the population values, σ^2 (sigma squared) and σ (sigma).

You might have noticed that the numerator for the sample variance and standard deviation in Table 8.1 is $n - 1$, while in the population formulas it is N. The reason for this is somewhat complicated, but the short version is that dividing by n tends to underestimate the variance and standard deviation in samples, so $n - 1$ is a correction for this. You should note, though, that in practical terms, dividing by $n - 1$ instead of n makes little difference in large samples. For small samples, however, it can make an important difference.

Not all samples are equally useful, however, from the perspective of statistical inference. What we are looking for is a sample that is representative of the population, one that "looks like" the population. The key to a good sample, generally, is that it should be large and randomly drawn from the population. A pure random sample is hard to achieve, but it is what we should strive for. The key benefit to this type of sample is that every unit in the population has an equal chance of being selected. If this is the case, then the sample should "look" a lot like the population.

SAMPLING ERROR

Even with a large, representative sample we can't expect that a given sample statistic is going to have the same value as the population parameter. There will always be some error—this is the nature of sampling. Because we are using a sample and not the population, the sample estimates will not be the same as the population parameter. Let's look at a decidedly non-social science illustration of this idea.

Imagine that you have 6000 colored ping pong balls (2000 yellow, 2000 red, and 2000 green), and you toss them all into a big bucket and give the bucket a good shake, so the yellow, red, and green balls are randomly dispersed in the bucket. Now, suppose you reach into the bucket and randomly pull out a sample of 600 ping pong balls. How many yellow, red, and green balls would you expect to find in your 600-ball sample? Would you expect to get exactly 200 yellow, 200 red, and 200 green balls, perfectly representing the color distribution in the full 6000-ball population? Odds are, it won't work out that way. It is highly unlikely that you will end up with the exact same color distribution in the sample as in the population. However, if the balls are randomly selected and there is no inherent bias (e.g., one color doesn't weigh more than the others, or something like that), you should end up with close to one-third yellow, one-third red, and one-third green.

This is the idea behind sampling error: Sample statistics, by their very nature, will differ population from parameters, but given certain characteristics (large, random samples), they should be close approximations of the population parameters. Because of this, when you see reports of statistical findings, such as the results of a public opinion poll, they are often reported along with a caveat, something like "a margin of error of plus or minus two percentage points." This is exactly where this chapter is eventually headed, measuring and taking into account the amount of sampling error when making inferences about the population.

Let's look at how this works in the context of some real-world political data. For the rest of this chapter, we will use county-level election returns from the 2020 presidential election, found in the `county20` dataset, to illustrate some principles of sampling and inference.[1] We will focus our attention on Biden's percentage of the two-party vote in 2020 (`county20$d2pty20`), starting with the following histogram and a few summary statistics.

```
#Histogram of county-level Biden vote
hist(county20$d2pty20, xlab = "Biden % of Two-Party Vote",
     ylab = "Number of Counties",
     main = "County-Level Votes for Biden in 2020 (All Counties)")

#Add vertical lines at the mean and median.
abline(v = mean(county20$d2pty20, na.rm = T), lwd = 2)
abline(v = median(county20$d2pty20, na.rm = T), lwd = 2, lty = 2)

#Add a legend
legend("topright", legend = c("Mean","Median"), lty = 1:2)
```

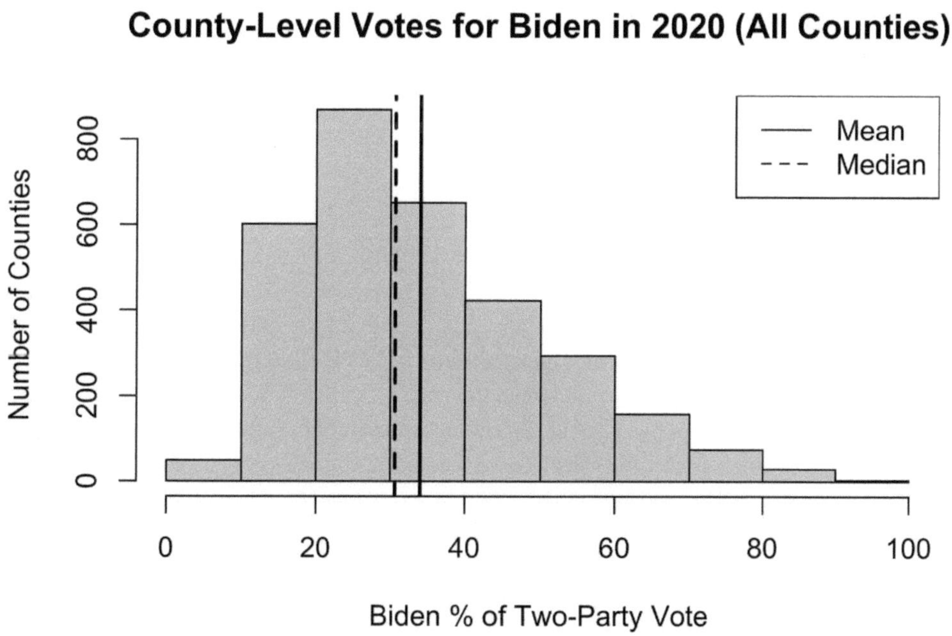

```
#Get summary stats and skewness
summary(county20$d2pty20)

   Min.  1st Qu.  Median    Mean  3rd Qu.    Max.
  3.114   21.327  30.640  34.044   43.396  94.467

Skew(county20$d2pty20, na.rm = T)
[1] 0.8091358
```

As you can see, this variable is somewhat skewed to the right, with the mean (34.04) a bit larger than the median (30.64), producing a skewness value of .81.[2] Since the histogram and statistics reported are based on the *population of counties*, not a sample of counties, $\mu = 34.04$.

To get a better idea of what we mean by *sampling error*, let's take a sample of 300 counties, randomly drawn from the population of counties, and calculate the mean Democratic percentage of the two-party vote in those counties. The R code that follows shows how you can do this. One thing I want to draw your attention to here is the `set.seed()` function. This tells R where to start the random selection process. It is not absolutely necessary to use this, but doing so provides reproducible results. This is important in this instance because, without it, the code would produce slightly different results every time it is run (and I would have to rewrite the text reporting the results over and over).

```
#Tell R where to start the process of making random selections
set.seed(250)

#draw a sample of 300 counties, using "d2pty20";
#store in "d2pty_300"
d2pty300 <- sample(county20$d2pty20, 300)

#Get sample stats
summary(d2pty300)

   Min.  1st Qu.  Median   Mean 3rd Qu.     Max.
  6.148   21.144  30.749  33.937  42.503  94.467
```

The new object created from this code `d2pty_300`, contains the values of `d2pty20` from 300 randomly selected counties. From this sample, we get $\bar{x} = 33.94$. As you can see, the sample mean is not equal to but is pretty close to the population value (34.04). This is what we expect from a large, randomly drawn sample. One of the keys to understanding sampling is that if we take another sample of 300 counties and calculate another mean from that sample, not only will it not equal the population value, but it will also be different from the first sample mean, as shown next.

Let's see.

```
#Different "set.seed" here because I want to show that the
#results from another sample are different.
set.seed(200)
```

```
#Draw a different sample of 300 counties
d2pty300b <- sample(county20$d2pty20, 300)

#Summary stats
summary(d2pty300b)

   Min.  1st Qu.  Median    Mean  3rd Qu.    Max.
  5.459   20.807  30.401  33.461   43.911  83.990
```

We should not be surprised that we get a different result here ($\bar{x} = 33.46$), since we are dealing with a different sample. Nor should we be surprised that this sample mean is close in value to the first sample mean (33.94) or that they are both close to the population value (34.04). That's the nature of sampling.

SAMPLING DISTRIBUTIONS

We could go on like this and calculate a lot of different sample means from a lot of different samples, and they would all take on different values. It is also likely that none of them would be exactly equal to the population value unless by chance, due to rounding. If we kept sampling like this, the *distribution of sample means* we get would constitute a sampling distribution. **Sampling distributions** are theoretical distributions representing what would happen if we took an infinite number of large, independent samples from the population. Figure 8.1 provides an illustration of a theoretical sampling distribution, where Biden's percentage of the two-party vote in the population of counties is 34.04.

FIGURE 8.1 ■ A Theoretical Sampling Distribution, *mu* = 34.04, Sample Size = 300

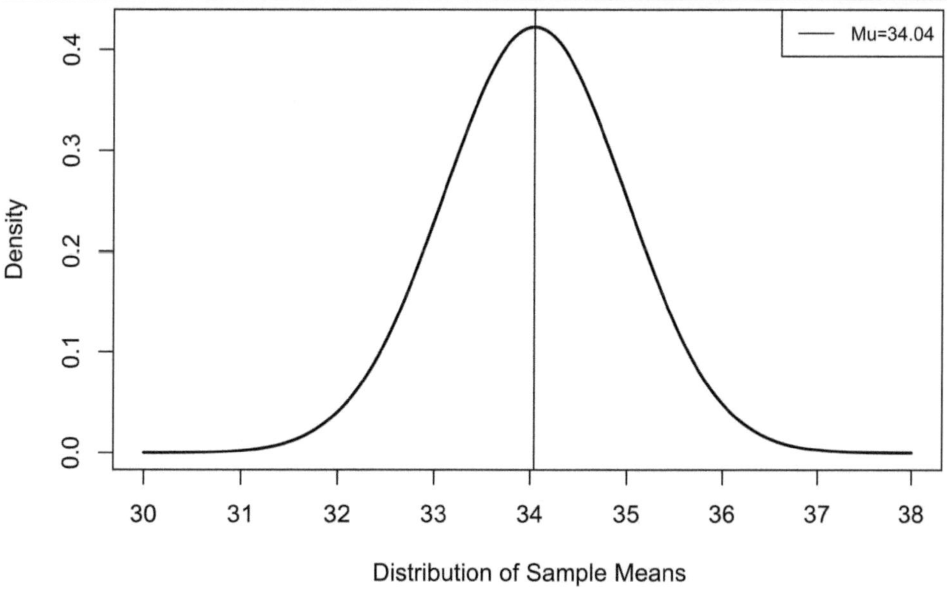

To be clear, this figure illustrates a theoretical distribution of sample means that we obtain if we took an infinite number of samples of 300 counties from the population of U.S. counties.

There are several things to note here: The mean of the distribution of sample means (sampling distribution) is equal to the population value (34.04); most of the sample means are relatively close to the population value, between about 32 and 36; and there are a just few means at the high and low ends of the horizontal axis.

This theoretical distribution reflects a number of important characteristics we can expect from a sampling distribution based on repeated large, random samples from a population with a mean equal to μ and a standard deviation equal to σ:

- It is normally distributed.
- Its mean (the mean of all sample means) equals the population value, μ.
- It has a standard deviation equal to $\frac{\sigma}{\sqrt{n}}$.

This idea, known as the **central limit theorem**, holds as long as the sample is relatively large ($n > 30$). What this means is that statistics (means, proportions, standard deviations, etc.) from large, randomly drawn samples are good approximations of underlying population parameters, as most sample estimates will be relatively close in value to the population parameters.

This makes sense, right? If we take multiple large, random samples from a population, some of the sample statistics should be a bit higher than the population parameter, and some will be a bit lower. And, of course, a few will be a lot higher and a few a lot lower than the population parameter, but most of them will be **clustered near the mean** (this is what gives the distribution its bell shape). By its very nature, *sampling produces statistics that are different from the population values, but the sample statistics should be relatively close to the population values.*

It is important to note that the central limit theorem holds for relatively small samples, and the shape of the sampling distribution does not depend on the distribution of the *empirical variable.* In other words, the variable being measured does not have to be normally distributed in order for the sampling distribution to be normally distributed. This is good, since very few empirical variables follow a normal distribution, including `county20$d2pty` (used here), which is positively skewed. Note, however, that in cases where there are extreme outliers, this rule may not hold.

Simulating the Sampling Distribution

In reality, we don't actually take repeated samples from the population. Think about it, if we had the population data, we would not need a sample, let alone a sampling distribution. Instead, sampling distributions are *theoretical distributions* with known characteristics. Still, we can take multiple samples from the population of counties to illustrate that the pattern shown in Figure 8.1 is a realistic expectation. We know from earlier that *in the population,* the mean county-level share of the two-party vote for Biden was 34.04% ($\mu = 34.04$), and the distribution is somewhat skewed to the right (skewness = .81), but with no extreme outliers. We can *simulate* a sampling distribution by taking repeated samples from the population of counties and then observing the shape of the distribution of means from those samples. The resulting distribution should start to look like the distribution presented in Figure 8.1, especially as the number of samples increases.

To do this, I had R take fifty different samples, each of which includes 50 counties, get the mean outcome from each sample, and save those means as a new object. Note that this is a small number of relatively small number of samples, but we should still see the distribution trending toward normal.

Let's look at the fifty separate samples means:

```
 [1] 31.24475 34.72631 34.84539 37.19550 34.83349 33.52246 35.85605 35.42220
 [9] 33.60402 32.97881 31.80286 30.33704 33.65232 33.28219 35.40609 35.97478
[17] 37.88739 34.55101 32.68982 30.96713 32.90207 37.19780 30.35538 33.33603
[25] 33.68002 29.96251 33.70846 35.77540 34.63077 35.51688 35.98887 34.35845
[33] 34.97154 38.55704 33.17959 36.86423 31.40136 30.17257 34.18038 35.64399
[41] 38.93687 30.61432 36.25869 32.81192 37.76805 33.14322 40.73194 34.50676
[49] 30.94900 35.40402
```

Here's how you read this output. Each of the values represents a single mean drawn from a sample of 50 counties. The first sample drawn had a mean of 31.24, the second sample 34.73, and so on, with the 50th sample having a mean of 35.4. Most of the sample means are fairly close to the population value (34.04), and a few are more distant. Looking at the summary statistics (following), we see that, on average, these fifty sample means balance out to an overall mean of 34.3, which is very close to the population value.

```
   Min.  1st Qu.   Median    Mean  3rd Qu.    Max.
  29.96    32.92    34.43   34.29    35.74   40.73
```

Figure 8.2 uses a density plot (solid line) of this sampling distribution, displayed alongside a normal distribution (dashed line), to get a sense of how closely the distribution fits the contours of a normal distribution. As you can see, with just 50 relatively small samples, the sampling distribution is beginning to take on the shape of a normal distribution.

So, let's see what happens when we create another sampling distribution but increase the number of samples to 500. In theory, this sampling distribution should resemble a normal distribution more closely. In the summary statistics, we see that the mean of this sampling distribution (33.98) is slightly closer to the population value (34.04) than in the previous example. Further, if you examine the density plot shown in Figure 8.3, you will see that the sampling distribution of 500 samples of 50 counties follows the contours of a normal distribution more closely, as should be the case. If we increased the number of samples to 1000 or 2000, the sampling distributions would grow even more similar to a normal distribution.

FIGURE 8.2 ■ Normal Distribution and Simulated Sampling Distribution, 50 Samples of 50 Counties

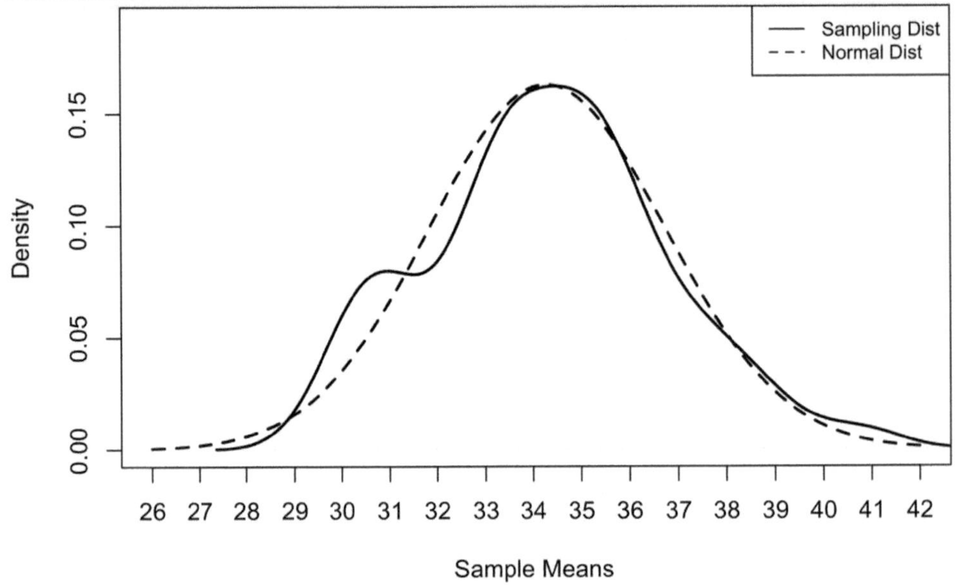

FIGURE 8.3 ■ Normal Distribution and Sampling Distribution, 500 Samples of 50 Counties

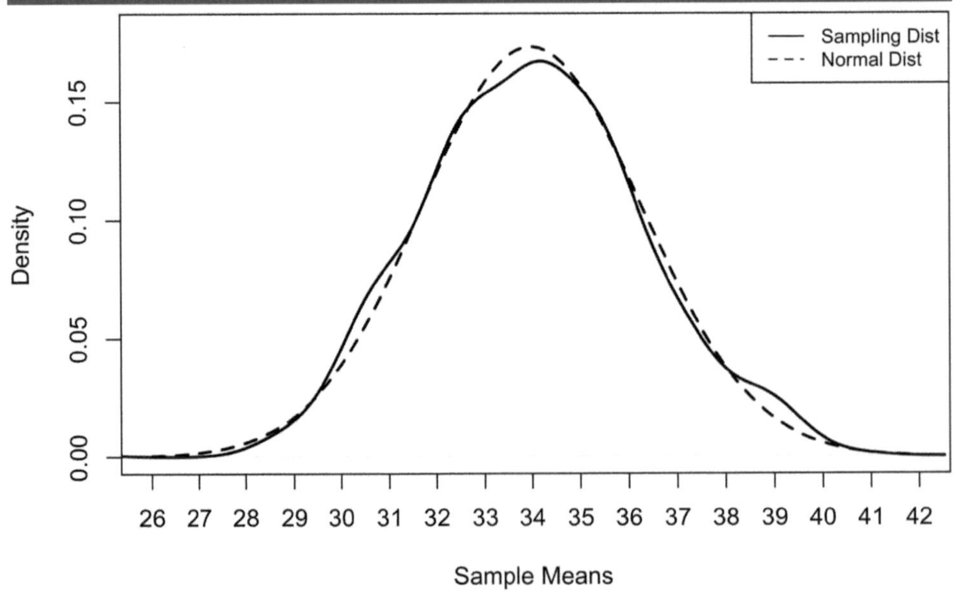

CONFIDENCE INTERVALS

In both cases, the simulations support one of the central tenets of the central limit theorem: With large, random samples, the sampling distribution will be nearly normal, and the mean of the sampling distribution equals μ. This is important because it is also means that most sample means are fairly close to the mean of the sampling distribution. In fact, because sampling distributions follow the normal distribution, we know that approximately 68% of all sample means will be within one standard deviation of the population value. We refer to the standard deviation of a sampling distribution as the **standard error**. In the preceding examples where the distributions represent collections of different sample means, this is referred to as the **standard error of the mean**. Remember this, the standard error is a measure of the standard deviation of the sampling distribution.

The formula for a standard error of the mean is

$$\sigma_{\bar{x}} = \frac{\sigma}{\sqrt{N}},$$

where σ is the standard deviation of the variable in the population, and N is the number of observations in the empirical sample.

Of course, this assumes that we know the population value, σ, which we do not. Fortunately, because of the central limit theorem, we do know of a good *estimate* of the population standard deviation, the sample standard deviation. So, we can substitute S (the sample standard deviation) for σ:

$$S_{\bar{x}} = \frac{S}{\sqrt{n}}$$

This formula is saying that the standard error of the mean is equal to the sample standard deviation divided by the square root of the sample size. Let's go ahead and calculate the standard error for a sample of just 100 counties (this makes the whole $\sqrt{N}$ business a lot easier). The observations for this sample are stored in a new object, `d2pty100`. First, just a few descriptive statistics, presented next. Of particular note here is that the mean from this sample (32.71) is again pretty close to μ, 34.04 (Isn't it nice how this works out?).

```
set.seed(251)

#draw a sample of d2pty20 from 100 counties
d2pty100 <- sample(county20$d2pty20, 100)

mean(d2pty100)
[1] 32.7098

sd(d2pty100)
[1] 17.11448
```

We can use the sample standard deviation (17.11) to calculate the standard error of the sampling distribution (based on the characteristics of this single sample of 100 counties).

```
se100 = 17.11/10
se100
[1] 1.711
```

Of course, we can also get this more easily.

```
#This function is in the "DescTools" package.
MeanSE(d2pty100)
[1] 1.711448
```

The mean from this sample is 32.71, and the standard error is 1.711. For right now, treating this as our only sample, 32.71 is our best guess for the population value. We can refer to this as the **point estimate**. But we know that this is probably not the population value because we would get different sample means if we took additional samples, and they can't all be equal to the population value. But we also know that most sample means, including ours, are going to be fairly close to μ.

Now, suppose we want to use this sample information to create a range of values that we are pretty confident includes the population parameter. We know that the sampling distribution is normally distributed and that the standard error is 1.711, so we can be confident that 68% of all sample means are within one standard error of the population value. Using our sample mean as the estimate of the population value (it's the best guess we have), we can calculate a 68% **confidence interval**:

$$c.i._{.68} = \overline{x} \pm z_{.68} * S_{\overline{x}}$$

Here we are saying that a 68% confidence interval ranges from the $\overline{x}$ *plus* and *minus* the value for z that give us 68% of the area under the curve around the mean times the standard error of the mean. In this case, the multiplication is easy because the *critical value* of z (the z-score for an area above and below the mean of about .68) is 1, so

$$c.i._{.68} = 32.71 \pm 1 * 1.711.$$

```
#Estimate lower limit of confidence interval
LL_68 = 32.71-1.711
LL_68
[1] 30.999

#Estimate upper limit of confidence interval


UL_68 = 32.71 + 1.711
UL_68
[1] 34.421
```

$$c.i._{.68} = 30.999 \text{ to } 34.421$$

The lower limit of the confidence interval (LL) is about 31, and the upper limit (UL) is 34.42, a narrow range of just 3.42 that does happen to include the population value, 34.04.

You can also use the `MeanCI` function to get a confidence interval around a sample mean.

```
#You need to specify the standard deviation and level of confidence
MeanCI(d2pty100, sd = 17.11, conf.level = .68, na.rm = T)

    mean   lwr.ci   upr.ci
32.70980 31.00828 34.41132
```

As you can see, the results are very close to those we calculated, with the differences no doubt due to rounding error.

So, rather than just assuming that the $\mu = 32.71$ based on just one sample mean, we can incorporate a bit of uncertainty that acknowledges the existence of sampling error. We can be 68% confident that this confidence interval includes the μ. In this case, since we know that the population value is 34.04, we can see that the confidence interval does incorporate the population value. But, usually, we do not know the population values (that's why we use samples!), and we have to trust that the confidence interval includes the μ. How much can we trust that this is the case? That depends on the confidence level specified in setting up the confidence interval. In this case, we can say that we are 68% certain that this confidence interval included the population value.

A good way to think of this is that if we were to calculate 68% confidence intervals across hundreds of samples, 68% of those intervals would include the population value. I demonstrate

FIGURE 8.4 ■ Fifty 68% Confidence Intervals, *n* = 100, *mu* = 34.04

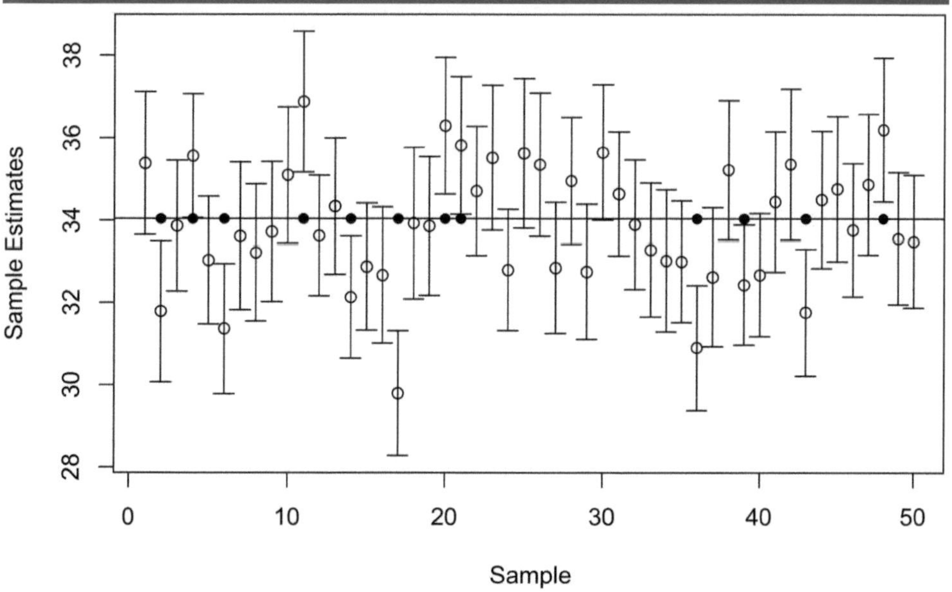

this idea next, in Figure 8.4, where I plot 50 different 68% confidence intervals around estimates of the Democratic percentage of the two-party vote, based on fifty samples of 100 counties. The expectation is that at least 34 (68%) of the intervals will include the value of μ, and up to 16 (32%) intervals might not include μ. The circles represent the point estimates, and the capped vertical lines represent the width limits of the confidence intervals:

Here you can see that most of the confidence intervals include μ (the horizontal line at 34.04), but there are 12 confidence intervals that do not overlap with μ, identified with the solid dots on the horizontal line at the value of μ. This is consistent with our expectation that there is at least a .68 probability that any given confidence interval constructed using a z-score of 1 will overlap with μ.

In reality, while a 68% confidence interval works well for demonstration purposes, we usually demand a bit more certainty. The standard practice in the social sciences is to use a 95% confidence interval. All we have to do to find the lower- and upper-limits of a 95% confidence interval is find the critical value for z that will give us 42.5% of the area under the curve between the mean and the z-score, leaving .025 of the area under the curve at each of the tails of the distribution. You can use a standard z-distribution table to find the critical value of z, or you can ask R to do it for you, using the `qnorm` function.

```
#qnorm gives the z-score for a specified area on the distribution tail.
#Specifying "lower.tail = F" instructs R to find the upper tail area.
qnorm(.025, lower.tail = F)
[1] 1.959964
```

The critical value for $z_{.95} = 1.96$. So, now we can substitute this into the equation we used earlier for the 68% confidence interval to obtain the 95% confidence interval:

$$c.i._{.95} = 32.71 \pm 1.96 * 1.711$$
$$c.i_{.95} = 32.71 \pm 3.35$$

```
#Estimate lower limit of confidence interval
LL_95 = 32.71-3.35
LL_95
[1] 29.36

#Estimate upper limit of confidence interval


UL_95 = 32.71 + 3.35
UL_95
[1] 36.06
```

$$c.i_{.95} = 29.36 \text{ to } 36.06$$

Check this using the `MeanCI` command.

```
MeanCI(d2pty100, sd = 17.11, conf.level = .95)

   mean  lwr.ci  upr.ci
32.7098 29.3563 36.0633
```

Now we can say that we are 95% confident that the interval between 29.36 and 36.06 includes the population mean. Technically, what we should say is that *95% of all confidence intervals based on z = 1.96 include the value for* μ, so we can be 95% certain that this confidence interval includes μ.

Sample Size and Confidence Limits

Note that the 95% confidence interval (almost 6.7 points) is wider than the 68% interval (about 3.4 points), because we are demanding a higher level of confidence. Suppose we want to narrow the width of the interval, but we do not want to sacrifice the level of confidence. What can we do about this? The answer lies in the formula for the standard error:

$$S_{\bar{x}} = \frac{S}{\sqrt{n}}$$

There is only one thing in this formula that we can manipulate, the sample size. If we took another sample, we would get a very similar standard deviation, something around 17.11. However, we might be able to affect the sample size, and as the sample size increases, the standard error of the mean decreases, as does the width of the confidence interval (holding level of confidence constant).

Let's look at this for a new, larger sample of 500 counties.

```
set.seed(251)
#draw a sample of d2pty20 from 500 counties
d2pty500 <- sample(county20$d2pty20, 500)

mean(d2pty500)
[1] 33.24397

sd(d2pty500)
[1] 15.8867

MeanSE(d2pty500)
[1] 0.7104747
```

Here, you can see that the mean (33.24) and standard deviation (15.89) are fairly close in value to those obtained from the smaller sample of 100 counties (32.71 and 17.11), but the standard error of the mean is much smaller (.71 compared to 1.71). This difference in standard error, produced by the larger sample size, results in a much narrower confidence interval, even though the level of confidence (95%) is the same.

```
MeanCI(d2pty500, sd = 15.89, conf.level = .95)

    mean   lwr.ci   upr.ci
33.24397 31.85118 34.63677
```

The width of the confidence interval (31.85 to 34.64) is now 2.78 points, compared to 6.7 points for the 100-county sample.

Let's take a closer look at how the width of the confidence interval responds to sample size, using data from the current example (Figure 8.5).

FIGURE 8.5 ■ Width of a 95% Confidence Interval at Different Sample Sizes (Std Dev. = 15.89)

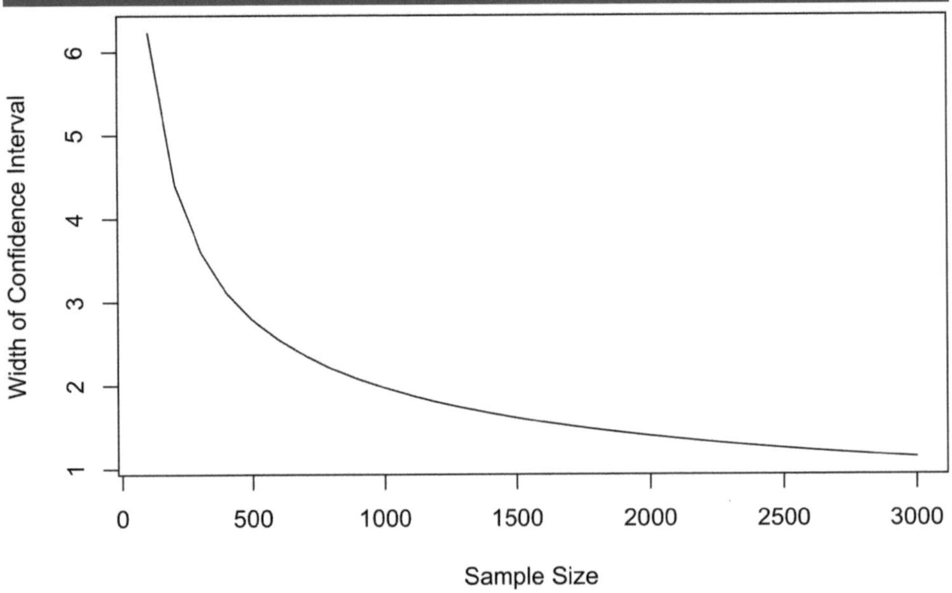

As you can see, as the sample size increases, the width of the confidence intervals decreases, but there are diminishing returns after a point. Moving from small samples of 100 or so to larger samples of 500 or so results in a steep drop in the width of the confidence interval. Moving from 500 to 1000 results is a smaller reduction in width, and moving from 1000 to 2000 results in an even smaller reduction in the width of the confidence interval.

This pattern has important implications for real-world research. Depending on how a researcher is collecting their data, they may be limited by the very real costs associated with increasing the size of a sample. If conducting a public opinion poll or recruiting experimental participants, for instance, each additional respondent costs money, and spending money on increasing the sample size inevitably means taking money away from some other part of the research enterprise.[3]

PROPORTIONS

Everything we have just seen regarding the distribution of sample means also applies to the distribution of sample proportions. It should, since a proportion is just the mean of a dichotomous variable scored 0 and 1. For example, with the same data we just used, we can focus on a dichotomous variable that indicates whether Biden won (1) or lost (0) in each county. The mean of this variable across all counties is the proportion of counties won by Biden.

```
#Create dichotomous indicator for counties won by Biden
demwin <- as.numeric(county20$d2pty20 > 50)
table(demwin)

demwin
   0    1
2595  557

mean(demwin, na.rm = T)
[1] 0.1767132
```

Biden won 557 counties and lost 2595, for a winning proportion of .1767. This is the population value (P).

Again, we can take multiple samples from this population and none of the proportions calculated from them may match the value of P exactly, but most of them should be fairly close in value *and* the mean of the sample proportions should equal the population value over infinite sampling.

Let's check this out for 500 samples of 50 counties each, stored in a new object, `sample_prop500`.

```
summary(sample_prop500)

  Min.  1st Qu.  Median    Mean  3rd Qu.    Max.
0.0200   0.1400  0.1800  0.1742   0.2000  0.3400
```

Here we see that the mean of the sampling distribution (.1742) is, as expected, very close to the population proportion (.1767), and the distribution has very little skew. The density plots in Figure 8.6 show that the shape of the sampling distribution mimics the shape for the normal distribution fairly closely. This is all very similar to what we saw with the earlier analysis using the mean Democratic share of the two-party vote.

Everything we learned about confidence intervals around the mean also applies to sample estimates of the proportion. A 95% confidence interval for a sample proportion is

$$c.i._{.95} = p \pm z_{.95} * S_p,$$

FIGURE 8.6 ■ Normal and Sampling Distribution (Proportion), 500 Samples of 50 Counties

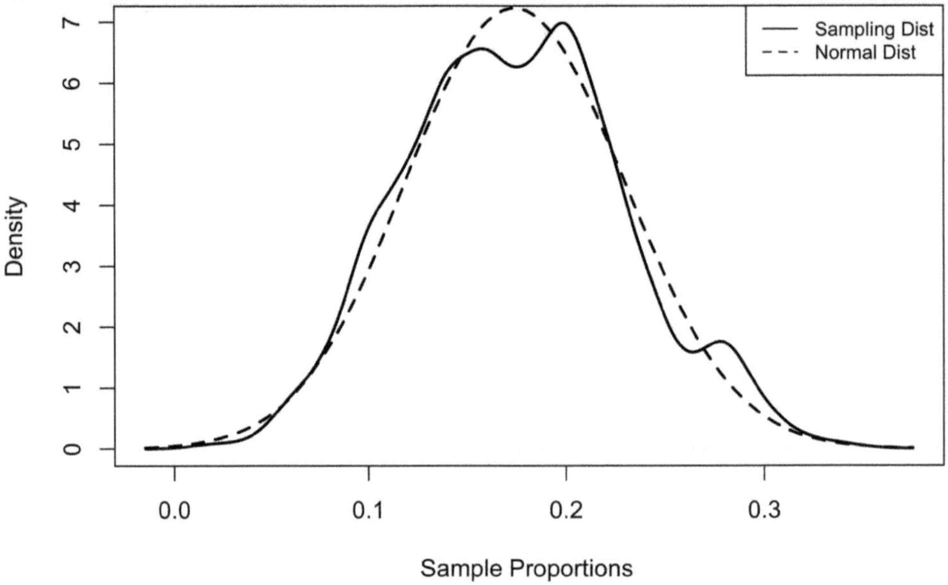

where the standard error of the proportion is calculated the same as the standard error of the mean: The standard deviation divided by the square root of the sample size.

Let's turn our attention to estimating a confidence interval for the proportion of counties won by Biden from a single sample with 300 observations (`demwin300`).

```
set.seed(251)
#Sample 300 counties for demwin
demwin300 <- sample(demwin, 300)
mean(demwin300)
[1] 0.1533333

sd(demwin300)
[1] 0.3609105
```

For the sample of 300 counties, the mean is .153, which is quite a bit lower that the known population value of .1767, and the standard deviation is .3609. To calculate a 95% confidence interval, we need to estimate the standard error.[4]

```
seprop300 = .3609/sqrt(300)
seprop300
[1] 0.02083657
```

We can now plug the standard error of the proportion into the confidence interval:

$$c.i._{.95} = .153 \pm 1.96 * .0208$$
$$c.i._{.95} = .153 \pm .041$$

```
#Estimate lower limit of confidence interval
LL_95 = .153-.041
LL_95
[1] 0.112

#Estimate upper limit of confidence interval

UL_95 = .153 + .041
UL_95
[1] 0.194
```

$$c.i._{.95} = .112 \text{ to } .194$$

Check this with `MeanCI`.

```
MeanCI(demwin300, sd = .3609, conf.level = .95)

     mean      lwr.ci     upr.ci
0.1533333 0.1124944 0.1941723
```

The confidence interval is .082 points wide, meaning that we are 95% confident that the interval between .112 and .194 includes the population value for this variable. If you want to put this in terms of percentages, we are 95% certain that Biden won between 11.2% and 19.4% of counties.

The "±" part of the confidence interval might sound familiar to you from media reports of polling results. This figure is sometimes referred to as the *margin of error*. When you hear the results of a public opinion poll reported on television and the news reader adds language like "plus or minus 3.3 percentage points," they are referring to the confidence interval (usually 95% confidence interval), except they tend to report percentage points rather than proportions.

So, for instance, a Fox News poll taken between September 12 and September 15, 2021, with a sample of 1002 respondents, reported President Biden's approval rating at 50%, with a margin of error of ±3.0.

If you do the math (go ahead, give it a shot!), based on $p = .50$ and $n = 1002$, this ±3.0 corresponds with the upper and lower limits of a 95% confidence interval ranging from .47 to .53. So we can say with a 95% level of confidence that at the time of the poll, President Biden's approval rating was between .47 (47%) and .53 (53%), according to this Fox News poll.

NEXT STEPS

Now that you have learned about some of the important principles of statistical inference (sampling, sampling error, and confidence intervals), you have all the tools you need to learn about hypothesis testing, which is taken up in the next chapter. In fact, it is easy to illustrate how you could use one important tool—confidence intervals—to test certain types of hypotheses. For instance, using the results of Fox News poll just reported, if someone stated that they thought Biden's approval rating was no higher than 45% at the time of the poll, you could tell them that you are 95% certain that they are wrong, since you have a 95% confidence interval for Biden's approval rating that ranges from 47% to 53%, which means that the likelihood of it being 45% or lower is very small. Hypothesis testing gets a bit more complicated than this, but this example captures the spirit of it very well: We use sample data to test ideas about the values of population parameters. On to hypothesis testing!

EXERCISES

Concepts and Calculations

1. A group of students on a college campus are interested in how much students spend on books and supplies in a typical semester. They interview a random sample of 300 students and find that the average semester expenditure is $350 and the standard deviation is $78.
   a. Are the results reported from an *empirical* distribution or a *sampling* distribution? Explain your answer.
   b. Calculate the standard error of the mean.
   c. Construct and interpret a 95% confidence interval around the mean amount of money students spend on books and supplies per semester.

2. In the same survey used in Question 1, students were asked if they were satisfied or dissatisfied with the university's response to the COVID-19 pandemic. Among the 300 students, 55% reported being satisfied. The administration hailed this finding as evidence that a majority of students support the actions they've taken in reaction to the pandemic. What do you think of this claim? Of course, as a bright college student in the midst of learning about political data analysis, you know that 55% is just a point estimate and you really need to construct a 95% confidence interval around this sample estimate before concluding that more than half the students approve of the administration's actions. So, let's get to it (Hint: this is a "proportion" problem):
   a. Calculate the standard error of the proportion. What does this represent?
   b. Construct and interpret a 95% confidence interval around the reported proportion of students who are satisfied with the administration's actions.
   c. Is the administration right in their claim that a majority of students support their actions related to the pandemic?

3. One of the data examples used in Chapter 4 combined five survey questions on LGBTQ rights into a single index ranging from 0 to 6. The resulting index has a mean of 3.4, a standard deviation of 1.56, and a sample size of 7,816.
   - Calculate the standard error of the mean and a 95% confidence interval for this variable.
   - What is the standard error of the mean if you assume that the sample size is 1,000?
   - What is the standard error of the mean if you assume that the sample size is 300?
   - Discuss how the magnitude of change in sample size is related to changes in the standard error in this example.

R Problems

For these problems, you should use load the `county20large` dataset to analyze the distribution of *internet access* across U.S. counties, using `county20large$internet`. This variable measures the percentage of households with broadband access from 2015 to 2019.

1. Describe the distribution of `county20large$internet`, using a histogram and the mean, median, and skewness statistics. Note that since these 3142 counties represent the population of counties, the mean of this variable is the population value (μ).

2. Use the code provided next to create a new object named `web250` that represents a sample of internet access in 250 counties, drawn from `county20large$internet`.

```
set.seed(251)
web250 <- sample(county20large$internet, 250)
```

3. Once you've generated your sample, describe the distribution of `web250` using a histogram and the sample mean. How does this distribution compare to population values you produced in Question 1?

4. Create a new object that represents a sample of internet access rates in 750 counties, drawn from `county20large$internet`. Name this object `web750`. Describe the distribution of `web750` using a histogram and the sample mean. How does this distribution compare to population values you produced in Question 1? Does it resemble the population more closely than the distribution of `web250` does? If so, in what ways?

5. Use `MeanSE` and `MeanCI` to produce the standard errors and 95% confidence intervals for both `web250` and `web750`. Use words to interpret the substantive meaning of both confidence intervals. How are the confidence intervals and standard errors in these two samples different from each other? How do you explain these differences?

9 HYPOTHESIS TESTING

GET READY

In this chapter, the concepts used in Chapters 7 and 8 are extended to focus more squarely on making statistical hypothesis testing. The focus here is on taking the abstract ideas that are the foundation for hypothesis testing and applying them to some concrete examples. The only thing you need to load in order to follow among is the `anes20.rda` dataset.

THE LOGIC OF HYPOTHESIS TESTING

When engaged in the process of hypothesis testing, we are essentially asking, "What is the probability that the statistic found in the sample could have come from a population in which it is equal to some other, specified, value?" As discussed in Chapter 8, social scientists want to know something about a population value but frequently are only able to work with sample data. We generally think the sample data represent the population fairly well, but we know that there will be some sampling error. In Chapter 8, we took this into account using confidence intervals around sample statistics. In this chapter, we apply some of the same logic to determine if the sample statistic is *different enough* from a hypothesized population parameter that we can be confident it did not occur just due to sampling error. (Come back and reread this paragraph when you are done with this chapter; it will make a lot more sense then.)

We generally consider two different types of hypotheses, the null and alternative (or research) hypotheses.

- **Null hypothesis** (H_0): This hypothesis is tested directly. It usually states that the population parameter (μ) is equal to some specific value, even if the sample statistic ($\bar{x}$) is a different value. The implication is that the difference between the sample statistic and the hypothesized population parameter is attributable to sampling error, not a real difference. We usually hope to reject the null hypothesis. This might sound strange now, but it will make more sense to you soon.
- **Alternative (research) hypothesis** (H_1): This is a substantive hypothesis that we think is true. Usually, the alternative hypothesis posits that the population parameter does not equal the value specified in H_0. We don't actually test this hypothesis directly.

Rather, we try to build a case for it by showing that the sample statistic is *different enough* from the population value hypothesized in H_0 that it is unlikely that the null hypothesis is correct.

We can use what we know about the z-distribution to test the validity of the null hypothesis by stating and testing hypotheses about specific values of population parameters. Consider the following problem:

A data analyst in the human resources department for a large metropolitan county is asked to evaluate the impact of a new method of documenting sick leave among county employees. The new policy is intended to cut down on the number of sick leave hours taken by workers. Last year, the **average** number of hours of sick leave taken by workers was **59.2** (about 7.4 days), a level determined to be too high. To evaluate if the new policy is working, the analyst took a sample of 100 workers at the end of one year under the new rules and found a sample mean of 54.8 hours (about 6.8 days) and a standard deviation of 15.38. The question is, does this sample mean represent a real change in sick leave use, or does it only reflect sampling error? To answer this, we need to determine how likely it is to get a sample mean of 54.8 from a population in which $\mu = 59.2$.

Using Confidence Intervals

As alluded to at the end of Chapter 8, you already know one way to test hypotheses about population parameters by using confidence intervals. In this case, the county data analyst can calculate the lower- and upper-limits of a 95% confidence interval around the sample mean (54.8) to see if it includes μ (59.2):

$$c.i._{.95} = 54.8 \pm 1.96(S_{\bar{x}})$$

$$S_{\bar{x}} = \frac{15.38}{\sqrt{100}} = 1.538$$

$$c.i._{.95} = 54.8 \pm 1.96(1.538)$$

$$c.i._{.95} = 54.8 \pm 3.01$$

$$51.78 \leq \mu \leq 57.81$$

From this sample of 100 employees, after one year of the new policy in place, the county analyst is able to estimate that there is a 95% chance that the interval between 51.78 and 57.81 includes μ, and the probability that μ is outside this range is less than .05. *Based on this alone, we can say there is less than a 5% chance that the number of hours of sick leave taken is the same that it was in the previous year.* In other words, there is a fairly high probability that fewer sick leave hours were used in the year after that policy change than in the previous year.

DIRECT HYPOTHESIS TESTS

We can be a bit more direct and precise by setting this up as a hypothesis test and then calculating the probability that the sample mean came from a population in which the null hypothesis is true. First, the null hypothesis.

$$H_0: \mu = 59.2$$

This is saying that there is no real difference between last year's mean number of sick days (μ) and the sample we've drawn from this year ($\bar{x}$). Even though the sample mean *looks* different from 59.2, the true population mean is 59.2, and the sample statistic is just a result of random sampling error. After all, if the population mean *is* equal to 59.2, any sample drawn from that population will produce a mean that is different from 59.2, due to sampling error. In other words, H_0 is saying that the new policy had no effect, even though the sample mean suggests otherwise.

Because the county analyst is interested in whether the new policy reduced the use of sick leave hours, the alternative hypothesis is

$$H_1: \mu < 59.2$$

Here, we are saying that the sample statistic is different enough from the hypothesized population value (59.2) that it is unlikely to be the result of random chance; that it represents real change, and the population value is less than 59.2.

Note here that we are not testing whether the number of sick days is equal to 54.8 (the sample mean). Instead, we are testing whether the average hours of sick leave taken this year is lower than the average number of sick days taken last year. The alternative hypothesis reflects what we really think is happening; it is what we're really interested in. We test the null hypothesis as a way of gathering evidence to support the alternative.

Sampling Distributions and Hypothesis Testing

To test the null hypothesis, we need to ask how likely it is that a sample mean of this magnitude (54.8) could be drawn from a population in which $\mu = 59.2$. We know that we would get lots of different mean outcomes if we took repeated samples from this population. We also know that most of them would be clustered near μ and a few would be relatively far away from μ at both ends of the distribution. All we have to do is estimate the probability of getting a sample mean of 54.8 from a population in which $\mu = 59.2$. If the probability of drawing $\bar{x}$ from μ is small enough, then we can reject H_0.

We can assess this probability by converting the difference between $\bar{x}$ and μ into a *z*-score and then use what we know about sampling distributions to calculate the probability of getting $\bar{x}$ from a population in which the null hypothesis is true (see Figure 9.1).

FIGURE 9.1 ■ The Logic of Hypothesis Testing

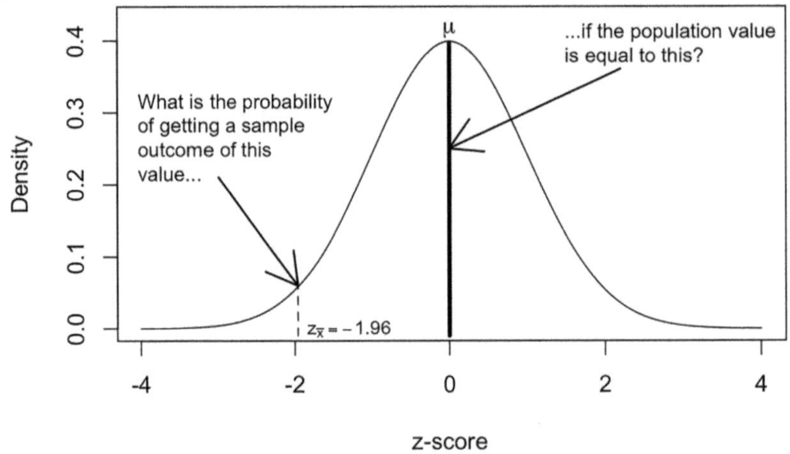

Note: Inspired by a figure on p. 164 of *Statistical Analysis* by S. K. Kachigan, 1986, Radius Press.[1]

The essential question in hypothesis testing is, "What is the probability of getting a particular sample outcome ($\bar{x}$) from a population in which the population value is equal to something else (μ)?" The process for answering that question is laid out in Figure 9.1, which illustrates a sampling distribution in which the *z*-score for a sample mean ($\bar{x}$) is -1.96. We can calculate the probability of getting a sample mean of this magnitude or lower ($z = -1.96$) by estimating the area under the curve to the left of -1.96. The area on the tail of the distribution used for hypothesis testing is referred to as the α (alpha) area. As shown next, the alpha area for $z = -1.96$ is .025. This result is also commonly referred to as the ***p*-value**, for probability value.

```
pnorm(-1.96)
[1] 0.0249979
```

With this information, we can say that the probability of drawing a sample mean with a *z*-score less than or equal to -1.96 from a population in which the null hypothesis is true ($\mu = 0$, in this case) is about .025. This can also be interpreted as saying that the probability that the observed sample outcome occurred by chance is about .025. This is pretty low, so we reject the null hypothesis. The smaller the *p*-value, the more confident you can be in rejecting the null hypothesis.

To reiterate, the critical step in moving from empirical data to hypothesis testing is to convert your sample statistic to a *z*-score. Once you've done that, you simply use what you know about sampling distributions to test the null hypothesis.

Critical Values

A common and fairly quick way to use the *z*-score in hypothesis testing is by comparing it to the critical value (*cv*) for *z*. The critical value is the *z*-score associated with the probability level

(p-value) required to reject the null hypothesis. To determine the critical value of z, we need to determine what the *probability threshold* is for rejecting the null hypothesis. In other words, how confident do you have to be that the observed sample outcome did not come from a population in which the null hypothesis is true before you can reject the null hypothesis? In the social sciences, it is fairly standard to consider any probability level lower than .05 sufficient for rejecting the null hypothesis. This probability level is also known as the *significance level.*

Typically, the critical value is the z-score that gives us .05 as the area on the tail (left in this case) of the normal distribution. Looking at the z-score table from Chapter 6 or using the `qnorm` function in R, we see that this is $z = -1.645$. The area beyond the critical value is referred to as the critical region and is sometimes also called the area of rejection: If the z-score falls in this region, the null hypothesis is rejected.

```
#Get the z-score for .05 area at the lower tail of the distribution

qnorm(.05, lower.tail = T)
[1] -1.644854
```

Once we have the cv, we can calculate the z-score for the difference between $\bar{x}$ and μ. The z-score will be positive if $\bar{x} - \mu > 0$ and negative if $\bar{x} - \mu < 0$. If $|z| > |z_{cv}|$, then we reject the null hypothesis.

So let's get back to the sick leave example:

- First, what's the critical value? −1.65 (Make sure you understand why this is the value.)
- What is the obtained value of z?

$$z = \frac{\bar{x} - \mu}{S_{\bar{x}}} = \frac{54.8 - 59.2}{1.538} = \frac{-4.4}{1.538} = -2.86$$

This z-score of −2.86 means that the sample mean (54.8) is 2.86 standard errors lower than the hypothesize population value (59.2).

- If the $|z|$ is greater than the $|cv|$, then reject H_0. If the $|z|$ is less than the critical value, then fail to reject $H_{0.}$

In this case, z (−2.86) is of much greater (absolute) magnitude than cv (−1.65), so we *reject the null hypothesis* and conclude that μ is probably less than 59.2. By rejecting the null hypothesis, we build a case for the alternative hypothesis, though we never test the alternative directly. Rejecting the null hypothesis means that we are saying that the difference is statistically significant, representing a real difference rather than random error.

p-values

We can be a bit more precise about the level of confidence in rejecting the null hypothesis (the level of statistical significance) by estimating the alpha area to the left of $z = -2.86$.

```
#Area under to curve to the left of -2.86

pnorm(-2.86)
[1] 0.002118205
```

This alpha area (or p-value) is close to zero, reinforcing the conclusion that we can confidently reject the null hypothesis. Check out Figure 9.2 as an illustration of how unlikely it is to get a sample mean of 54.8 (thick solid line) from a population in which $\mu = 59.2$ (thin solid line), based on our sample statistics. Remember, the area to the left of the critical value (dashed line) is the critical region, equal to .05 of the area under the curve, and the sample mean is far to the left of this point.

One useful way to think about this p-value (and others like it) is that if we took 1000 samples of 100 workers from a population in which $\mu = 59.2$ and calculated the mean hours of sick leave taken for each sample, only two samples would give you a result equal to or less than 54.8, simply due to sampling error. In other words, there is a 2/1000 chance that the sample mean was the result of random variation instead of representing a real difference from the hypothesized value.

FIGURE 9.2 ■ An Illustration of Key Concepts in Hypothesis Testing

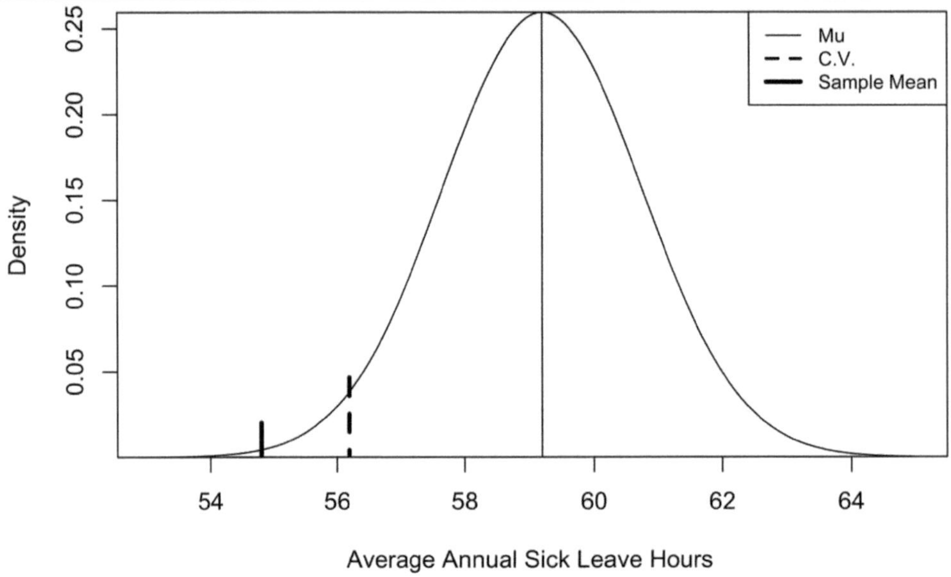

One Tail or Two?

Note that we were explicitly testing a one-tailed hypothesis in the preceding example. We were saying that we expect a decrease in the number of sick days used, due to the new policy. But suppose someone wanted to argue that there was a loophole in the new policy that might make it easier for people to take sick days. These sorts of unintended consequences almost always occur with new policies. Given that it could go either way (μ could be higher or lower than 59.2), the county data analyst who is evaluating this policy might decide to test a two-tailed hypothesis: that the new policy could create a difference in sick day use, and that difference could be positive or it could be negative:

$$H_1 : \mu \neq 59.2$$

The process for testing two-tailed hypotheses is exactly the same, except that we use a larger critical value because even though the α area is the same (.05), we must now split it between two tails of the distribution. Again, this is because we are not sure if the policy will increase or decrease sick leave. When the alternative hypothesis does not specify a direction, the two-tailed test is most appropriate.

The Figure 9.3 illustrates the difference in critical values for one- and two-tailed hypothesis tests. Since we are splitting .05 between the two tails, the *cv* for a two-tailed test is now the *z*-score that gives us .025 as the area at the tails of the distribution. Using the `qnorm` function in R (following), we see that this is $z = -1.96$, which we take as ± 1.96 for a two-tailed test critical value ($p = .05$).

```
#Z-score for .025 area at one tail of the distribution

qnorm(.025)
[1] -1.959964
```

If we obtain a *z*-score (positive or negative) that is larger in absolute magnitude than this, we reject H_0. Using a two-tailed test requires a larger *z*-score, making it slightly harder to reject the

FIGURE 9.3 ■ Critical Values for One- and Two-Tailed Tests

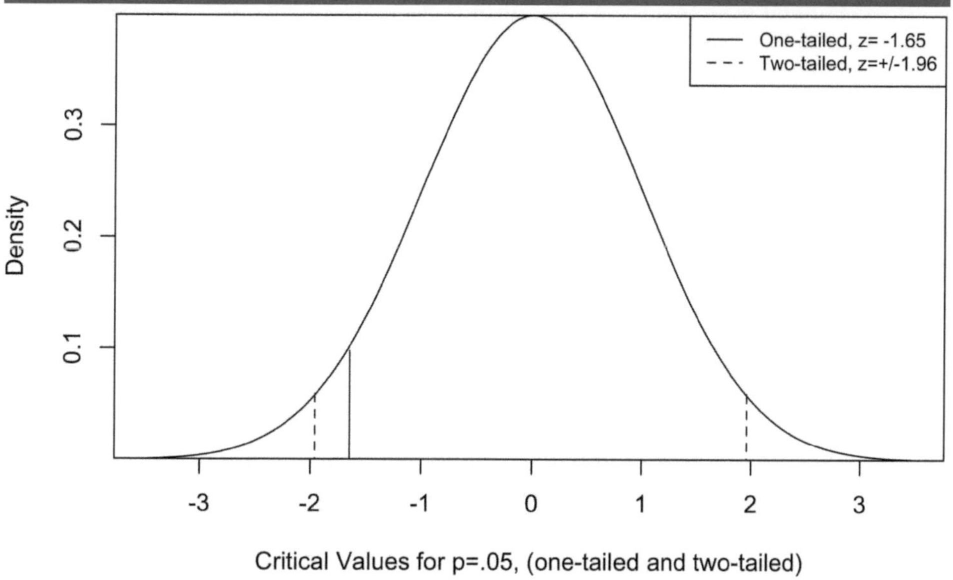

null hypothesis. However, since the z-score in the sick leave example was −2.86, we would still reject H_0 under a two-tailed test.

In truth, the choice between a one- or two-tailed test rarely makes a difference in rejecting or failing to reject the null hypothesis. The choice matters most when the p-value from a one-tailed test is greater than .025 but less than .05 (just barely into the rejection area), in which case it would be greater than .05 in a two-tailed test. It is worth scrutinizing findings from one-tailed tests that are just barely statistically significant to see if a two-tailed test would be more appropriate. Because it provides a more conservative basis for rejecting the null hypothesis, researchers often choose to report two-tailed significance levels even when a one-tailed test could be justified. Many statistical programs, including R, report two-tailed p-values by default.

T-DISTRIBUTION

Thus far, we have focused on using z-scores and the z-distribution for testing hypotheses and constructing confidence intervals. Another distribution available to us is the t-distribution. The t-distribution has an important advantage over the z-distribution: It does not assume that we know the population standard error. This is very important because we rarely know the population standard error. In other words, the t-distribution assumes that we are using an *estimate* of the standard error. As shown in Chapter 8, the estimate of the standard error of the mean is

$$S_{\bar{x}} = \frac{S}{\sqrt{N}}.$$

$S_{\bar{x}}$ is our best guess for $\sigma_{\bar{x}}$, but it is based on a sample statistic, so it does involve some level of error.

To account for the fact that we are estimating the standard error with sample data rather than the population, the t-distribution is somewhat flatter (see Figure 9.4) than the z-distribution. Comparing the two distributions, you can see that they are both perfectly symmetric but that the t-distribution is a bit more squat and has slightly fatter tails. This means that the critical value for a given level of significance will be larger in magnitude for a t-score than for a z-score. This difference is especially noticeable for small samples and virtually disappears for samples greater than 100, at which point the t-distribution becomes almost indistinguishable from the z-distribution (see Figure 9.5).

Now, here's the fun part—the t-score is calculated the same way as the z-score:

$$t = \frac{\bar{x} - \mu}{S_{\bar{x}}}$$

We use the t-score and the t-distribution in the same way and for the same purposes that we use the z-score: choose a desired level of significance (p-value), find the critical value for the t-score, calculate the t-score, compare the t-score to the critical value, and make a decision about the null hypothesis.

FIGURE 9.4 ■ Comparison of Normal and *t*-distributions

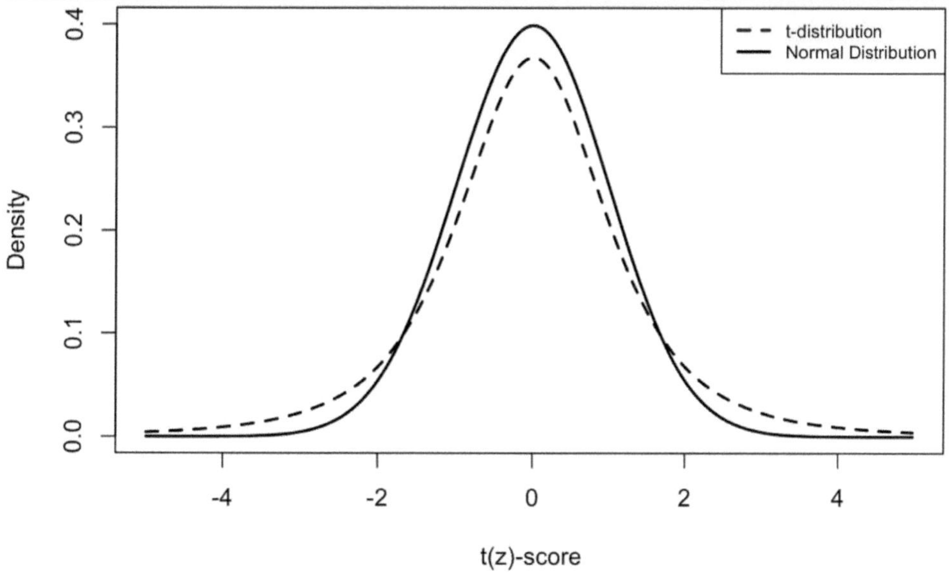

FIGURE 9.5 ■ Degrees of Freedom and Resemblance of *t*-distribution to the Normal Distribution

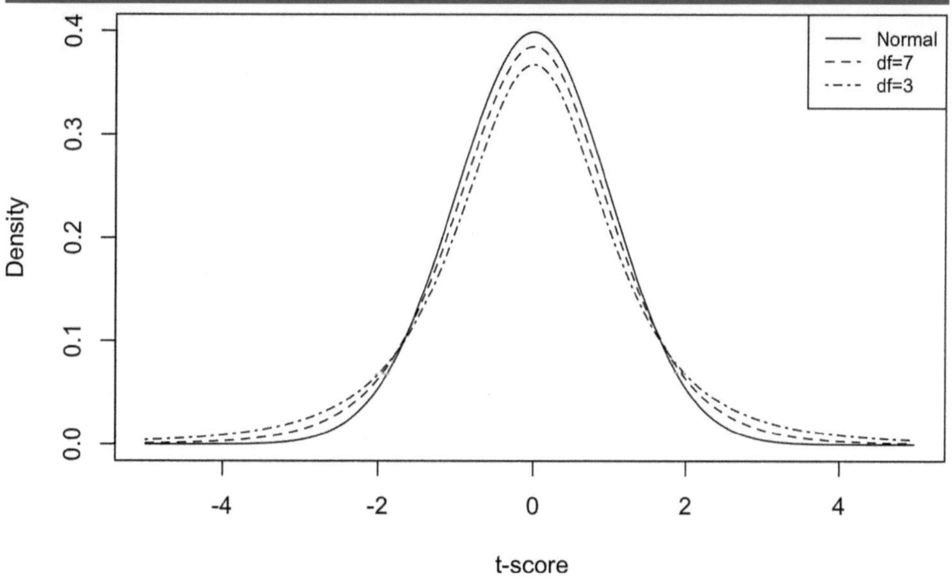

While everything here looks about the same as the process for hypothesis testing with *z*-scores, determining the critical value for a *t*-distribution is somewhat different and depends

upon sample size. This is because we have to consider something called **degrees of freedom (*df*)**. Taking degrees of freedom into account addresses an issue discussed in Chapter 8, that sample data tend to slightly underestimate the variance and that this underestimation is a bigger problem with small samples. For testing hypotheses about a single mean, degrees of freedom equal

$$df = n - 1.$$

So for the sick leave example used previously,

$$df = 100 - 1 = 99.$$

You can see the impact of sample size (through degrees of freedom) on the shape of the *t*-distribution in Figure 9.5: As sample size and degrees of freedom increase, the *t*-distribution grows more and more similar to the normal distribution. At $df = 100$ (not shown here) the *t*-distribution is virtually indistinguishable from the *z*-distribution.

There are two different methods you can use to find the critical value of *t* for a given level of degrees of freedom. You can go "old school" and look it up in a *t*-distribution table (Table 9.1),[2] or you can ask R to figure it out for you. It's easier to rely on R for this, but there is some benefit to going old school at least once. In particular, it helps reinforce how degrees of freedom, significance levels, and critical values fit together. You should follow along.

TABLE 9.1 ■ *t*-score Critical Values at Different *p*-values and Degrees of Freedom

	p(one-tail)	0.10	0.05	0.025	0.01	0.005
	p(two-tails)	0.20	0.10	0.05	0.02	0.010
df	10	1.372	1.812	2.228	2.764	3.169
	11	1.363	1.796	2.201	2.718	3.106
	12	1.356	1.782	2.179	2.681	3.055
	13	1.350	1.771	2.160	2.650	3.012
	14	1.345	1.761	2.145	2.624	2.977
	15	1.341	1.753	2.131	2.602	2.947
	16	1.337	1.746	2.120	2.583	2.921
	17	1.333	1.740	2.110	2.567	2.898
	18	1.330	1.734	2.101	2.552	2.878
	19	1.328	1.729	2.093	2.539	2.861
	20	1.325	1.725	2.086	2.528	2.845

p(one-tail)	0.10	0.05	0.025	0.01	0.005
p(two-tails)	0.20	0.10	0.05	0.02	0.010
21	1.323	1.721	2.080	2.518	2.831
22	1.321	1.717	2.074	2.508	2.819
23	1.319	1.714	2.069	2.500	2.807
24	1.318	1.711	2.064	2.492	2.797
25	1.316	1.708	2.060	2.485	2.787
26	1.315	1.706	2.056	2.479	2.779
27	1.314	1.703	2.052	2.473	2.771
28	1.313	1.701	2.048	2.467	2.763
29	1.311	1.699	2.045	2.462	2.756
30	1.310	1.697	2.042	2.457	2.750
40	1.303	1.684	2.021	2.423	2.704
50	1.299	1.676	2.009	2.403	2.678
60	1.296	1.671	2.000	2.390	2.660
70	1.294	1.667	1.994	2.381	2.648
80	1.292	1.664	1.990	2.374	2.639
90	1.291	1.662	1.987	2.368	2.632
100	1.290	1.660	1.984	2.364	2.626
120	1.289	1.658	1.980	2.358	2.617
∞	1.282	1.645	1.960	2.326	2.576

The first step is to decide if you are using a one-tailed or two-tailed test and then decide what the desired level of significance (*p*-value) is. For instance, for the sick leave policy example, we can assume a one-tailed test with a .05 level of significance. The relevant column of the table is found by going across the top row of *p*-values to the column headed by .05. Then, scan down the column until we find the point where it intersects with the appropriate degree of freedom row. In this example, $df = 99$, but there is no listing of $df = 99$ in the table so we will err on the side of caution and use the next lowest value, 90. The .05 one-tailed level of significance column intersects with the $df = 90$ row at $t = 1.662$, so -1.662 is the critical value of t in the sick leave example. Note that it is only slightly different than the *cv* for z that we used in the sick leave calculations, -1.65. This is because the sample size is relatively large (in statistical terms) and the t-distribution closely approximates the z-distribution for large samples. So, in this case, the z- and t-distributions lead to the same outcome, we decide to reject H_0.

Alternatively, we could ask R to provide this information using the `qt` function. For this, you need to declare the desired *p*-value and specify the degrees of freedom, and R reports the critical value.

```
#Calculate t-score for .05 at the lower tail, with df = 99
#The command is: qt(alpha, df)

qt(.05, 99)
[1] -1.660391
```

By default, `qt()` provides the critical values for a specified alpha area at the **lower tail** of the distribution (hence, −1.66). To find the *t*-score associated with an alpha area at the right (upper) tail of the distribution, just add `lower.tail = F` to the command.

```
#Specifying "lower.tail = F" instructs R to find the upper tail area.

qt(.05, 99, lower.tail = F)
[1] 1.660391
```

For a two-tailed test, you need to cut the alpha area in half, because you are splitting .05 between two tails.

```
#Calculate t-score for .025 at one tail, with df = 99

qt(.025, 99)
[1] -1.984217
```

Here, R reports a critical value of −1.984, which we take as ±1.984 for a two-tailed test from a sample with $df = 99$. Again, this is slightly larger than the critical value for a *z*-score (1.96). If you used the *t*-score table to do this the old-school way, you would find the critical value is $t = 1.99$, for $df = 90$. The results from using the `qt` function are more accurate than from using the *t*-table since you are able to specify the correct degrees of freedom.

Whether using a one- or two-tailed test, the conclusion for the sick leave example is unaffected: the *t*-score obtained from the sample (−2.68) is in the critical region, so reject H_0.

We can also get a bit more precise estimate of the probability of getting a sample mean of 54.8 from a population in which $\mu = 59.2$ by using the `pt()` function to get the area under the curve to the left of $t = -2.86$.

```
pt(-2.86, df = 99)
[1] 0.002583714
```

Note that this result is very similar to what we obtained when using the *z*-distribution (.002118). To get the area under the curve to the right of a positive *t*-score, add `lower.tail = F` to the command.

```
#Specifying "lower.tail = F" instructs R to find the area to the right of
#the t-score

pt(2.86, df = 99, lower.tail = F)
[1] 0.002583714
```

For a two-tailed test using the t-distribution, we double this to find a p-value equal to .005167.

PROPORTIONS

As discussed in Chapter 8, the logic of hypothesis testing about mean values also applies to proportions. For example, in the sick leave example, instead of testing whether $\mu = 59.2$, we could test a hypothesis regarding the proportion of employees who take a certain number of sick days. Let's suppose that in the year before the new policy went into effect, 50% of employees took at least 7 sick days. If the new policy has an impact, then the proportion of employees taking at least 7 days of sick leave during the year after the change in policy should be lower than .50. In the sample of 100 employees used, the proportion of employees taking at least 7 sick days was .41. In this case, the null and alternative hypotheses are the following:

$$H_0: P = .50$$
$$H_1: P < .50$$

To review, in the previous example, to test the null hypothesis, we established a desired level of statistical significance (.05), determined the critical value for the t-score (−1.66), calculated the t-statistic, and compared it the critical value. There are a couple of differences, however, when working with hypotheses about the population value of proportions.

Because we can calculate the *population* standard deviation based on the hypothesized value of P (.5), we can use the z-distribution rather than the t-distribution to test the null hypothesis. This is kind of a confusing point. Because the hypothesis is telling us what the population proportion is, it is also, by default, telling us the value of the population standard deviation. If the population value is .50, then the *population* standard deviation is $\sqrt{P(1 - P)}$ and the standard error can be calculated as follows:

$$S_p = \sqrt{\frac{P(1 - P)}{n}} = \sqrt{\frac{.5(.5))}{100}} = .05$$

Since this is the based the hypothesized population standard deviation, we use a z-score, calculated as before.

$$z = \frac{p - P}{S_p}$$

Using the data from the problem, this gives us the following:

$$z = \frac{p - P}{S_p} = \frac{.41 - .5}{.05} = \frac{-.09}{.05} = -1.8$$

We know from before that the critical value for a one-tailed test using the z-distribution is −1.65. Since this z-score is larger (in absolute terms) than the critical value, we can reject the null hypothesis and conclude that the proportion of employees using at least 7 days of sick leave per year is lower than it was in the year before the new sick leave policy went into effect.

Again, we can be a bit more specific about the p-value.

```
pnorm(-1.8)
[1] 0.03593032
```

Here are a couple of things to think about with this finding. First, while the p-value is lower than .05, it is not much lower. In this case, if you took 1000 samples of 100 workers from a population in which $P = .50$ and calculated the proportion who took 7 or more sick days, approximately 36 of those samples would produce a proportion equal to .41 or lower, just due to sampling error. This still means that the probability of getting this sample finding from a population in which the null hypothesis was true is pretty small (.03593), so we should be comfortable rejecting the null hypothesis. But what if there were good reasons to use a two-tailed test? Would we still reject the null hypothesis? No, because the critical value for a two-tailed test (−1.96) would be larger in absolute terms than the z-score, and the p-value would be .07186. These findings stand in contrast to those from the analysis of the *average number* of sick days taken, where the p-values for both one- and two-tailed tests were well below the .05 cut-off level.

One of the take-home messages from this example is that confidence in research findings is sometimes fragile, since "significance" can be a function of how you frame the hypothesis test (one- or two-tailed test?) or how you measure your outcomes (average hours of sick days taken or proportion who take a certain number of sick days). For this reason, it is always a good idea to be mindful of how the choices you make might influence your findings.

TYPES OF ERROR

In both the difference in means and difference in proportions tests we just discussed, we concluded that there was decrease in the use of sick leave in the year following the implementation of the new sick leave policy. Based on the p-values from the hypothesis tests, we rejected the null hypotheses, concluding that there was a small enough probability of getting the sample findings from a population in which H_0 was true that it was unlikely to be true.

But what if we are wrong? What if there actually was no change in sick leave usage in the year following the implementation of the new policy but we concluded there was? Is this possible? Yes, it is possible to make errors like this when testing hypotheses. Generally, we think of these types of errors as **Type I** and **Type II** errors, as described in Figure 9.6.

In the current example, we rejected H_0 and concluded, based on sample findings, that there was a significant decline in sick leave usage in the year following the implementation of a new sick leave policy. If, in the population, the null hypothesis were actually true (there was no change in sick leave use), then we would be making a Type I error, rejecting H_0 when it is true. This is also referred to as an alpha (α) error, because the probability of committing a Type I error is equal to the p-value used to decide if the null hypothesis should be rejected. In both cases, we used a .05 threshold for rejection, but we also know from the p-values that the probability of making a Type I error is .0359 for the difference in proportions test and .0052 for the difference in means test. Given this, we know that the probability of making a Type I error is low.

FIGURE 9.6 ■ Potential Errors in Hypothesis Testing

Research Decision	Reality: H_0: True	Reality: H_0: False
Reject H_0	Type I Error (alpha)	No Error
Fail to Reject H_0	No Error	Type II Error (beta)

The other type of error that can occur in hypothesis testing is a Type II, or beta (β), error, where you fail to reject H_0 even though it is false. Suppose in the sick leave example, we had decided that we wanted to be very certain before rejecting H_0, so we decided to use a *p*-value of .01 as the threshold for rejection instead of .05. In that case, we would have failed to reject H_0 in the difference in proportions test since the *p*-value there was .0359. By requiring a smaller *p*-value, we decreased the probability of a Type I error but increased the likelihood of committing a Type II error, concluding that the sick leave policy had no impact when in fact it did. This illustrates that the two types of errors are inversely related to each other: As the probability of Type I error decreases, perhaps by requiring smaller *p*-values before rejecting H_0, the probability of a Type II error increases.

One way to minimize both Type I and Type II errors is to increase the sample size. If you are able to control the size of the sample, moving from small to reasonably large samples pays dividends, as shown in Chapter 8.

T-TEST IN R

Let's say you are looking at data on public perceptions of the presidential candidates in 2020 and you have a sense that people had mixed feelings about Democratic nominee, Joe Biden, going into the election. This leads you to expect that his average rating on the 0 to 100 feeling thermometer scale from the ANES was probably about 50. You decide to test this directly with the `anes20` dataset.

The null hypothesis is

$$H_0: \mu = 50.$$

Because there are good arguments for expecting the mean to be either higher or lower than 50, the alternative hypothesis is two-tailed:

$$H_1: \mu \neq 50$$

First, you get the sample mean.

```
#Get the sample mean for Biden's feeling thermometer rating
mean(anes20$V202143, na.rm = T)
[1] 53.41213
```

Here, you see that the mean feeling thermometer rating for Biden in the fall of 2020 was 53.41. This is higher than what you thought it would be (50), but you know that it's possible to get a sample outcome of 53.41 from a population in which the mean is actually 50, so you need to do a *t*-test to rule out sampling error as a reason for the difference.

In R, the command for a one-sample two-tailed *t*-test is relatively simple; you just have to specify the variable of interest and the value of μ under the null hypothesis.

```
#Use 't.test' and specify the variable and mu
t.test(anes20$V202143, mu = 50)

    One Sample t-test

data:  anes20$V202143
t = 8.1805, df = 7368, p-value = 3.303e-16
alternative hypothesis: true mean is not equal to 50
95 percent confidence interval:
 52.59448 54.22978
sample estimates:
mean of x
 53.41213
```

These results are pretty conclusive, the *t*-score is 8.2 and the *p*-value is very close to 0.[3] Also, if it makes more sense for you to think of this in terms of a confidence interval, the 95% confidence interval ranges from about 52.6 to 54.2, which does not include 50. You should reject the null hypothesis and conclude instead that Biden's feeling thermometer rating in the fall of 2020 was greater than 50.

Even though Joe Biden's feeling thermometer rating was greater than 50, from a substantive perspective, it is important to note that a score of 53 does not mean Biden was wildly popular, just that his rating was greater than 50. This highlights an important thing to remember: Statistically significant results are not necessarily substantively large results. This point is addressed at greater length in the next several chapters where we explore measures of substantive importance that can be used to complement measures of statistical significance.

NEXT STEPS

The last three chapters have given you a foundation in the principles and mechanics of sampling, statistical inference, and hypothesis testing. Everything you have learned thus far is interesting and important in its own right, but what is most exciting is that it prepares you for testing hypotheses about outcomes of a dependent variable across two or more categories of an independent variable. In other words, you now have the tools necessary to begin looking at relationships among variables. We take this up in the next chapter by looking at differences in outcomes across two groups. Following that, we test hypotheses about outcomes across multiple groups in Chapters 11 through 13. In each of the next several chapters, we continue to focus on methods of statistical inference, exploring alternative ways to evaluate statistical significance. At the same time, we also introduce the idea of evaluating the *strength* of relationships by focusing on measures of *effect size*. Both of these concepts—statistical significance and effect size—continue to play an important role in the remainder of the book.

EXERCISES

Concepts and Calculations

1. The survey of 300 college students introduced in the end-of-chapter exercises in Chapter 8 found that the average semester expenditure was \$350 with a standard deviation of \$78. At the same time, campus administration has done an audit of required course materials and claims that the average cost of books and supplies for a single semester should be no more than \$340. In other words, the administration is saying the population value is \$340.
   a. State a null and alternative hypothesis to test the administration's claim. Did you use a one- or two-tailed alternative hypothesis? Explain your choice.
   b. Test the null hypothesis and discuss the findings. Show all calculations.

2. The same survey reports that among the 300 students, 55% reported being satisfied with the university's response to the COVID-19 pandemic. The administration hailed this finding as evidence that a majority of students support the course they've taken in reaction to the pandemic. (Hint: this is a "proportion" problem.)
   a. State a null and alternative hypothesis to test the administration's claim. Did you use a one- or two-tailed alternative hypothesis? Explain your choice.
   b. Test the null hypothesis and discuss the findings. Show all calculations.

3. Determine whether the null hypothesis should be rejected for the following pairs of t-scores and critical values:
   a. $t = 1.99$, $cv = 1.96$
   b. $t = 1.64$, $cv = 1.65$
   c. $t = -2.50$, $cv = -1.96$
   d. $t = 1.55$, $cv = 1.65$
   e. $t = -1.85$, $cv = -1.96$

4. A researcher at the county elections office was interested in whether voter turnout in the county had changed significantly since the last election, when it was 59%. They were impatient and didn't want to tabulate all of the precincts, so they took a random sample of 100 precincts and found that turnout in the sample of precincts was 53%. The t-score for the difference is -2.02. and the p-value was .023.

a. In this example, is the research testing a one- or two-tailed hypothesis?
b. What is the value of μ?
c. Does the evidence here support rejecting the null hypothesis that there is no change in turnout?
d. What is the probability of committing a Type I error in this example?
e. Suppose the researcher decides to use a .01 instead of .05 threshold for rejecting the null hypothesis. Does this affect the probability of making Type I and Type II errors? Explain.

R Problems

For the first three questions, use the feeling thermometers for Donald Trump (`anes20$V202144`), liberals (`anes20$V202161`), and conservatives (`anes20$V202164`).

1. Using descriptive statistics and either a histogram, boxplot, or density plot, describe the central tendency and distribution of each feeling thermometer.

2. Use the `t.test` function to test the null hypotheses that the mean for each of these variables in the population is equal to 50. State the null and alternative hypotheses and interpret the findings from the *t*-test.

3. Use the `pt()` function to calculate the *p*-value (area at the tail) for each of the following *t*-scores (assume one-tailed tests):
   a. $t = 1.45$, $df = 49$
   b. $t = 2.11$, $df = 30$
   c. $t = -.69$, $df = 200$
   d. $t = -1.45$, $df = 100$

4. What are the *p*-values for each of the *t*-scores listed in Problem 4, if you assume a two-tailed test?

5. Treat the *t*-scores from Problem 4 as *z*-scores and use the `pnorm()` to calculate the *p*-values. List the *p*-values and comment on the differences between the *p*-values associated with the *t*- and *z*-scores. Why are some closer in value than others?

10 HYPOTHESIS TESTING WITH TWO GROUPS

GET READY

This chapter extends the discussion of hypothesis testing to include the comparison of means and proportions across population subgroups. To follow along in R, you should load the `anes20.rda` dataset. We will be using a lot of techniques from separate packages, so also make sure you attach the following libraries: `DescTools`, `Hmisc`, `gplots`, `descr`, and `effectsize`.

TESTING HYPOTHESES ABOUT TWO MEANS

A common use of hypothesis testing involves examining the difference between two sample means. For instance, instead of testing hypotheses about the level of support for a political candidate or some group in the population as a whole, it is often more interesting to speculate about differences in support across population subgroups, such as men and women, whites and people of color, city dwellers and suburbanites, religious and secular voters, and so forth. When comparing subgroups like these, we are really asking if variables such as sex, race, place of residence, and religiosity influence or are related to differences in the dependent variable.

Let's begin looking at sex-based differences in political attitudes, commonly referred to as the *gender gap*. We'll start with a somewhat obvious dependent variable for testing the presence of a gender gap in political attitudes, the feeling thermometer rating for feminists. As a quick reminder, the feeling thermometers in the American National Election Study ask people to rate how they feel about certain groups and individuals on a 0 (negative, "cool" feelings) to 100 (positive, "warm" feelings) scale. Which group do you think is likely to rate feminists the highest, men or women?[1] Although both women and men can be feminists (or anti-feminists), the connection between feminism and the fight for the rights of women leads quite reasonably to the expectation that women support feminists at higher levels than men do.

Before shedding light on this with data, let's rename the ANES measures for respondent sex (`anes20$V201600`) and the feminist feeling thermometer (`anes20$V202160`) so they are a bit easier to use in subsequent commands.

```
#Create new respondent sex variable
anes20$Rsex <- factor(anes20$V201600)
#Assign category labels
levels(anes20$Rsex) <- c("Male", "Female")
#Create new feminist feeling thermometer variable
anes20$femFT <- anes20$V202160
```

Generating Subgroup Means

There are a couple of relatively simple ways to examine subgroup means, in this case, the mean levels of support for feminists among men and women. First, you can use the `aggregate` function to get the subgroup means or many other statistics. Since we are thinking of subgroup analysis as a way of testing hypotheses, it is useful to think of the format of this function as `aggregate(dependent, by = list(independent), FUN = stat_you_want)`. In this case, we are telling R to generate the mean outcomes of the dependent variable for different categories of the independent variable.

For sex-based differences on the feminist feeling thermometer:

```
#Store the mean Feminist FT, by sex, in a new object
agg_femFT <- aggregate(anes20$femFT, by = list(anes20$Rsex),
                        FUN = mean, na.rm = TRUE)
#List the results of the aggregate command
agg_femFT

    Group.1      x
1      Male  54.54
2    Female  62.65
```

What this table shows is that the average feeling thermometer for feminists was 62.55 for women and 54.44 for men, a difference of about eight points on a scale from 0 to 100. So, it looks like there is a difference in attitudes toward feminists, with women viewing them more positively than men. At the same time, it is important to note that both groups, on average, have positive feelings toward feminists.

Another helpful R function we can use to compare the means of these two groups is `compmeans`, which produces subgroup means *and* standard deviations, as well as a boxplot that shows the distribution of the dependent variable for each value of the independent variable. The format for this command is `compmeans(dependent, independent)`. You should also include axis labels and other plot commands (see next) since a boxplot is included by default (you can suppress it with `plot = F`). When you run this command, you will get a warning message telling you that there are missing data. Don't worry about this for now unless the number of missing cases seems relatively large compared to the sample size.

```
#List the dependent variable first, then the independent variable.
#Add graph commands
compmeans(anes20$femFT, anes20$Rsex,
          xlab = "Sex",
          ylab = "Feminist Feeling Thermometer",
          main = "Feminist Feeling Thermometer, by Sex")
```

```
Mean value of "POST: Feeling thermometer: feminists" according to "anes20$Rsex"
           Mean        N Std.   Dev.
Male       54.54    3321        26.11
Female     62.55    3952        26.75
Total      58.89    7273        26.76
```

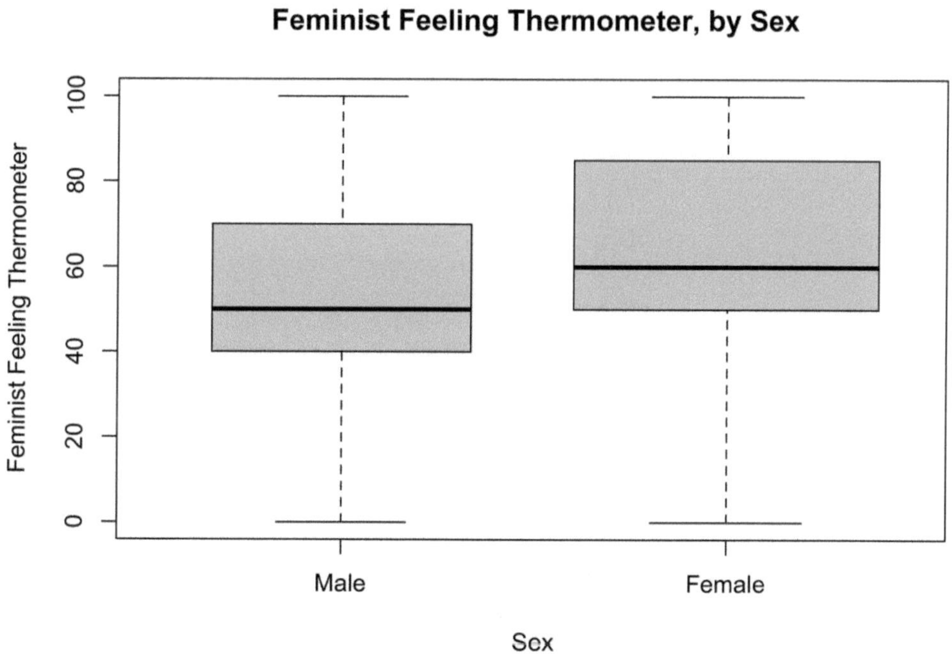

First, note that the subgroup means are the same as those produced using the `aggregate` command. The key difference in the numeric output is that you also get information on the standard deviation in both subgroups, as well as the mean and standard deviation for the full sample.

In addition, the boxplot produced by the `compmeans` command provides a visualization of the two subgroup distributions side-by-side, giving us a chance to see how similar or dissimilar they are. Remember, the boxplots do not show the differences in means, but they do show differences in medians and interquartile ranges, both of which can be indicative of group-based differences in outcomes on the dependent variable. The side-by-side distributions in the boxplot do appear to be different, with the distribution of outcomes concentrated a bit more at the high end of the feeling thermometer among women than among men.

So, it *looks* like there is a gender gap in attitudes toward feminists, with women rating feminists about eight points (8.01) higher than men rate them. But there is a potential problem with this conclusion. The main problem is that we are using sample data and we know from sampling theory that *it is possible to find a difference in the sample data even if there really is no difference in the population*. The question we need to answer is whether the difference we observe in this sample is *large enough* that we can reject the possibility that there is no difference between the two groups in the population. In other words, we need to expand the logic of the hypothesis test developed earlier to incorporate differences between two sample means.

HYPOTHESIS TESTING WITH TWO MEANS

When comparing means, the language of a hypothesis test changes just a bit.

$H_0: \mu_1 = \mu_2$ There is no relationship. The means are equal in the population.

Alternative hypotheses state that there is a relationship in the population:

$H_1: \mu_1 \neq \mu_2$ The means differ in the population (two tailed).

$H_1: \mu_1 < \mu_2$ There is a negative difference in the population (one tailed).

$H_1: \mu_1 > \mu_2$ There is a positive difference in the population (one tailed).

The logic of hypothesis testing for mean differences is very much the same as that for a single mean: If we observe a difference between two sample subgroup means, we must ask if there is really a difference between these groups in the population or if the sample difference is due to random variation. In other words, is the difference large enough that we can attribute it to something other than sampling error? If so, then we can reject the null hypothesis.

The difference between the two sample means ($\bar{x}_1 - \bar{x}_2$) is a sample statistic, so the sampling distribution for the difference between the two groups has the same properties as the sampling distribution for a single mean: If the sample difference comes from a large, random sample, the sampling distribution will follow a normal curve and the mean will equal the difference between the two subgroups in the population ($\mu_1 - \mu_2$).

A Theoretical Example

Figure 10.1 extends the example used in Chapter 9 to illustrate the logic of hypothesis testing for mean differences. If H_0 is true and there is no difference between the two groups in the population, how likely is it that we would get a sample difference of the magnitude of $\bar{x}_1 - \bar{x}_2$?

To answer this, we need to transform $\bar{x}_1 - \bar{x}_2$ from a raw score for the difference into a standard score (*t*- or *z*-score). Since we are working with a sample, we focus on *t*-scores in this application. Recall, though, that the calculation for a *t*-score is the same as for a *z*-score:

$$t = \frac{(\bar{x}_1 - \bar{x}_2) - (\mu_1 - \mu_2)}{S_{\bar{x}_1 - \bar{x}_2}}$$

However, since we always assume that $\mu_1 - \mu_2 = 0$, we are asking if the sample finding is different from 0, and the equation becomes

$$t = \frac{(\bar{x}_1 - \bar{x}_2)}{S_{\bar{x}_1 - \bar{x}_2}}.$$

So, as we did with a single mean, we divide the raw score difference by the *standard error of the sampling distribution* to convert the raw difference into a *t*-score. The standard error of the difference is a function of the variance in both subgroups, along with the sample sizes. Since we do not know the population variances, we rely on sample variances to estimate them:

$$S_{\bar{x}_1 - \bar{x}_2} = \sqrt{\frac{S_1^2}{n_1} + \frac{S_2^2}{n_2}},$$

FIGURE 10.1 ■ The Logic of Hypothesis Testing

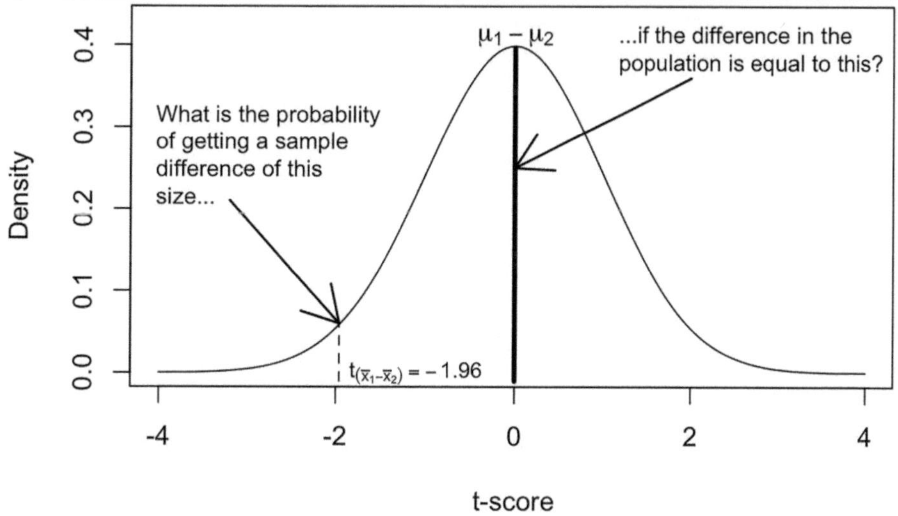

Note: Inspired by a figure on p. 164 of *Statistical Analysis* by S. K. Kachigan, 1986, Radius Press.[2]

where S_1^2 and S_2^2 are the sample variances of the subgroups.

The standard error represents the standard deviation of the sampling distribution that would result from repeated (large, random) samples from which we would calculate differences between the group means. The *t*-score expresses the raw difference between the two groups $(\bar{x}_1 - \bar{x}_2)$ relative to the standard error of the sampling distribution.

In Figure 10.1, the *t*-score for the difference is -1.96, which means that the sample difference is 1.96 standard errors below the hypothesized population difference. What is the probability of getting a *t*-score of this magnitude if there is really no difference between the two groups in the population? That probability is equal to the area to the left of $t = -1.96$ (the same for both *t*- and *z*-scores with a large sample), which is .025. So, with a one-tailed hypothesis, we would reject H_0 because there is less than a .05 probability that it is true.

Returning to the Empirical Example

Okay, now let's apply this to the sex-based differences in feeling thermometer rating for feminists. The null hypothesis, of course, states that there is no difference between the groups:

$$H_0: \mu_W = \mu_M$$

What about the alternative hypothesis? Do we expect the mean for women to be higher or lower than that of men? Since high values signify more positive evaluations, I anticipate that the mean for women is higher than the mean for men:

$$H_1: \mu_W > \mu_M$$

We can use the same process we used to test hypotheses about single means:

1. Choose a *p*-value (α area) for determining the level of statistical significance required for rejecting H_0 (usually .05).

2. Find the critical value of *t* associated with α. This depends on the degrees of freedom.

   When testing the difference between two means, degrees of freedom is equal to $n - 2$, reflecting the fact that we are using information from two means instead of one. In this case, the degrees of freedom is a very large number ($7273 - 2 = 7271$), and the critical value for a one-tailed test is -1.645, essentially the same as if we were using the *z*-distribution. Recall that as the sample size increases, the *t*-distribution grows increasingly similar to the *z*-distribution.

```
#Get critical values (t) for p = .05, df = 7271
qt(.05, 7271)
[1] -1.645
```

3. Calculate the *t*-score from the sample data.

4. Compare *t*-score to the critical value. If $|t| > |cv|$, then reject H_0; if $|t| < |cv|$, then fail to reject H_0.

Calculating the *t*-score

First, we'll plug the appropriate numbers into the *t*-score formula to illustrate a bit more concretely how we arrive at the *t*-score for the difference. All of the input for these calculations are taken from the `compmeans` results:

$$t = \frac{54.54 - 62.55}{\sqrt{\frac{26.11^2}{3321} + \frac{26.75^2}{3952}}} = \frac{-8.01}{.6216} = -12.89$$

You may have noticed that I subtracted the mean for women from the mean for men in the numerator, leading to a negative *t*-score. The reason for this is that the R function for conducting *t-tests* subtracts the second value (the mean for women) it encounters from the first value (the mean for men) by default, so the calculation is set up to reflect what we should expect to find when using R. The negative value makes sense in the context of our expectations, since it means that the value for women is higher than the value for men.

In this case, the *t*-score far exceeds the critical value (−1.645), so we *reject the null hypothesis* and conclude that there is a gender gap in evaluations of feminists on the feeling thermometer scale, with women providing higher ratings of feminists than those provided by men.

t-test in R

The R command for conducting a *t*-test (`t.test`) is straightforward and easy to use. The format is `t.test(dependent ~ independent)`. The ~ symbol is used in this and other functions to signal that you are using a formula that specifies a dependent and independent variable.

```
#use t.test to get t-score for Feminist FT by Sex
t.test(anes20$femFT ~ anes20$Rsex)


        Welch Two Sample t-test


data:  anes20$femFT by anes20$Rsex
t = -13, df = 7111, p-value <2e-16
alternative hypothesis: true difference in means between group Male
and group Female is not equal to 0
95 percent confidence interval:
   -9.231   -6.794
sample estimates:
   mean in group Male mean in group Female
                54.54                62.55
```

There are a few important things to pick up on here. First, the reported *t*-score (-13) is very close to the one we calculated (−12.89), the difference due to rounding in the R output. Second, the reported *p*-value is $2e$ - 16. Recall from earlier that scientific notation like this is used as a shortcut when the actual numbers have several digits. In this case, the notation means that the *p*-value is less than .0000000000000002. Since R uses a two-tailed test by default, this is the total area under the curve at the two tails combined. This means that there is virtually no chance the null hypothesis is true. Of course, for a one-tailed test, we still reject the null hypothesis since the *p*-value is even lower.

The `t.test` output also provides a 95% confidence interval around the sample estimate of the difference between the two groups. The way to interpret this is that you can be 95% certain that interval between -9.231 and -6.794 includes the population value. Importantly, since the confidence interval does not include 0 (no difference), we can also use this as a basis for rejecting the null hypothesis. Finally, you probably noticed that the reported degrees of freedom (7111) is different than what I previously calculated (7271). This is because one of the assumptions underlying *t*-tests is that the variances in the two groups are the same. If they are not, then some corrections need to be made, including adjustments to the degrees of freedom. The Welch's two-sample test used by R does not assume that the two sample variances are the same and, by default, always makes the correction. In this case, with such a large sample, the findings are not really affected by the correction other than the degrees of freedom. This makes the Welch's two-sample test a slightly more conservative test, which I see as a virtue.

You can also run a one-tailed test (specifying `alternative = "less"` so the direction fits our expectations) and assume equal variances, as I do next. As you can see, the results are virtually identical, other than the degrees of freedom and the confidence interval.

```
#t.test with one-tailed test and equal variance
t.test(anes20$femFT ~ anes20$Rsex, alternative = "less", var.equal = T)

       Two Sample t-test

data:  anes20$femFT by anes20$Rsex
t = -13, df = 7271, p-value <2e-16
alternative hypothesis: true difference in means between group Male
and group Female is less than 0
95 percent confidence interval:
       -Inf -6.987
sample estimates:
   mean in group Male mean in group Female
                54.54                62.55
```

Let's take a quick look at another application of the gender gap but this time using a less obvious dependent variable, the feeling thermometer for "Big Business" (`anes20$V202163`). While this variable is not as directly connected to gender-related issues, it is reasonable to expect that female respondents are less supportive of big business than are male respondents, if for no other reason than that women tend to be more liberal than men. Again, though, this connection is not as obvious as it was in the previous example.

```
#Create new object, "anes20$busFT
anes20$busFT <- anes20$V202163
#T-test for sex-based differences in 'busFT'
t.test(anes20$busFT ~ anes20$Rsex)

    Welch Two Sample t-test

data:  anes20$busFT by anes20$Rsex
t = 2.1, df = 7015, p-value = 0.04
alternative hypothesis: true difference in means between group Male
and group Female is not equal to 0
95 percent confidence interval:
    0.07143 2.16569
sample estimates:
    mean in group Male mean in group Female
                 48.41                47.29
```

First, there is a statistically significant difference in support for big business between male and female respondents. The *t*-score is 2.1, and the *p*-value is .04 (less than .05), so we reject the null hypothesis. The average rating among male respondents is 48.41, and the average among female respondents is 47.29, a difference of just 1.12. The 95% confidence interval for this difference ranges from .07143 to 2.16569. As expected, given the *p*-value, this confidence interval does not include 0. There is a relationship between sex and support for big business.

Statistical Significance Versus Effect Size

The example of attitudes toward big business illustrate an important issue related to statistical significance: Sometimes, statistically significant findings represent relatively small substantive effects. In this case, we have a statistically significant difference between two groups, but that difference is only 1.12 on a dependent variable that is scaled from 0 to 100. Yes, male and female respondents hold different attitudes toward big business, but just barely different!

This finding highlights an important issue: Statistically significant differences are different from 0, but that doesn't mean they are large differences. Let's put this finding in the context of the sample size and other results. Two important factors influence the size of the *t*-score, the magnitude of the effect and the size of the sample. As a consequence, when the sample size is very large, as in this case ($n > 7200$), the standard error for the difference between two groups is so small that sometimes even relatively trivial subgroup differences are statistically significant: The relationship exists, but it is of little consequence.

We can appreciate this by comparing this result to the earlier example using the feminist feeling thermometer. The difference between men and women on the feminist feeling thermometer was 8.01, more than seven times the size of the difference in ratings of big business, yet both findings are statistically significant. This is an important issue in statistical analysis, as it is often the case that the focus on statistical significance leaves substantive importance unattended.

What this discussion points to is the need to complement the findings related to statistical significance with a measure of the *size of the effect*. One such statistic that is used a lot in conjunction with *t*-tests is Cohen's *D*. The most direct way to calculate *D* is to express the difference between the two group means relative to the size of the pooled standard deviation.[3]

$$D = \frac{\overline{x}_1 - \overline{x}_2}{S_{\text{pooled}}}$$

All of this information can be obtained for both examples from the following `compmeans` output.

```
#Get means and standard deviations using 'compmeans'
compmeans(anes20$femFT, anes20$Rsex, plot = F)

Mean value of "POST: Feeling thermometer: feminists" according to "anes20$Rsex"
          Mean      N Std.    Dev.
Male      54.54  3321         26.11
Female    62.55  3952         26.75
Total     58.89  7273         26.76

compmeans(anes20$busFT, anes20$Rsex, plot = F)

Mean value of "POST: Feeling thermometer: big business" according to
"anes20$Rsex"
          Mean      N Std.    Dev.
Male      48.41  3337         23.01
Female    47.29  3953         22.38
Total     47.80  7290         22.67
```

The values of Cohen's *D* for the feminist and big business feeling thermometer tests are calculated next (using the overall standard deviation as a conservative estimate of the pooled standard deviation).

```
#Cohen's D for Feminist FT
(54.54-62.55)/26.76
[1] -0.2993

#Cohen's D for Big Business FT
(48.41-47.29)/22.67
[1] 0.0494
```

Let's check our work with the R command for getting Cohen's *D*.

```
#Get Cohens D for impact of sex on 'femFT' and 'busFT'
cohens_d(anes20$femFT ~ anes20$Rsex)

Cohen's d |          95% CI
--------------------------
-0.30     | [-0.35, -0.26]

- Estimated using pooled SD.

cohens_d(anes20$busFT ~ anes20$Rsex)

Cohen's d |          95% CI
------------------------
0.05      |   [0.00, 0.10]

- Estimated using pooled SD.
```

As expected, the *effect size* is much greater for the gender gap on the feminist feeling thermometer than it is for the big business feeling thermometer. Notice also that the R output shows a negative effect for the feminist feeling thermometer model and a positive effect for the big business model, reflecting the direction of the difference between group means.

Although we have shown that the impact of respondent sex is much greater when looking at the feminist feeling thermometer than when using the big business feeling thermometer, this still doesn't tell us if the effect is strong or weak, other than in comparison to the meager impact in the case of the big business feeling thermometer. In absolute terms, how strong is this effect? Does $D = -.30$ indicate a strong or weak effect on its own? The table that follows provides some conventional guidelines for evaluating the substantive meaning of Cohen's D values.

TABLE 10.1 ■ Cohen's *D* and Effect Size Interpretations

Cohen's *D*	Effect Size
.1	
.2	Small
.3	
.4	
.5	Medium
.6	
.7	
.8	Large

Note that these are all positive values, but the finding of $D = -.3$ should be treated the same as $D = .3$. Using the guidelines in Table 10.1, it is fair to describe the impact of respondent sex on the feminist feeling thermometer rating as small to medium while the impact on big business ratings is tiny at best.

DIFFERENCE IN PROPORTIONS

Finally, we can also extend difference in means hypothesis testing to differences between group proportions. For instance, suppose I'm interested in the gender gap in abortion attitudes, a topic that is frequently assumed to be an important issue on which men and women disagree. The ANES has a few variables measuring abortion attitudes, including one that it has used for the past few decades. Respondents are asked which of the listed positions best matches with their position on abortion.

> "1. By law, abortion should never be permitted"
>
> "2. The law should permit abortion only in case of rape, incest, or when the woman's life is in danger"
>
> "3. The law should permit abortion other than for rape/incest/danger to woman but only after need clearly established"
>
> "4. By law, a woman should always be able to obtain an abortion as a matter of personal choice"
>
> "Other"

One thing we can do with this variable is focus on whether respondents think abortion should never be permitted (the first category) and create a new variable distinguishing between those who do and do not think abortion should be banned. To do this, I relabeled the categories and then created a numeric dichotomous variable scored 1 for those who think abortion should never be permitted and 0 for all other responses.

```
#Create abortion attitude variable
anes20$banAb <- factor(anes20$V201336)
#Change levels to create two-category variable
levels(anes20$banAb) <- c("Illegal", "Other", "Other", "Other", "Other")
#Create numeric indicator for "Illegal"
anes20$banAb.n <- as.numeric(anes20$banAb == "Illegal")
```

The mean of this variable is the proportion who think abortion should never be permitted. Based on conventional wisdom and on the gender gaps reported in earlier in this chapter, the expectation is that the proportion of women who think abortion should never be permitted is lower than the proportion of men who support this position.

Since these means are actually proportions,

$$H_0: P_W = P_{M.}$$

$$H_1: P_W < P_{M.}$$

Let's see what the sample statistics tell us about the sex-based difference in the proportion who support banning all abortions. In this case, we suppress the boxplot because it is not a useful tool for examining variation in dichotomous outcomes (Go ahead and generate a boxplot if you want to see what I mean).

```
#Generate means (proportions), by sex
compmeans(anes20$banAb.n, anes20$Rsex, plot = F)

Mean value of "anes20$banAb.n" according to "anes20$Rsex"
           Mean       N Std.     Dev.
Male     0.1041  3728          0.3054
Female   0.1074  4422          0.3097
Total    0.1059  8150          0.3077
```

Two things stand out from this table. First, banning abortions completely is not a popular position; only 10.59% of respondents support this position. Second, there doesn't seem to be much real difference between men and women on this issue (just .0033), and women are ever-so-slightly more likely than men to take this position.

Still, even with a difference so small, the question for us is whether this represents a real difference in the population or is due to random error. Given the sample size, even small differences could be statistically significant. We know that the data do not support the original alternative hypothesis ($P_W < P_M$), but it is still interesting to know if the difference between men and women is statistically significant. To figure this out, we need to estimate the probability of getting a sample difference of this magnitude from a population in which there is no difference between the two groups. Of course, since the mean difference is opposite of what we originally expect, we should use a two-tailed test ($P_W \neq P_M$).

So, we need to go through the process again of calculating a *t*-score for the difference between the two groups and compare it to the critical value (1.96 for a two-tailed test). The formula should look very familiar to you, as it is the same as that for the difference between two means (because the sample proportions are means of dichotomous variables):

$$t = \frac{(p_1 - p_2) - (P_1 - P_2)}{S_{p_1 - p_2}} = \frac{(p_1 - p_2)}{S_{p_1 - p_2}} = \frac{.1041 - .1074}{S_{p_1 - p_2}}$$

Fair enough, this all looks good. We just divide the difference between the two sample proportions by the standard error of the difference. It gets a bit more complicated when calculating the standard error of the difference:

$$S_{p_1-p_2} = \sqrt{p_u(1-p_u)} * \sqrt{\frac{N_1+N_2}{N_1N_2}}$$

Here, p_u is the estimate of the population proportion (P), which we can get from the `compmeans` table (.1059), and N_1 and N_2, which you can also get from the `compmeans` table, are the sample sizes for p_1 and p_2, respectively. The proportion in the full sample supporting a ban on abortions is .1059, so $S_{p_1-p_2}$ is as follows.

```
#Calculate the standard error for the difference
sqrt(.1059*(1-.1059))*sqrt((3728 + 4422)/(3728*4422))
[1] 0.006842
```

$$S_{p_1-p_2} = .006842$$

$$t = \frac{.0033}{.006842} = .4823.$$

Okay, now let's see what R tells us.

```
#Test for difference in banAb.n, by sex
t.test(anes20$banAb.n ~ anes20$Rsex)


    Welch Two Sample t-test


data:   anes20$banAb.n by anes20$Rsex
t = -0.49, df = 7952, p-value = 0.6
alternative hypothesis: true difference in means between group Male
and group Female is not equal to 0
95 percent confidence interval:
  -0.01674  0.01006
sample estimates:
  mean in group Male mean in group Female
              0.1041               0.1074
```

It looks like our calculations were just about spot on. There is no significant relationship between respondent sex and supporting a ban on abortions. None whatsoever. The *t*-score is only -.49, far smaller than a critical value of -1.96, and the reported *p*-value is .626, meaning there is a pretty good chance of drawing a sample difference of this magnitude from a population in which there is no real difference. This is also reflected in the confidence interval for the difference (-.0167 to .01), which includes the value of 0 (no difference). So, we fail to reject the null hypothesis.

What about conventional wisdom? Doesn't everyone know that there is a huge gender gap on abortion? Sometimes, *conventional wisdom meets data*, and conventional wisdom loses. Results similar to the one presented are not unusual in quantitative studies of public opinion on this issue. Sometimes there is no gender gap, and sometimes there is a gap, but it tends to be a small one. For instance, if we focus on the other end of the original abortion variable and create a dichotomous variable indicating those who think abortion generally should be available as a matter of choice, we find a significant gender gap.

```
#Create "choice" variable
anes20$choice <- factor(anes20$V201336)
#Change levels to create two-category variable
levels(anes20$choice) <- c("Other", "Other", "Other", "Choice by Law", "Other")
#Create numeric indicator for "Choice by law"
anes20$choice.n <- as.numeric(anes20$choice == "Choice by Law")

#Test for differences in "choice.n", by sex
t.test(anes20$choice.n ~ anes20$Rsex)


   Welch Two Sample t-test


data:  anes20$choice.n by anes20$Rsex
t = -4.5, df = 7923, p-value = 0.000008
alternative hypothesis: true difference in means between group Male
and group Female is not equal to 0
95 percent confidence interval:
-0.07134  -0.02782
sample estimates:
 mean in group Male mean in group Female
            0.4622               0.5118
```

Here, we see that there is a statistically significant difference between male and female respondents on this position, with about 46% of men and 51% of women favoring abortion availability as a matter of choice. But the substantive difference is not very large: A bit less than half the male respondents and just more than half the female respondents support this position. We can confirm the limited effect size with `Cohen's` *h*, an alternative to Cohen's *D*, which should be used when comparing proportions (Cohen's *D* will give you the same value in this case).

```
#Get Cohens D for impact of sex on "choice.n"
cohens_h(anes20$choice.n, anes20$Rsex)

Cohen's h  |        95% CI
------------------------
0.10       |[0.06, 0.14]
```

Yes, $h = .10$ confirms that, though this is a statistically significant relationship, it is not a very strong one. This gives us one non-significant finding on abortion attitudes and one finding that is statistically significant but weak. This combination of findings is in keeping with research in this area. We will return to this issue in Chapter 13, where we will utilize information from all categories of the abortion variable to provide a more thorough evaluation of the relationship between sex and abortion attitudes.

PLOTTING MEAN DIFFERENCES

As you saw earlier in the chapter, you can get boxplot comparison of the two groups with the `compmeans` command. This mode of comparison is useful for getting a sense of where the middle of the distributions are located, as well as how much the distributions for the two groups overlap. What's missing from a side-by-side boxplot graph, however, is a graphic comparison of the means themselves. One thing we can do to remedy this is add markers to the boxplots to identify the group means, as shown next for the gender gap in the feminist feeling thermometer.

Boxplots With Means

First, you create a boxplot (`boxplot(dependent ~ independent)`), with appropriate axis labels. Then, you add a shape that identifies the mean outcomes for each group, using the `points` function to identify the mean values. In this case, I added an *X* for each mean outcome using `pch = 4` and doubled the size of the markers with `cex = 2`. Since I had already created and saved the mean values earlier using the `aggregate` function, I was able to use that saved object (`agg_femFT`) to create the points. If you do not have the means saved but know what they are, you can also add them separately to the points command, as shown in the alternative code following. Fiddle around a bit to the values for `pch` and `cex` to see how they change the graph.

```
#Create a boxplot
boxplot(anes20$femFT ~ anes20$Rsex,
        xlab = "Respondent Sex",
        ylab = "Feminist Feeling Thermometer")
#Add points to the plot reflecting mean values
#Use below, or points(c("55.54", "62.55"), pch = 4, cex = 2)
points(agg_femFT, pch = 4, cex = 2)
```

One of the nice things about the addition of the means to this graph is that it reinforces how small the difference in mean outcomes is, even though it is statistically significant.

Bar Charts

It is also good to explore alternatives to the boxplots that focus more exclusively on the mean values. You are already familiar with one popular alternative, the bar chart. In this case, we want the height of the bars to represent the mean levels of the dependent variable for each of

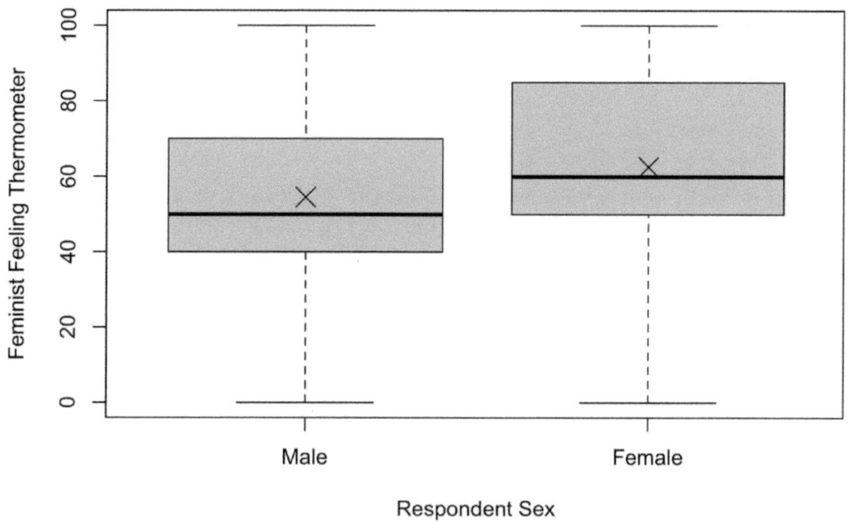

the two subgroups. Once again, we can use the results from the `aggregate()` command. R stored this object as a data.frame (`agg_femFT`) with two variables, `x`, the group means, and `Group.1`, the group labels. We use this information to create a bar chart using the format `barplot(dependent ~ independent)` and add some labels.

```
#Use 'barplot' to show the mean outcomes of "femFT" by sex
barplot(agg_femFT$x ~ agg_femFT$Group.1,
        xlab = "Sex of Respondent",
        ylab = "Mean Feminist Feeling Thermometer")
```

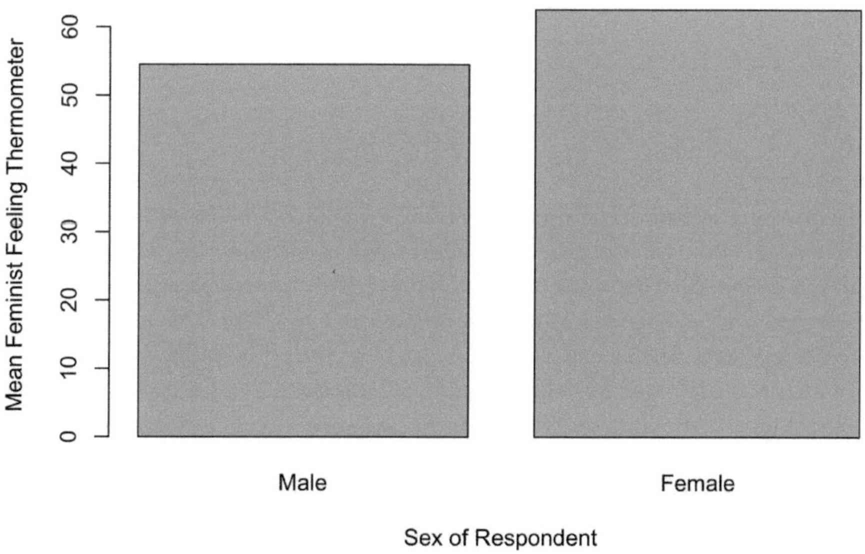

This bar plot does a nice job of showing the difference between the two groups while also communicating that difference in the context of the scale of the dependent variable. It shows that there is a difference but a somewhat modest one.

Means Plot

Another option for graphing the mean differences is `plotmeans`, a function found in the `gplots` package. The structure of this command is straight forward: `plotmeans(dependent ~ independent)`. Let's take a close look at the means plot for sex-based differences in the feminist feeling thermometer.

```
#Use 'plotmeans' to show the mean outcomes of "femFT" by sex
plotmeans(anes20$femFT ~ anes20$Rsex,
          n.label = F, #Do not include the number of observations
          ylab = "Mean Feminist Feeling Thermometer",
          xlab = "Respondent Sex")
```

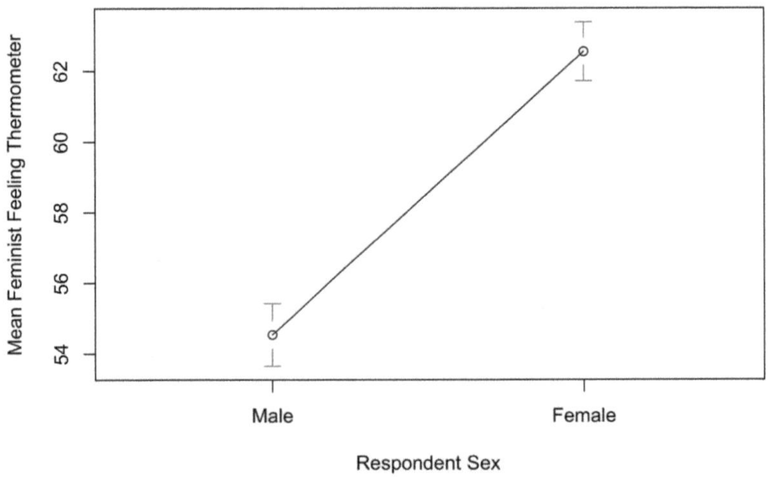

What you see here are two small circles representing the mean outcomes on the dependent variable for each of the two independent variable categories and error bars (vertical lines within endcaps) representing the confidence intervals around each of the two subgroup means. As you can see, there appears to be a substantial difference between the two groups. This is represented by the vertical distance between the group means (the circles) but also by the fact that neither of the confidence intervals overlaps with the other group mean. If the confidence interval of one group overlaps with the mean of the other, then the two groups are not statistically different from each other.

I like using `plotmeans` as a graphing tool, but one drawback is that it can create the impression that the difference between the two groups is more meaningful than it is. Note in this graph that the span of the *y*-axis is only as wide as the lowest and highest confidence limits. There is nothing technically wrong with this. However, since the scale of the dependent variable is from 0 to 100, restricting the view to these narrow limits can make it seem that there is a more important difference between the two groups than there actually is. This is why it is important to measure the effect size with something like Cohen's D and to pay attention to the scale of the dependent variable when evaluating the size of the subgroup differences. It is worth pointing out that this "scale" problem was not a problem for the boxplot or the bar chart.

WHAT'S NEXT?

It is quite common to make comparisons between two groups as we have done in this chapter. However, we are frequently interested in comparisons across more than just two groups. For instance, if you think back to some of the other group characteristics mentioned at the beginning of this chapter—race and ethnicity, place of residence, and religiosity—it is easy to see how we could compare several subgroups at the same time. In the case of race and ethnicity, while the dominant comparison tends to be between whites and people of color, it is probably more useful to take full advantage of the data and compare outcomes among several groups—whites, Blacks, Hispanics, Asian -Americans and Pacific Islanders, and other identities. While t-tests play a role in these types of comparisons, a more appropriate method is analysis of variance (ANOVA), a statistical technique we take up in the next chapter.

EXERCISES

Concepts and Calculations

1. The means plot that follows illustrates the mean feeling feminist thermometer rating for two groups of respondents, those aged 18 to 49 and those aged 50 and older. Based just on this graph, does there appear to be relationship between age and support for feminists? Justify and explain your answer.

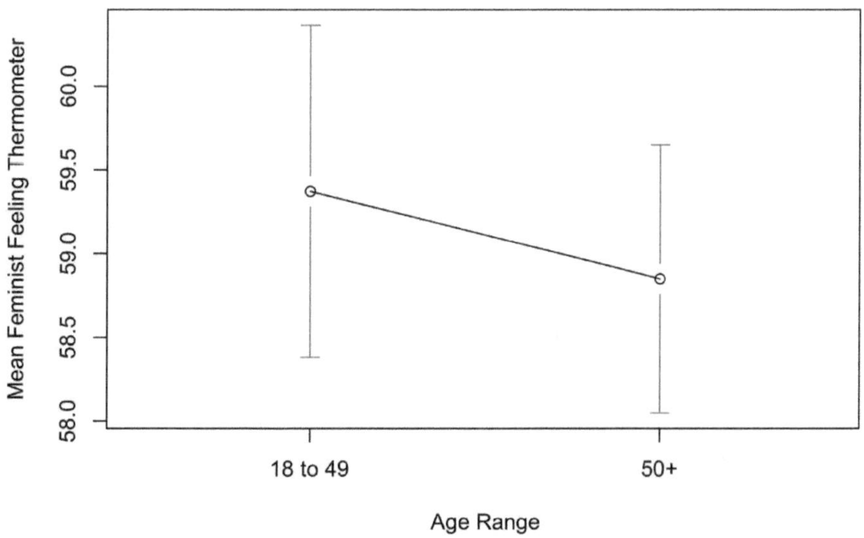

2. In response to the student survey that was used for the exercises in earlier chapters, a potential donor wants to provide the campus bookstore with gift certificates as a way of defraying the cost of books and supplies. In consultation with the student government leaders, the donor decides to prioritize first and second year students for this program because they think that upper-class students spend less than others on books and supplies. Before finalizing the decision, the student government wants to test whether there really is a difference in the spending patterns of the two groups of students.
   a. What are the null and alternative hypotheses for this problem? Explain.
   b. Using the following listed data, test the null hypothesis and summarize your findings for the student government. Is there a significant relationship between class standing and expenditures? Is the relationship strong?
   c. Based on these findings, should the donor prioritize first- and second-year students for assistance? Be sure to go beyond just reciting the statistics when you answer this question.

Expenditures on Books and Supply, by Class Status

	1st & 2nd Year	Upper Class
Mean	$358	$340
Std. Dev	77	79
n	165	135

3. On a number of cultural, demographic, and political measures, the states in the American South stand out as somewhat different from the rest of the country. The results shown here summarize the mean levels of several different variables in Southern and non-Southern states, along with the pooled standard deviation for those variables. Use this information to calculate the difference between the means of the two groups of states for each variable, along with Cohen's *D*. For which variable is the impact of region the strongest? For which variable is the impact of region the weakest? Do any of these differences surprise you?

Comparison of Southern and Non-Southern States

Variable	South Mean	Non-South Mean	Pooled SD
Gallons of Beer Per Capita	32.8	32.1	5.4
Congregations per 10k	16.9	12.1	4.9
Diabetes %	12.6	9.7	1.4
Gun Deaths per 100k	16.6	11.8	4.5
Tax Burden	8.8	9.6	1.3

4. A researcher is interested in examining the relationship between veteran status, measured with a dichotomous variable (0 = Never served, 1 = Veteran or currently serving), and support for the military, measured with a 0 to 100 military feeling thermometer. Convert each of the following statements of research expectations into formal alternative hypotheses that can be used to test for mean differences in support for the military between veterans and non-veterans. Also, what is the null hypothesis?
   - I expect the level of support for the military to be higher among veterans than among non-veterans.
   - I expect that there is a difference in support for the military between veterans and non-veterans.
   - I expect the level of support for the military to be lower among veterans than among non-veterans.

R Problems

For these problems, use the `county20large` dataset to examine how county-level educational attainment is related to COVID-19 cases per100k population. You need to load the following libraries: `dplyr`, `Hmisc`, `gplots`, `descr`, `effectsize`.

1. The first thing you need to do is take a sample of 500 counties from the `counties20large` dataset and store that sample in a new dataset, `covid500`, using the following command.

```
set.seed(1234)
#create a sample of 500 rows of data from the county20large data set
covid500 <- sample_n(county20large, 500)
```

The `sample_n` command samples rows of data from the dataset, so we now have 500 randomly selected counties with all data on all of the variables in the dataset. The dependent variable in this assignment is `covid500$cases100k_sept821` (cumulative COVID-19 cases per 100,000 people, up to September 8, 2021), and the independent variable is a two-category version of `covid500$postgrad`, the percentage of adults in the county with a post-graduate degree. The expectation is that case rates are lower in counties with relatively high levels of education than in other counties.

2. Transform `covid500$postgrad` into a two-category variable with a roughly equal number of counties in each category. Store this variable in a new object named `covid500$postgrad2` and label the categories "Low Education" and "High Education". The generic format is `data$newvariable <- cut2(data$oldvariable, g = # of groups)`. If you are unclear about how to do this, go back and take a quick look at the variable transformation section of Chapter 4 for a refresher. Produce a frequency table for `covid500$postgrad2` to check on the transformation.

3. State a null and alternative hypothesis for this pair of variables.

4. Use the `compmeans` command to estimate the level of COVID-19 cases per 100k in low- and high-education counties. Describe the results. What do the data and boxplot tell you? Make sure to use clear, intuitive labels for the boxplot and make specific references to the group means.

5. Conduct a *t*-test for the difference in COVID-19 rates between low- and high-education counties. Interpret the results.

6. Add a means plot (`plotmeans` command) and Cohen's *D* (`cohens_d` command) and discuss what additional insights they provide.

11 HYPOTHESIS TESTING WITH MULTIPLE GROUPS (ANOVA)

GET READY

This chapter extends the examination of mean differences to include comparisons among several subgroup means. Examining differences between two subgroup means is an important and useful method of hypothesis testing in the social and natural sciences. It is not, however, without limitations. For example, in many cases, independent variables have more than just two categories or can be collapsed from continuous values into several discrete categories. To follow along in R, you should load the `countries2` dataset and attach the libraries for the following packages: `descr`, `DescTools`, `dplyr`, `gplots`, `effectsize`, and `Hmisc`.

INTERNET ACCESS AS AN INDICATOR OF DEVELOPMENT

Suppose you are interested in studying indicators of development around the world. One thing you might use for this purpose is access to the internet. By now, access to the internet is not considered a luxury but has come to be thought of as an important part of infrastructure, similar to roads, bridges, and more basic forms of communication. The variable `internet` in the `countries2` dataset provides an estimate of the country-level percentage of households that have regular access to the internet. Before delving into the factors that might explain differences in internet access, we should get acquainted with its distribution, using a histogram and some basic descriptive statistics. Let's start with a histogram.

```
hist(countries2$internet, xlab = "% Internet Access",
     ylab = "Number of Countries", main = "")
```

Here we see a fairly even distribution from the low to the middle part of the scale and then a spike in the number of countries in the 70% to 90% access range. The distribution looks a bit left-skewed, though not too severely. We can also get some descriptive statistics to help round out the picture.

```
#Get some descriptive statistics for internet access
summary(countries2$internet, na.rm = T)

   Min.  1st Qu.  Median  Mean  3rd Qu.  Max.  NA's
    0.0     27.6    58.4  54.2     79.6  99.7     1

Skew(countries2$internet, na.rm = T)
[1] -0.2114
```

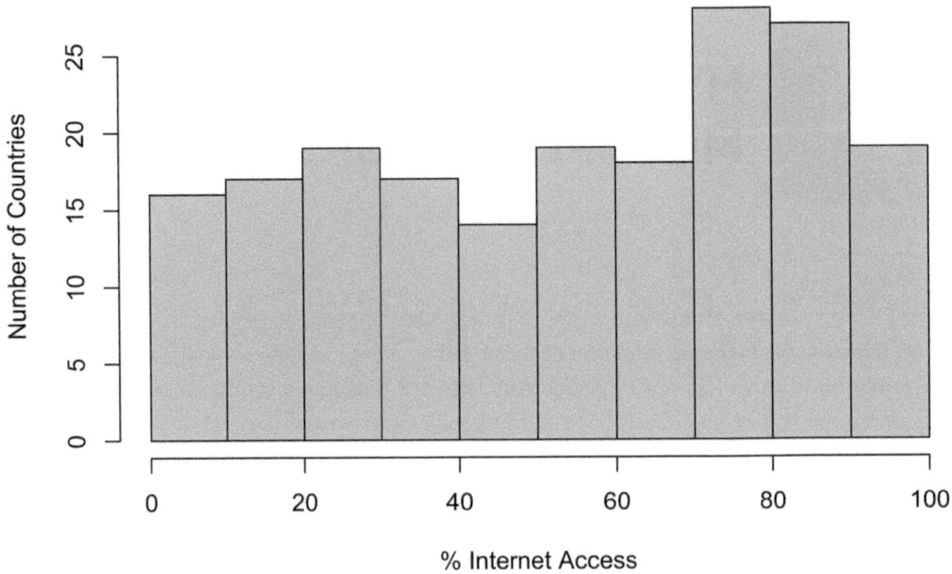

The mean, median, and skew statistics confirm the initial impression of relatively modest negative skewness. Both the histogram and descriptive statistics also show that there is a lot of variation in internet access around the world, with the bottom quarter of countries having access levels ranging from about 0% to about 28% access, and the top quarter of countries having access levels ranging from about 80% to almost 100% access. As with many such cross-national comparisons, there are clearly some "haves" and "have nots" when it comes to internet access.

THE RELATIONSHIP BETWEEN WEALTH AND INTERNET ACCESS

If internet access is an indicator of development, then it makes sense that knowing the level of wealth in countries should help explain variation in access to the internet across countries. We test this idea using GDP per capita (`countries2$gdp_pc`) as a measure of wealth. If our intuition is right, we should see on average higher levels of internet access as per capita GDP increases.

Comparing Two Means

As a first step and as a bridge to demonstrating the benefit of using multiple categories of the independent variable, we explore the impact of GDP on internet access using a simple *t*-test for the difference between two groups of countries, those in the top and bottom halves of the distribution of GDP. Let's create a new, two-category variable measuring high and low GDP, and then let's look at a boxplot and the mean differences. The `cut2` command tells R to slice the original variable into two equal-size pieces (g = 2) and assign it to a new object, `countries2$gdp2`.

```
#Create two-category GDP variable
countries2$gdp2 <- cut2(countries2$gdp_pc, g = 2)

#Assign levels to "gdp2"
levels(countries2$gdp2) <- c("Low", "High")
```

Now, let's use compmeans to examine the mean differences in internet access between low and high GDP countries. Note that I added an "X" to each box to reflect the mean of the dependent variable within the columns, as in Chapter 10, except this time specifying the mean values as they are listed in the compmeans results.

```
#Evaluate mean internet access, by level of gdp
compmeans(countries2$internet, countries2$gdp2,
          xlab = "GDP per Capita",
          ylab = "Number of Countries")

#Add points to the plot, reflecting mean values of internet access
#"pch" sets the marker type, "cex" sets the marker size
points(c(33.39, 76.41), pch = 4, cex = 2)

Mean value of "countries2$internet" according to "countries2$gdp2"
             Mean       N Std.    Dev.
Low          33.39     92         19.36
High         76.41     91         16.28
Total        54.78    183         27.99
```

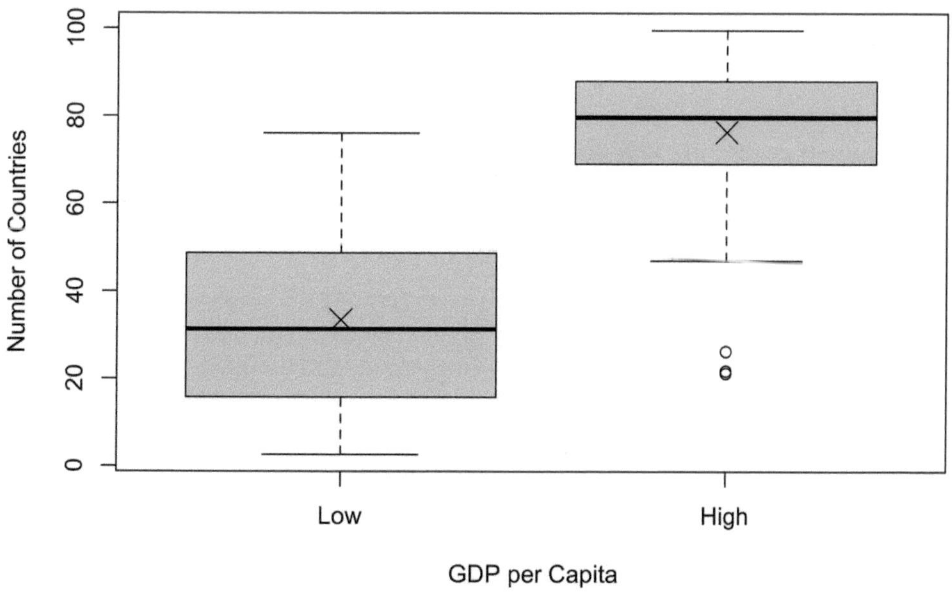

In this graph, we see what appears to be strong evidence that GDP per capita is related to internet access: On average, 33.4% of households in countries in the bottom half of GDP have internet access, compared to 76.4% in countries in the top half of the distribution of per capita GDP. The boxplot provides a dramatic illustration of this difference, with the vast majority of low GDP countries also having low levels of internet access, and the vast majority of high GDP countries having high levels of access. We can do a *t*-test and get the Cohens *D* value just to confirm that this is indeed a statistically significant and strong relationship.

```
#Test for the impact of "gdp2" on internet access
t.test(countries2$internet ~ countries2$gdp2, var.equal = T)

    Two Sample t-test

data: countries2$internet by countries2$gdp2
t = -16, df = 181, p-value <2e-16
alternative hypothesis: true difference in means between group Low and
group High is not equal to 0
95 percent confidence interval:
 -48.24 -37.80
sample estimates:
 mean in group Low mean in group High
             33.39              76.41

#Get Cohen's D for the impact of "gdp2" on internet access
cohens_d(countries2$internet ~ countries2$gdp2)

Cohen's d  |           95% CI
---------------------------
-2.40      | [-2.78, -2.02]

- Estimated using pooled SD.
```

These tests confirm the expectations, with a *t*-score of –16.3, a *p*-value close to 0, and Cohen's $D = -2.40$ (a very strong relationship).[1]

Comparing Multiple Means

This is all as expected, but we can do a better job of capturing the relationship between GDP and internet access by taking fuller advantage of the variation in per capita GDP. Because we created and used just two categories of GDP, we were treating all countries in the "Low" GDP category as if they are the same (low on GDP), and all countries in the "High" GDP category as if they are the same (high on GDP), even though there is a lot of variation within these categories. The range in outcomes within the "Low" GDP group is from \$725 to \$13,078, and in the "High" GDP category, it is from \$13,654 to \$114,482. Certainly, we should expect that the variation in GDP within the high and low categories should be related to difference in internet access within those categories and that relying on such overly broad categories as "High" and "Low" represents a loss of important information.

We can take greater advantage of the variation in per capita GDP by dividing the countries into four quartiles instead of two halves.

```
#Create a four-category GDP variable
countries2$gdp4 <- cut2(countries2$gdp_pc, g = 4)

#Assign levels
levels(countries2$gdp4) <- c("Q1", "Q2", "Q3", "Q4")
```

Again, we start the analysis by looking at the differences in mean levels of internet access across levels of GDP per capita, using the `compmeans` function.

```
#Evaluate mean internet access, by level of gdp
compmeans(countries2$internet, countries2$gdp4,
          xlab = "GDP per Capita (Quartile)",
          ylab = "% Households with Internet Access")

Mean value of "countries2$internet" according to "countries2$gdp4"
        Mean    N Std.    Dev.
Q1     19.39   46       11.427
Q2     47.38   46       15.073
Q3     65.91   45       15.097
Q4     86.68   46        9.441
Total  54.78  183       27.994

#Add points to the plot reflecting mean values of internet access
points(c(19.39, 47.38, 65.91, 86.68), pch = 4, cex = 2)
```

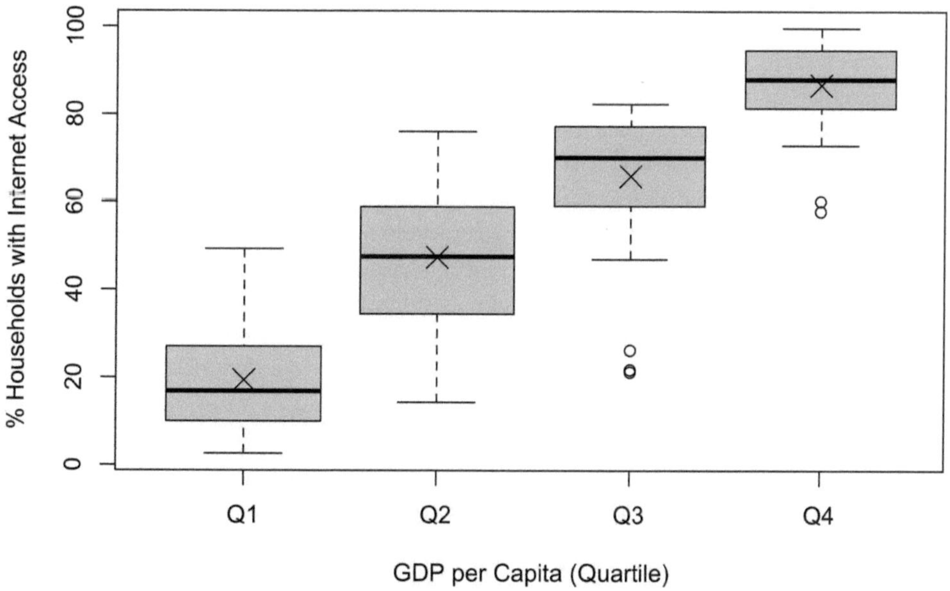

The mean level of internet access increases from about 19.4% for countries in the first quartile to 47.4% in the second quartile, to 65.9% for the third quartile, to 86.7% in the top quartile. Although there is a steady increase in internet access as GDP per capita increases, there is also considerable variation in internet access within each category of GDP. Because of this variation, there is overlap in the internet access distributions across levels of GDP. There are some countries in the lowest quartile of GDP that have greater access to the internet than some countries in the second and third quartiles; some in the second quartile with higher levels of access than some in the third and fourth quartiles, and so on.

On its face, this looks like a strong pattern, but the key question is whether the differences in internet access across levels of GDP are great enough, relative to the level of variation within categories, that we can determine that there is a statistically significant relationship between the two variables.

ANALYSIS OF VARIANCE

Using analysis of variance (ANOVA), we are able to judge whether there are *significant differences in the mean value of the dependent variable across categories of the independent variable.* This significance is a function of both the *magnitude of the mean differences between categories* and the *amount of variation within categories*. If there are relatively small differences in the mean level of the dependent variable across categories of the independent variable or relatively large variances in the dependent variable within categories of the independent variable, the differences in the data may be due to random error.

Just as with the *t*-test, we use ANOVA to test a null hypothesis. The null hypothesis in ANOVA is that the mean value of the dependent variable does not vary across the levels of the independent variable:

$$H_0: \mu_1 = \mu_2 = \mu_3 = \mu_4$$

In the example of internet access and GDP, the null hypothesis would be that the average level of internet access is unrelated to the level of GDP. Of course, we will observe some differences in the level of internet access across different levels of GDP. It's only natural that there will be some differences. The question is whether these differences are substantial enough that we can conclude they are real differences and not just apparent differences that are due to random variation. When the null hypothesis is true, we expect to see relatively small differences in the mean of the dependent variable across values of the independent variable.

Technically, the alternative hypothesis states that at least one of the mean differences is statistically significant, but it is often more casually stated, something like this:

$$H_1: \text{The means are not equal.}$$

In the current case, this means that the mean level of internet access varies across levels of GDP per capita. Even though we usually have expectations regarding the direction of the relationship (high GDP associated with high level of internet access), the alternative hypothesis does not state a direction, just that there are some differences in mean levels of the dependent variable across levels of the independent variable.

Now, how do we test this? As with *t*-tests, it is the null hypothesis that is tested directly. I'll describe the ideas underlying ANOVA in intuitive terms first, and then we will move on to some formulas and technical details.

Recall that with the difference in means test we calculated a *t*-score that was based on the size of the mean difference between two groups divided by the standard error of the difference. The mean difference represented the impact of the independent variable, and the standard error reflected the amount of variation in the data. If the mean difference was relatively large or the standard error relatively small, then the *t*-score would usually be larger than the critical value and we could conclude that there was a significant difference between the means.

We do something very similar to this with ANOVA. We use an estimate of the overall variation in means between categories of the independent variable (later, we call this "mean square between") and divide that by an estimate of the amount of variation around those means within categories (later, we call this "mean squared error"). Mean squared between represents the size of the effect, and mean squared error represents the amount of variation. The resulting statistic is called an *F*-ratio and can be compared to a critical value of *F* to determine if the variation between categories (magnitude of differences) is *large enough* relative to the total variation within categories that we can conclude that the differences we see are not due to random error.

Important Concepts/Statistics

To fully grasp what is going on with ANOVA, you need to understand the concept of variation in the dependent variable and break it down into its component parts.

Sum of squares total (*SST*) is the total variation in y, the dependent variable; it is very similar to the variance, except that we do not divide through by the $n - 1$. We express each observation of y (dependent variable) as a deviation from the mean of y ($\bar{y}$) and square the deviations. Then, sum the squared deviations to get the sum of squares total:

$$SST = \sum(y_i - \bar{y})^2$$

Let's calculate this in R for the dependent variable, `countries2$internet`.[2]

```
#Generate mean of "countries2$internet"
mn_internet = mean(countries2$internet[countries2$gd p_pc! = "NA"],
                   na.rm = T)
#Generate squared deviation from "mn_internet"
dev_sq = (countries2$internet-mn_internet)^2
#Calculate sst (sum of squared deviations)
sst <- sum(dev_sq[countries2$gdp_pc! = "NA"], na.rm = T)
sst
[1] 142629
```

$$SST = 142629.5$$

This number represents the total variation in y. The SST can be decomposed into two parts, the variation in y *between categories* of the independent variable (the variation in the means reported in `compmeans`) and the amount of variation in y *within categories* of the independent variable (in spirit, this is reflected in the separate standard deviations in the `compmeans` results). The variation between categories is referred to as sum of squares between (*SSB*) and the variation within categories is referred to as sum of squares within (*SSW*).

$$SST = SSB + SSW$$

Sum of squares within is the sum of the variance in y within categories of the independent variable:

$$SSW = \sum (y_i - \bar{y}_k)^2,$$

where $\bar{y}_k$ is the mean of y within a category of the independent variable.

Here we subtract the mean of y for each category (k) from the value of y for each observation in the respective category, square those differences, and sum all of the squared deviations. It should be clear that *SSW* summarizes the variation around the means of the dependent variable within each category of the independent variable, similar to variation seen in the side-by-side distributions shown in the boxplot previously presented.

Sum of squares between summarizes the variation in mean values of y across categories of the independent variable:

$$SSB = \sum N_k (\bar{y}_k - \bar{y})^2,$$

where $\bar{y}_k$ is the mean of y within a given category of x, and N_k is the number of cases in the category k.

Here we subtract the overall mean of the dependent variable from the mean of the dependent variable in each category (k) of the independent variable, square that difference, and then multiply it times the number of observations in the category. We then sum this across all categories. This provides us with a good sense of how much the mean of y differs across categories of x, weighted by the number of observations in each category. You can think of *SSB* as representing the impact of the independent variable on the dependent variable. If y varies only slightly across categories of the independent variable, then *SSB* will be small; if it varies a great deal, then *SSB* will be larger.

Using information from the `compmeans` command, we can calculate the *SSB* (Table 11.1). In this case, the sum of squares between = 112490.26.

Since we now know *SST* and *SSB*, we can calculate *SSW*:

$$SSW = SST - SSB$$

$$SSW = 142629.5 - 112490.26 = 30139.24$$

TABLE 11.1 ■ Calculations for Sum of Squares Between

Sum of Squares Between for Internet Access, by GDP					
	Mean Internet Access	Difference From Overall Mean	Difference^2	N	Sum of Difference^2
Bottom Quarter	19.395	-35.385	1252.10	46	57596.52
2nd Quarter	47.38	-7.4	54.76	46	2518.96
3rd Quarter	65.912	11.132	123.92	45	5576.46
Top Quarter	86.676	31.896	1017.35	46	46798.32
Overall	54.78			183	112490.26

Degrees of freedom for SSB and SSW (dfb and dfw, respectively) need to be estimated to calculate the F-ratio.

For SSW, $dfw = n - k\ (183 - 4 = 179)$;

for SSB, $dfb = k - 1\ (4 - 1 = 3)$,

where n = total number of observations; k = number of categories in x.

The degrees of freedom can be used to standardize SSB and SSW, so we can evaluate the variation between groups in the context of the variation within groups. Here, we calculate the mean squared error[3] (MSE), based on SSW, and mean squared between (MSB), based on the SSB:

Mean squared error (MSE) =

$$\frac{SSW}{dfw} = \frac{30139.24}{179} = 168.4.$$

Mean square between (MSB) =

$$\frac{SSB}{dfb} = \frac{112490.26}{3} = 37496.8.$$

As described earlier, you can think of the MSB as summarizing the differences in mean outcomes of the dependent variable across categories of the independent variable and the MSE as summarizing the amount of variation around those means. As with t- and z-tests, we need to compare the magnitude of the differences across groups to the amount of variation in the dependent variable.

This brings us to the ***F*-ratio**, which is the actual statistic used to determine the significance of the relationship:

$$F = \frac{MSB}{MSE}$$

In this case,

$$F = \frac{37496.8}{168.4} = 222.7.$$

Similar to the *t*- and *z*-scores, if the value obtained is greater the critical value of *F* (the value that would give you an alpha area of .05 or less), then you can reject H_0. We can use R to find the critical value for *F*, using the `qf` function. To do this, we need to specify the desired *p*-value (.05), the degrees of freedom between (3), the degrees of freedom within (174), and that we want the *p*-value for the upper-tail of the distribution.

```
#Get critical value of F for p = .05, dfb = 3, dfw = 179

qf(.05, 3, 179, lower.tail = FALSE)
[1] 2.655
```

In this case, the critical value for *dfw* = 179; *dfb* = 3 is 2.66. The shape of the *F*-distribution is different from the *t*- and *z*-distributions: It is one-tailed (hence the non-directional alternative hypothesis), and the precise shape is a function of the degrees of freedom. Figure 11.1 illustrates the shape of the *f*-distribution and the critical value of *F* for the current example, with *dfw* = 179, *dfb* = 3, and a *p*-value of .05.

FIGURE 11.1 ■ *F*-distribution and Critical Value for *dfw* = 179, *dfb* = 3, *p* = .05

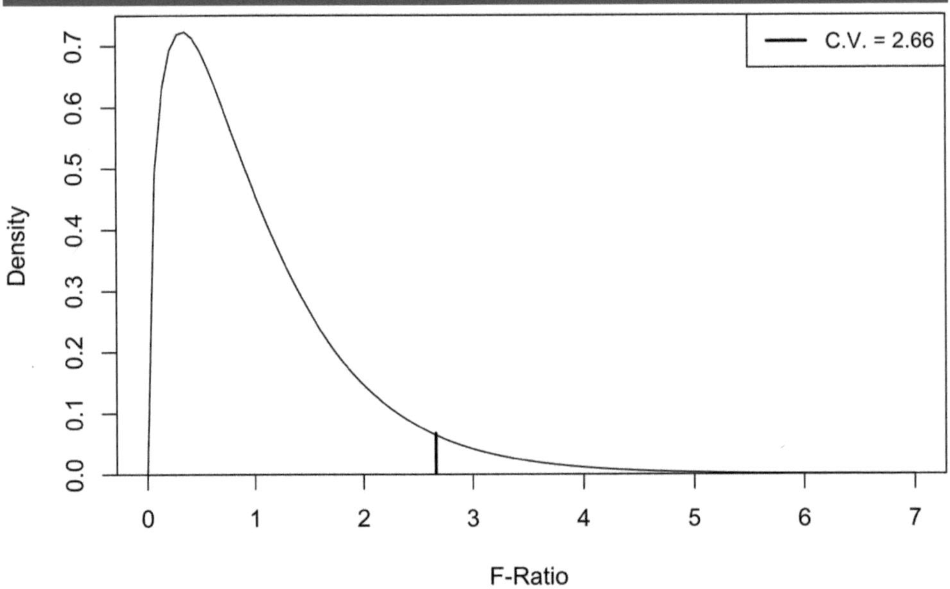

In this example, any obtained *F*-ratio greater than 2.66 provides a basis for rejecting the null hypothesis, as it would indicate less than a .05 probability of getting a difference in the means pattern of this magnitude or greater from a population in which there were no difference in the group means. The obtained value ($F = 222.7$) far exceeds the critical value, so we reject H_0.

ANOVA IN R

The ANOVA function in R is very simple: `aov(dependent variable ~ independent variable)`. When using this function, the results of the analysis should be stored in a new object, which is labeled `fit_gdp` in this particular case.

```
#ANOVA--internet access, by level of gdp
#Store in 'fit_gdp'

fit_gdp <- aov(countries2$internet ~ countries2$gdp4)
```

We can use the `summary` command to view the information stored in `fit _ gdp`.

```
#Show results of ANOVA
summary(fit_gdp)

                 Df  Sum Sq Mean Sq F value Pr( > F)
countries2$gdp4   3  112490   37497     223 < 2e-16 ***
Residuals       179   30140     168
---
Signif. codes: 0 '***' 0.001 '**' 0.01 '*' 0.05 '.' 0.1 ' ' 1
12 observations deleted due to missingness
```

Here's how to interpret these results. The first row, labeled with the independent variable name, is the "between" results, and the second row, labeled "Residuals", is the "within" results. `Df` represents degrees of freedom ($dfb = 3$, $dfw = 179$) for the ANOVA model. *Mean square between* is the first number under `Mean Sq`, and *mean squared error* is the second number. The *f*-statistic is $\frac{MSB}{MSE}$ and is found in the column headed by `F value`. The *p*-value associated with the *F*-statistic is in the column under `Pr( > F)` and is *the probability that the pattern of mean differences found in the data could come from a population in which there is no difference in internet access across the four categories of the independent variable.* Based on either the reported *p*-value (< 2e - 16) or using the three asterisks as a guide, this probability is very close to 0.

We conclude from this ANOVA model that differences in GDP are related to differences in internet access across countries. We reject the null hypothesis.

A Closer Look at Group Means

The *p*-value from the ANOVA model tells us there is a statistically significant relationship, but it does not tell us much about how important the individual group differences are. It is possible to have a significant *F*-ratio but still have fairly small differences between the group means, with only one of the group differences being statistically significant. To fill out the picture a bit, we need to take a closer look at a comparison of the group means to each other. We can see which group differences are statistically significant, using the `TukeyHSD` command (*HSD* stands for honest significant difference).

```
#Examine the individual mean differences
TukeyHSD(fit_gdp)

  Tukey multiple comparisons of means
    95% family-wise confidence level

Fit: aov(formula = countries2$internet ~ countries2$gdp4)
$`countries2$gdp4`
          diff    lwr    upr  p adj
Q2-Q1    27.99  20.97  35.00      0
Q3-Q1    46.52  39.46  53.57      0
Q4-Q1    67.28  60.26  74.30      0
Q3-Q2    18.53  11.48  25.59      0
Q4-Q2    39.30  32.28  46.31      0
Q4-Q3    20.76  13.71  27.82      0
```

This output compares the mean level of internet access across all combinations of the four groups of countries, producing six comparisons. For instance, the first row compares countries in the second quartile to countries in the first quartile. The first number in that row is the difference in internet access between the two groups (27.99), the next two numbers are the confidence interval limits for that difference (20.97 to 35.00), and the last number is the *p*-value for the difference (0). Looking at this information, it is clear the significant *f*-value is not due to just one or two significant group differences but that all groups are statistically different from each other.

EFFECT SIZE

Similar to Cohen's *D*, which was used to assess the strength of the relationship in a *t*-test, we can use `eta-squared` (η^2) to assess the strength of the relationship between two variables when the independent variable has more than two categories. Eta-squared measures the share of the total variance in the dependent variable (*SST*) that is attributable to variation in the dependent variable across categories of the independent variable (*SSB*):

$$\eta^2 = \frac{SSB}{SST} = \frac{112490}{142629} = .79$$

Now, let's check the calculations using R.

```
#Get eta-squared (measure of effect size) using 'fit_gdp'
eta_squared(fit_gdp)

For one-way between subjects designs, partial eta squared is equivalent
to eta squared.
Returning eta squared.

# Effect Size for ANOVA

Parameter        | Eta2  |        95% CI
----------------------------------------
countries2$gdp4  | 0.79  |  [0.75, 1.00]

- One-sided CIs: upper bound fixed at [1.00].
```

One of the things I like a lot about eta-squared is that it has a very intuitive interpretation. Eta-squared is bound by 0 and 1, where 0 means that independent variable does not account for any variation in the dependent variable and 1 means the independent variable accounts for all of the variation in the dependent variable. In this case, we can say that differences in per capita GDP explain about 79% of the variation in internet access across countries. This represents a strong relationship between these two variables. There are no strict guidelines for what constitutes strong versus weak relationships based on eta-squared, but here is my take.

TABLE 11.2 ■ Eta-Squared and Effect Size Interpretations

Eta-Squared	Effect Size
.1	
.2	Weak
.3	
.4	Moderate
.5	
.6	Strong
.7	
.8	Very Strong

The *F*-ratio tells us if there is a relationship between two variables but doesn't really tell us much about the substance of the relationship. Utilizing the size of the mean differences in the `compmeans` results, the boxplot, the `TukeyHSD` comparisons, and Cohen's *D* together provides a lot of additional useful information about what's going on "under the hood" of a significant *F*-test.

Plotting Multiple Means

In addition to the boxplot shown earlier, you can also use bar charts, and means plots to visualize differences across multiple groups, just as we did in Chapter 10. For the bar chart, you need to create a new object, `agg_internet`, using the aggregate command. Then, you can use the group names and group means from that object to create the plot.

```
#Use 'aggregate' to get group means
agg_internet <- aggregate(countries2$internet,
        by = list(countries2$gdp4), FUN = (mean), na.rm = TRUE)

#Get a barplot using information from 'agg_internet'
barplot(agg_internet$x ~ agg_internet$Group.1,
        xlab = "GDP Per Capita Quartiles",
        ylab = "% Households with Internet Access")
```

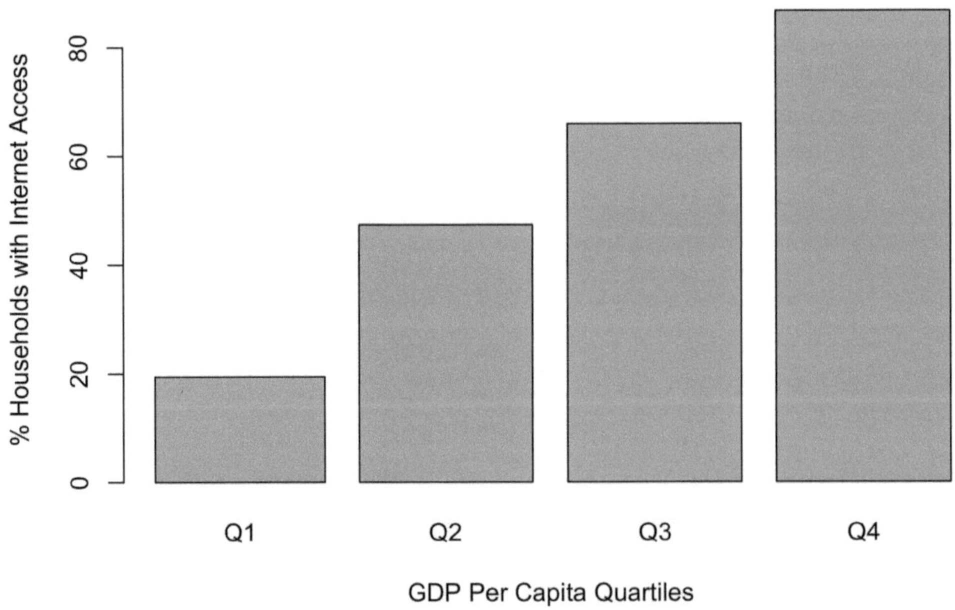

You can also plot the means and confidence intervals using the `plotmeans` command. In this example, R has trouble showing the confidence intervals because they are so small, which may result in the following warning message: *zero-length arrow is of indeterminate angle and so skipped.* Don't worry about this message, as the plot is still created and you can see the mean outcomes (small circles) very clearly.

```
#Use 'plotmeans' to view relationship
plotmeans(countries2$internet ~ countries2$gdp4,
          n.label = F,
          xlab = "GDP Per Capita Quartiles",
        ylab = "% Households with Internet Access")
```

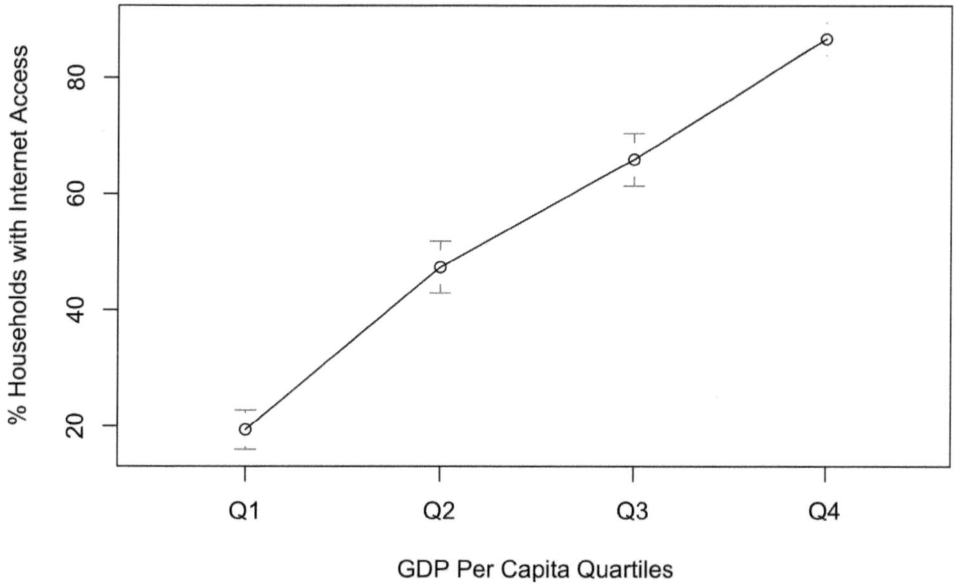

Regardless of which method you use—boxplot (earlier in the chapter), barchart, or means plot—the data visualizations for this relationship reinforce the ANOVA findings: There is a strong relationship between per capita GDP and internet access around the world. This buttresses the argument that internet access is a good indicator of economic development.

Population Size and Internet Access

Let's take a look at another example, using the same dependent variable, but focusing on how internet access is influenced by population size. I can see plausible arguments that smaller countries might be expected to have greater access to the internet, and I can see arguments for why larger countries should be expected to have greater internet access. For right now, though, let's assume that we think the mean level of internet access varies across levels of population size, though we are not sure how.

First, let's create the four-category population variable.

```
#Create four-category measure of population size
countries2$pop4 <- cut2(countries2$pop, g = 4)

#Assign levels
levels(countries2$pop4) <- c ("Q1", "Q2", "Q3", "Q4")
```

Now, let's look at some preliminary evidence, using compmeans.

```
#Examine internet access, by pop4
compmeans(countries2$internet, countries2$pop4,
          xlab = "Population size (Quartile)",
          ylab = "% Internet Access")

Mean value of "countries2$internet" according to "countries2$pop4"
        Mean    N Std.   Dev.
Q1      58.44   48       26.85
Q2      57.39   49       29.90
Q3      46.50   49       30.87
Q4      54.53   48       26.54
Total   54.19  194       28.79

#Add markers for group means
points(c(58.44, 57.39, 46.5, 54.5), pch = 4, cex = 2)
```

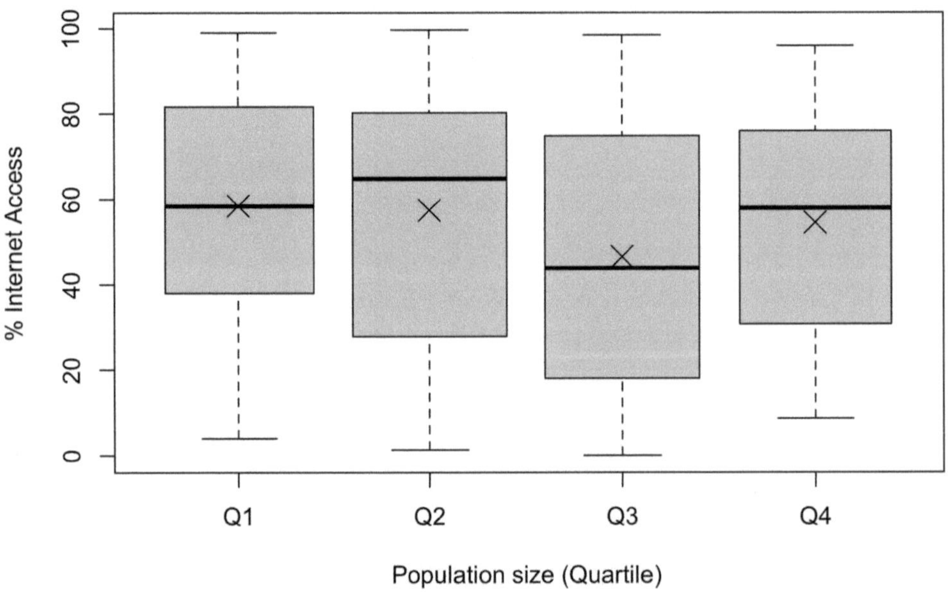

Hmmm. This is interesting. The mean outcomes across all groups are fairly similar to each other and not terribly different from the overall mean (54.2%). In addition, all of the within-group distributions overlap quite a bit with all others. So, we have what looks like very little between-group variation and a lot of within-group variation. ANOVA can sort this out and tell us if there is a statistically significant relationship between country size and internet access. Let's first determine the critical value for rejecting the null hypothesis ($H_0: \mu_1 = \mu_2 = \mu_3 = \mu_4$).

```
#Get critical value of F for p = .05, dfb = 3, dfw = 190

qf(.05, 3, 190, lower.tail = FALSE)
[1] 2.652
```

Now, let check with the ANOVA results.

```
#Run ANOVA for impact of population size on internet access
fit_pop <- aov(countries2$internet ~ countries2$pop4)

#View results
summary(fit_pop)

                  Df  Sum Sq Mean Sq F value Pr( > F)
countries2$pop4    3    4275    1425    1.74    0.16
Residuals        190  155651     819
1 observation deleted due to missingness
```

Okay, well, that's pretty definitive. The obtained *f*-ratio (1.74) is well below the critical value (2.65), and the *p*-value associated with the *F*-ratio is .16, meaning that the probability of getting an *F*-ratio of 1.74 from a population in which the null hypothesis is true is .16. This is far greater than the standard .05 cutoff point, so we *fail to reject* the null hypothesis. There is no evidence here that internet access is affected by population size.

Technically, there is no reason to check the `TukeyHSD` results because a non-significant *F*-test means that none of the mean differences are statistically significant. However, those results are shown next just to reinforce this point.

```
#Examine individual mean differences
TukeyHSD(fit_pop)

  Tukey multiple comparisons of means
    95% family-wise confidence level

Fit: aov(formula = countries2$internet ~ countries2$pop4)

$`countries2$pop4`
          diff      lwr    upr   p adj
Q2-Q1   -1.057  -16.123 14.008  0.9979
Q3-Q1  -11.947  -27.012  3.119  0.1718
Q4-Q1   -3.914  -19.057 11.229  0.9083
Q3-Q2  -10.889  -25.877  4.098  0.2387
Q4-Q2   -2.857  -17.922 12.209  0.9609
Q4-Q3    8.033   -7.033 23.099  0.5122
```

Still nothing here. None of the six mean differences are statistically significant. There is no evidence here to suggest that population size has any impact on internet access.

CONNECTING THE *T*-SCORE AND *F*-RATIO

If you are having trouble relating to ANOVA, it might be helpful to reinforce that the *F*-ratio is providing the same type of information you get from a *t*-score. Both tell you if the magnitude of the mean differences found in the data is large enough, relative to variation in the dependent variable, that you can be confident that it did not occur by chance. One way to appreciate how similar they are is to compare the *f*-ratio and *t*-score when an independent variable has only two categories. At the beginning of this chapter, we used a two-category measure of GDP to look at how GDP is related to internet access. The *t*-score for the difference in this example was −16.276. Although we wouldn't normally use ANOVA when the independent variable has only two categories, there is nothing technically wrong with doing so, so let's examine this same relationship using ANOVA.

```
#ANOVA with a two-category independent variable
fit_gdp2 <- aov(countries2$internet ~ countries2$gdp2)
summary(fit_gdp2)

                     Df Sum Sq Mean Sq F value Pr( > F)
countries2$gdp21      1  84669   84669     264 < 2e-16 ***
Residuals           181  57960     320
---
Signif. codes: 0 '***' 0.001 '**' 0.01 '*' 0.05 '.' 0.1 ' ' 1
12 observations deleted due to missingness
```

Here, again, the results point to a statistically significant relationship between these two variables. Of particular interest here is that the *F*-ratio is 264.4 (before rounding), while the *t*-score was −16.261 (before rounding). What do these two numbers have in common? It turns out, a lot. Just to reinforce the fact that *F* and *t* are giving you the same information, note that the square of the *t*-score (16.261^2) equals 264.42, almost exactly the value of the *F*-ratio (the difference due to rounding). With a two-category independent variable, $F = t^2$, the *f*-ratio and the *t*-score are doing exactly the same thing, telling us if the observed relationship between two variables is significantly different from what we would expect to find if there were no relationship between the two variables.

NEXT STEPS

ANOVA and *t*-tests are great tools to use when evaluating variables that influence the outcomes of quantitative dependent variables. However, we are often interested in explaining outcomes on ordinal and nominal dependent variables. For instance, when working with public opinion data, it is very common to be interested in political and social attitudes measured on ordinal scales with relatively few, discrete categories. In such cases, mean-based statistics are not appropriate. Instead, we can make use of other tools, notably crosstabs, which you have already seen (Chapter 7), ordinal measures of association (effect size), and a measure of statistical significance more suited to ordinal and nominal-level data (chi-square). The next two chapters cover these topics in some detail, helping to round out your introduction to hypothesis testing.

EXERCISES

Concepts and Calculations

1. The following table represents the results of two different experiments. In both cases, a group of political scientists were trying to determine if voter turnout could be influenced by different types of voter contacting (phone calls, knocking on doors, or fliers sent in the mail). One experiment took place in City A, and the other in City B. The table presents the mean level of turnout and the standard deviation for participants in each contact group in each city. After running an ANOVA test to see if there were significant differences in the mean turnout levels across the three voter contacting groups, the researchers found a significant *F*-ratio in one city but not the other. In which city do you think they found the significant effect? Explain your answer. (Hint: you don't need to calculate anything.)

City A			City B		
	Mean	Standard Deviation		Mean	Standard Deviation
Phone	55.7	7.5	Phone	57.7	4.3
Door Knock	60.3	6.9	Door Knock	62.3	3.7
Mail	53.2	7.3	Mail	55.2	4.1

2. When testing to see if poverty rates in U.S. counties are related to internet access across counties, a researcher divided counties into four roughly equal-sized groups according to their poverty level and then used ANOVA to see if poverty had a significant impact on internet access. The critical value for *F* in this case is 2.61, the $MSB = 27417$, and the $MSE = 52$. Calculate the *F*-ratio and decide if there is a significant relationship between poverty rates and internet access across counties. Explain your conclusion.

3. A recent study examined the relationship between race and ethnicity and the feeling thermometer rating (0 to 100) for Big Business, using data from the 2020 American National Election Study. The results are presented next. Summarize the findings, paying particular attention to statistical significance and effect size. Do any findings stand out as contradictory or surprising?

```
                    Df  Sum Sq Mean Sq F value Pr( > F)
anes20$raceth.5      4   16717    4179    8.16 0.0000015 ***
Residuals         7259 3716102     512
---
Signif. codes: 0 '***' 0.001 '**' 0.01 '*' 0.05 '.' 0.1 ' ' 1
```

```
1016 observations deleted due to missingness

# Effect Size for ANOVA

Parameter          |      Eta2 |       95%  CI
-----------------------------------------------
anes20$raceth.5    | 4.48e-03 | [0.00, 1.00]

- One-sided CIs: upper bound fixed at [1.00].
```

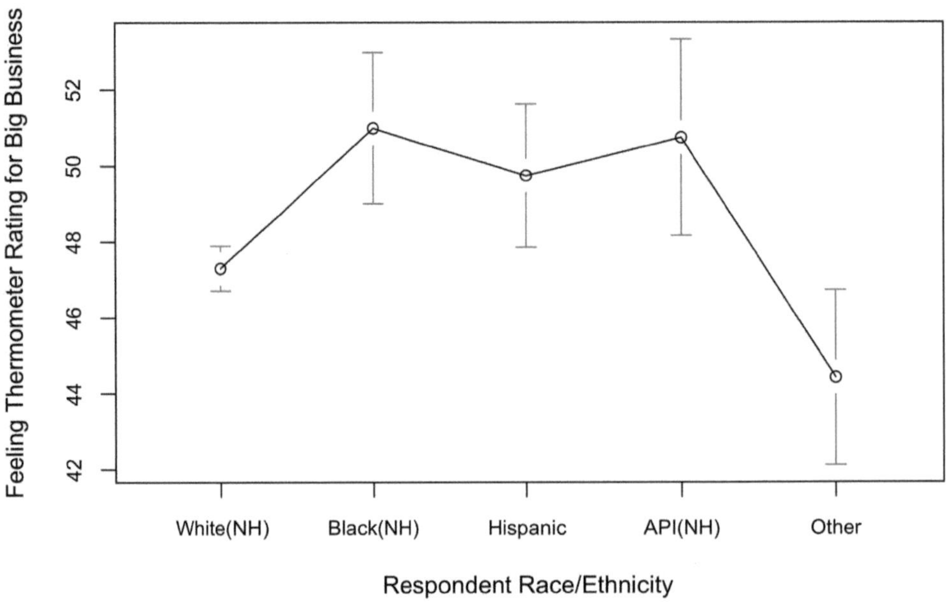

4. The following table presents several pairs of *F*-ratios and critical values for *F*. For each pair, decided if the null hypothesis should be rejected or not rejected.

Critical Value	F-Ratio	Reject H_0 ?
4.59	23.69	
3.78	3.79	
6.43	6.40	
2.76	5.36	
4.34	3.75	

5. In Chapter 10, the *t*-score for the relationship between sex and the feminist feeling thermometer rating was -12.89. I repeated this analysis but used an ANOVA model this time. What *F*-ratio do you think I obtained for this relationship?

R Problems

1. I tried to run an ANOVA model to examine the relationship between GDP per capita (gdp4) and infant mortality (infant_mort) using the countries2 dataset, but I can't get the following commands to work. What am I doing wrong? Correct the code and show the output. Hint: there are three errors.

```
fit_infant <- aov(countries2$Infant_mort,countries2$gdp4)
summary(infant_mort)
```

The remaining questions expand upon the R assignment from Chapter 10. First, create the sample of 500 counties, as you did in Chapter 10, using the code that follows.

```
set.seed(1234)
#create a sample of 500 rows of data from the county20large data set
covid500 <- sample_n(county20large, 500)
```

2. We will stick with covid500$cases100k_sept821 as the dependent variable and covid500$postgrad as the independent variable. Before getting started with the analysis, describe the distribution of the dependent variable using either a histogram, a density plot, or a boxplot. Supplement the discussion of the graph with relevant descriptive statistics.
3. Using cut2, convert the independent variable (covid500$postgrad) into a new independent variable with four categories.[4] Name the new variable covid500$postgrad4, use levels to assign intuitive labels to the categories, and generate a frequency table to check that this was done correctly.
4. State the null and alternative hypotheses for using ANOVA to test the differences in levels of COVID-19 cases across categories of educational attainment.
5. Use compmeans and the boxplot that comes with it to illustrate the apparent differences in levels of COVID-19 cases across the four categories of educational attainment. Use words to describe the apparent effects, referencing both the reported means and the pattern in the boxplot.
6. Use aov to produce an ANOVA model and check to see if there is a significant relationship between the COVID-19 cases and the level of education in the counties. Comment on statistical significance overall as well as between specific groups. You will need the TukeyHSD function for the discussion of specific differences.
7. Use eta-squared to assess the strength of the relationship.
8. What do you think of the findings? Are they stronger or weaker than you expected? What other variables do you think might be related to county-level COVID-19 cases?

12 HYPOTHESIS TESTING WITH NON-NUMERIC VARIABLES (CROSSTABS)

GET READY

This chapter explores ways to analyze relationships between two variables when the dependent variable is non-numeric, either ordinal or nominal in nature. To follow along in R, you should load the `anes20` dataset and make sure to attach the libraries for the `descr`, `dplyr`, and `Hmisc` packages.

CROSSTABS

ANOVA is a useful tool for examining relationships between variables, especially when the dependent variable is numeric. However, suppose we are interested in using data from the 2020 ANES survey to study how various independent variables are related to religiosity (dependent variable), measured as the importance of religion in one's life. We can't do this with ANOVA because it relies on differences in the *average* outcome of the dependent variable across categories of the independent variable, and the dependent variable in this case (importance of religion) is a three-category ordinal variable. Many of the variables that interest us—especially if we are using public opinion surveys—are measured at the nominal or ordinal level, so ANOVA is not an appropriate method in these cases.

Instead, we can use a **crosstab** (for cross-tabulation), also known as a **contingency table** (I use both "crosstab" and "contingency table" interchangeably to refer to this technique). A crosstab is nothing more than a joint frequency distribution that simultaneously displays the outcomes of two variables. You were introduced to crosstabs in Chapter 7, where one was used to demonstrate conditional probabilities.

Regional Differences in Religiosity

We begin the analysis of religiosity by examining regional differences in the importance of religion. Traditionally, the American South and parts of the Midwest are thought of as the "Bible Belt," so we expect to see distinctive outcomes across the four broadly defined regions in the ANES data (`anes20$V203003`): Northeast, Midwest, South, and West.

Before looking at a crosstab for region and religiosity, we need to collapse and reorder categories of religiosity from a five-category variable (`anes20$V201433`) that measures religious importance on a scale ranging from "1. Extremely important" to "5. Not important at all" into a three-category variable that ranges from "Low" to "Moderate" to "High" (`anes20$relig_import`).

```
#Create religious importance variable
anes20$relig_imp <- anes20$V201433

#Recode Religious Importance to three categories
levels(anes20$relig_imp) <- c("High", "High", "Moderate",  "Low", "Low")

#Change order to Low-Moderate-High
anes20$relig_imp <- ordered(anes20$relig_imp, levels = c("Low",
                                              "Moderate",  "High"))
```

Now, let's look at the table that follows, which illustrates the relationship between region (anes20$V203003) and religiosity (anes20$relig_imp). To review some of the basic elements of a crosstab covered earlier, this table contains outcome information for the independent and dependent variables. Each row represents a value of the dependent variable (religiosity), and each column a value of the independent variable (region). Each of the interior cells in the table represent the intersection of a given row and column or the joint frequency of outcomes on the independent and dependent variables. At the bottom of each column and the end of each row, are the row and column totals, also known as the marginal frequencies. These totals are the category frequencies for the dependent and independent variables, and the overall sample size (bottom right corner) is restricted to the 8249 people who answered both survey questions.

```
#Get a crosstab, list the dependent variable first
crosstab(anes20$relig_imp, anes20$V203003,
         prop.c = T, #Add column percentages
         plot = F)

   Cell Contents
|-------------------------|
|                   Count |
|          Column Percent |
|-------------------------|

=========================================================================
                    SAMPLE: Census region
anes20$relig_imp  1. Northeast  2. Midwest  3. South  4. West   Total
-------------------------------------------------------------------------
Low                        549         649       811      782    2791
                         39.4%       32.6%     26.4%    43.5%
-------------------------------------------------------------------------
Moderate                   320         385       548      343    1596
                         23.0%       19.3%     17.9%    19.1%
-------------------------------------------------------------------------
```

High	525	956	1710	671	3862
	37.7%	48.0%	55.7%	37.4%	
Total	1394	1990	3069	1796	8249
	16.9%	24.1%	37.2%	21.8%	

This crosstab includes two important pieces of information. The raw frequencies and the column percentages for each cell. The raw frequencies play a very important role in calculating a number of different summary statistics for the crosstab, but they are hard to interpret on their own because the column totals are not equal. For instance, 549 people from the Northeast attached a low level of importance to religion, which is fewer than the 649 people from the Midwest who are low on the religious importance scale. Does this mean that people from the Midwest are more likely to attach a low level of importance to religion than people from the Northeast? It's hard to tell by the cell frequencies alone because the base—or the total number of people in each of the regional categories—is different for each column.

Instead of relying on raw frequencies, we need to standardize the frequencies by their base (the column totals). This gives us column proportions/percentages, which can be used to make judgments about the relative levels of religiosity across different regional groups. Column percentages adjust each cell to the same metric, making the cell contents comparable.

Now, it is easier to judge the relationship between these two variables. The percentages represent the percentage within each column that reported each of the outcomes of the row variable. We can look *within the rows* and *across the columns* to assess how the likelihood of being in one of the religious importance categories is affected by being in one of the region categories.

Interpreting the Crosstab

Is there a relationship between region and level of religious importance? How can we tell? This depends on whether the column percentages change much from one column to the next as you scan within the row categories. If the column percentages in the "Low" religious importance row are constant (do not change) across categories of region, this could be taken as evidence that region is not related to religiosity. If we see a pattern of no substantial changes in the column percentages within multiple categories of the dependent variable, then there is a strong indication that there is not a relationship between the two variables. But this is not the case in the table. Instead, it seems that the outcomes of the dependent variable depend upon the outcomes of the independent variable.

The crosstab for region and religiosity shows what appears to be important regional differences in self-identified level of religiosity. The column percentages for the "Low" religious importance row range from a low of 26.4% for respondents from the South, to 32.6% for Midwesterners, to 39.4% for Northeastern respondents, to 43.5% for respondents from the

West. The polar opposite of this pattern is found in the "High" religious importance row, with Midwestern (48%) and Southern (55.7%) respondents having the highest column percentages, and Northeastern (37.7%) and Western (37.4%) respondents having the lowest. Generally, this tells us that respondent level of religiosity varies across regions of the country. More specifically, this tells us that Midwestern and Southern respondents assign more importance to religion than do Northeastern and Western respondents.

It is easy to see why crosstabs are also referred to as contingency tables because the outcome of the dependent variable is contingent upon the value of the independent variable. This should sound familiar to you, as the column percentages are giving the same information as conditional probabilities.

Mosaic Plots

One problem with reading crosstabs and column percentages is that there is a lot of information in the table and sometimes it can be difficult to process it all. Even in the simple table just presented, there are 12 different column percentages to process. You may recall that this was one of the limitations with frequency tables, especially for variables with several categories—the numbers themselves can be a bit difficult for some people to sort out. Just as barplots and histograms can be useful tools to complement frequency tables, **mosaic plots** can be useful summary graphs for crosstabs (you can get the mosaic plot by adding `plot = T` to the crosstab function).

In the mosaic plot in Figure 12.1, the horizontal width of each column reflects the relative sample size for each category of the independent variable, and the vertical height of the

FIGURE 12.1 ■ Example of a Mosaic Plot

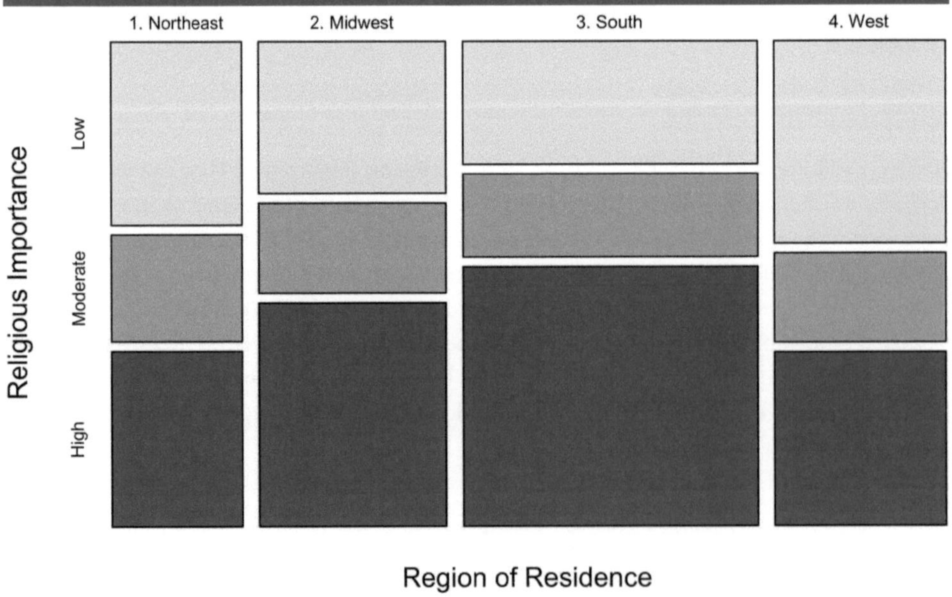

differently shaded boxes within the columns reflects the magnitude of the column percentages. Usually, the best thing to do is focus on how the height of the differently shaded segments (boxes in each row) changes as you scan from left to right. For instance, the height of the light gray bars (low religious importance) is greatest for the Western and Northeastern columns and the shortest for the Southern and Midwestern columns as you scan across regions, while the opposite pattern holds for the darkest column segments (high religious importance). These changes correspond to changes in the column percentages and reinforce the findings from the crosstab. Does the mosaic plot clarify things for you? If so, make sure to take advantage of its availability. If not, it's worth checking in with your instructor for help making sense of this useful tool.

SAMPLING ERROR

It appears that there is a relationship between region and religiosity. However, since these data are from a sample, we do not know if we can infer that there is a relationship in the population. As you know by now, if we took another sample and used the same two variables, we would get a different table with somewhat different percentages in the cells. Presumably, a table from a second sample would look a lot like the table from the first sample, but there would be differences. The same would be true of any successive samples we might take. And, of course, if there is no relationship between these two variables in the population, some number of samples drawn from that population would still feature differences in the column percentages, perhaps of the magnitude found in our sample, just due to sampling error. What we need to figure out is if the relationship in the table produced by this sample is significantly different from what would be expected if there were no relationship between these two variables in the population.

Think about what this table would look like if there were no relationship between these two variables. In that case, we would expect to see fairly constant percentages as we look from across the columns. We don't necessarily expect to see exactly the same percentages across the columns; after all, we should see some differences due to sampling error alone. However, the differences in column percentages would be rather small if there were no relationship between the two variables.

But what should we expect those column percentages to be? Since, for the overall sample, we find that about 34% of the cases are in the low importance row (2781 total in that row, divided by 8249 total sample size in the table), we should see about 34% or very slight deviations from that in this row across the columns, if there is no relationship. And the same is true for the other column percentages within rows; they should be close to the row total percentages (19% for "Moderate", 47% for "High").

When evaluating a mosaic plot, if there is no relationship between the independent and dependent variables, the height of the shaded boxes within each row should not vary across columns. In the plot just shown, the height of the boxes does vary across columns—but more so in the first row and last row than in the middle row.

HYPOTHESIS TESTING WITH CROSSTABS (CHI-SQUARE)

The crosstab just shown appears to deviate somewhat from what we would expect to see if there were no relationship between these two variables but not dramatically. What we need is a test of statistical significance that tells us if the differences that we observe across the columns are large enough that we can conclude that they did not occur due to sampling error; something like a *t*-test or an *F*-ratio. Fortunately, we have just that type of statistic available to us.

Chi-square (χ^2) is a statistic that can be used to judge the statistical significance of a bivariate table; it compares the observed frequencies in the cells to the frequencies that would be expected if there were no relationship between the two variables. If the differences between the observed and expected outcomes are large, χ^2 should be relatively large; when the differences are rather small, χ^2 should be relatively small.

The process for using χ^2 as a test of statistical significance is the same as that which underlies the *F*-ratio and *t*-score: We calculate a χ^2 statistic and compare it to a critical value (*cv*) for χ^2. If χ^2 is greater than the critical value, then we can reject the null hypothesis; if χ^2 is less than the critical value, we fail to reject the null hypothesis and conclude that there is no relationship between the two variables (in the parlance of chi-square, we would say the two variables are "statistically independent").

The null and alternative hypotheses for a χ^2 test are the following:

- H_0: No relationship. *The two variables are statistically independent.*
- H_1: There is a relationship. *The two variables are not statistically independent.*

Note that H_1 specifies a non-directional hypothesis; this is always the case with χ^2.

To test H_0, we calculate the χ^2 statistic:

$$\chi^2 = \sum \frac{(O - E)^2}{E},$$

where

- O = observed frequency for each cell;
- E = expected frequency for each cell; that is, the frequency we expect if there is no relationship (H_0 is true).

Chi-square is a table-level statistic that is based on summing information from each cell of the table. In essence, χ^2 reflects the extent to which the outcomes in each cell (the observed frequency) deviate from what we would expect to see in the cell if the null hypothesis were true (expected frequency). If the null hypothesis is true, we would expect to see very small differences between the observed and expected frequencies.

Raw frequencies (instead of column percentages) are used to calculate χ^2 because column percentages treat each column and each cell with equal weight, even though there are important differences in the relative contribution of the columns and cells to the overall sample size. For instance, the total number of cases in the "Northeast" column (1394) accounts for only 16.9% of total respondents in the table, so giving its column percentages the same weight as those for the "South" column, which has 37.2% of all cases, would over-represent the importance of the first column to the pattern in the table.

Let's use the table of raw frequencies that follows to walk through the calculation of the chi-square contribution of upper-left cell in the table (Low, Northeast). The observed frequency (the frequency produced by the sample) in this cell is 549. From the preceding discussion, we know that for the sample overall, about 34% (actually 33.83%) of respondents are in the first row of the dependent variable, so we expect to find about 34% of all respondents from the Northeast in this column. Because we are using raw frequencies instead of percentages, we need to calculate what 33.83% of the total number of observations in the first column is to get the expected frequency. There are 1394 respondents in this column, so the expected frequency for this cell is $.3383 * 1394 = 471.59$.

```
#Crosstab with raw frequencies
crosstab(anes20$relig_imp, anes20$V203003, plot = F)

   Cell Contents
|-------------------------|
|                   Count |
|-------------------------|

=========================================================================
                    SAMPLE: Census region
anes20$relig_imp     1. Northeast  2. Midwest  3. South  4. West   Total
-------------------------------------------------------------------------
Low                           549         649       811      782    2791
-------------------------------------------------------------------------
Moderate                      320         385       548      343    1596
-------------------------------------------------------------------------
High                          525         956      1710      671    3862
-------------------------------------------------------------------------
Total                        1394        1990      3069     1796    8249
=========================================================================
```

A shortcut for calculating expected frequency for any cell is

$$E_{c,r} = \frac{f_c * f_r}{n},$$

where f_c is the column total for a given cell, f_r if the row total, and n is the total sample size. For the upper-left cell in the crosstab, $(1394 * 2791)/8249 = 471.65$, virtually the same as the expected value using the first method of calculation. So, the **observed frequency** for the upper-left cell of the table is 549, and the **expected frequency** is 471.65.

To estimate this cell's contribution to the χ^2 statistic for the entire table,

$$\chi_c^2 = \frac{(549 - 471.65)^2}{471.65} = \frac{77.37^2}{471.65} = \frac{5983.02}{471.65} = 12.685.$$

To get the value of χ^2 for the whole table, we need to calculate the expected frequency for each cell, use that to estimate the cell contributions to the overall value of chi-square, and sum up all of the cell-specific values.

The following table shows the observed and expected frequencies (first two rows) as well as the cell contribution to overall χ^2 for each cell of the table (third row).

```
#Get expected frequencies and cell chi-square contributions
crosstab(anes20$relig_imp, anes20$V203003,
         expected = T, #Add expected frequency to each cell
         prop.chisq = T, #Total contribution of each cell
         plot = F)

   Cell Contents
|-------------------------|
|                   Count |
|         Expected Values |
| Chi-square contribution |
|-------------------------|

=========================================================================
                   SAMPLE: Census region
anes20$relig_imp   1. Northeast  2. Midwest  3. South  4. West     Total
-------------------------------------------------------------------------
Low                        549         649       811      782      2791
                         471.7       673.3    1038.4    607.7
                        12.685       0.877    49.790   50.015
-------------------------------------------------------------------------
Moderate                   320         385       548      343      1596
                         269.7       385.0     593.8    347.5
                         9.378       0.000     3.530    0.058
-------------------------------------------------------------------------
High                       525         956      1710      671      3862
                         652.6       931.7    1436.8    840.8
                        24.963       0.635    51.932   34.308
-------------------------------------------------------------------------
Total                     1394        1990      3069     1796      8249
=========================================================================
```

First, note that the overall contribution of the upper-left cell to the table χ^2 is 12.685, the same as the calculation we just made. As you can see, there are generally substantial differences between expected and observed frequencies, except for the second (Midwest) column, again indicating that there is not a lot of support for the null hypothesis. If we were to work through all the calculations and sum up all the individual cell contributions to χ^2, as shown next, we come up with $\chi^2 = 238.171$.

```
#Sum the individual cell contributions to shi-square
12.685 + .877 + 49.79 + 50.015 + 9.378 + 3.530 + .058 + 24.963 + .635
   + 51.932 + 34.308
[1] 238.2
```

So, what does this mean? Is this a relatively large value for χ^2? This is not a *t*-score or *F*-ratio, so we should not judge the number based on those standards. But still, this seems like a pretty big number. Does this mean the relationship in the table is statistically significant?

One important thing to understand is that the value of χ^2 for any given table is a function of three things: the strength of the relationship, the sample size, and the size of the table. Of course, the strength of the relationship and the sample size both affect the level of significance for other statistical tests, such as the *t*-score and *f*-ratio. But it might not be immediately clear why the size of the table matters. Suppose we had two tables, a 2x2 table and a 4x4 table, and that the relationships in the two tables were of the same magnitude and the sample sizes were also the same. Odds are that the chi-square for the 4x4 table would be larger than the one for the 2x2 table, simply due to the fact that it has more cells (16 vs. 4). The chi-square statistic is calculated by summing all the individual cell contributions, four for the 2x2 table and sixteen for the 4x4 table. The more cells, the more individual cell contributions to the table chi-square value. Given equally strong relationships and equal sample sizes, a larger table will tend to generate a larger chi-square. This means that we need to account for the size of the table before concluding whether the value of χ^2 is statistically significant. Similar to the *t*-score and *F*-ratio, we do this by calculating the degrees of freedom:

$$df_{\chi^2} = (r - 1)(c - 1),$$

where r = number of rows and c = number of columns in the table. This is a way of taking into account the size of the table.

The *df* for the table above is (3-1)(4-1) = 6.

We can look up the critical value for χ^2 with 6 degrees of freedom, using the following chi-square table (Table 12.1).[1] The values across the top of the χ^2 distribution table are the desired levels of significance (area under the curve to the right of the specified critical value), usually .05, and the values down the first column are the degrees of freedom. To find the *cv* of χ^2 for this table, we need to look for the intersection of the column headed by .05 (the desired level of significance) and the row headed by 6 (the degrees of freedom). That value is 12.59.

We can also get the critical value for χ^2 from the `qchisq` function in R, which requires you to specify the preferred *p*-value (.05), the degrees of freedom (6), and the upper end of the distribution (lower tail = *F*).

TABLE 12.1 ■ Chi-square Critical Values

Upper Tail		0.5	0.4	0.2	0.1	0.05	0.025	0.01	0.005
df	1	0.455	0.708	1.642	2.706	3.841	5.024	6.635	7.879
	2	1.386	1.833	3.219	4.605	5.991	7.378	9.210	10.597
	3	2.366	2.946	4.642	6.251	7.815	9.348	11.345	12.838
	4	3.357	4.045	5.989	7.779	9.488	11.143	13.277	14.860
	5	4.352	5.132	7.289	9.236	11.070	12.833	15.086	16.750
	6	5.348	6.211	8.558	10.645	12.592	14.449	16.812	18.548
	7	6.346	7.283	9.803	12.017	14.067	16.013	18.475	20.278
	8	7.344	8.351	11.030	13.362	15.507	17.535	20.090	21.955
	9	8.343	9.414	12.242	14.684	16.919	19.023	21.666	23.589
	10	9.342	10.473	13.442	15.987	18.307	20.483	23.209	25.188
	11	10.341	11.530	14.631	17.275	19.675	21.920	24.725	26.757
	12	11.340	12.584	15.812	18.549	21.026	23.337	26.217	28.300
	13	12.340	13.636	16.985	19.812	22.362	24.736	27.688	29.819
	14	13.339	14.685	18.151	21.064	23.685	26.119	29.141	31.319
	15	14.339	15.733	19.311	22.307	24.996	27.488	30.578	32.801
	16	15.339	16.780	20.465	23.542	26.296	28.845	32.000	34.267
	17	16.338	17.824	21.615	24.769	27.587	30.191	33.409	35.718
	18	17.338	18.868	22.760	25.989	28.869	31.526	34.805	37.156
	19	18.338	19.910	23.900	27.204	30.144	32.852	36.191	38.582
	20	19.337	20.951	25.038	28.412	31.410	34.170	37.566	39.997
	25	24.337	26.143	30.675	34.382	37.652	40.646	44.314	46.928
	30	29.336	31.316	36.250	40.256	43.773	46.979	50.892	53.672
	35	34.336	36.475	41.778	46.059	49.802	53.203	57.342	60.275
	40	39.335	41.622	47.269	51.805	55.758	59.342	63.691	66.766

Note: Code for creating this table was adapted from *How to Build a Chi-Square Distribution Table in R* by Alex Knorre, 2016, Stack Overflow, https://stackoverflow.com/questions/44198167/how-to-build-a-chi-square-distribution-table-in-r. CC BY-SA 3.0, https://creativecommons.org/licenses/by-sa/3.0/.

```
#Critical value of chi-square, p = .05, df = 8
qchisq(.05, 6, lower.tail = F)
[1] 12.59
```

This function confirms that the critical value for χ^2 is 12.59. Figure 12.2 illustrates the location of the chi-square value of 12.59 in a chi-square distribution when *df* = 8. The area under the curve to the right of the critical value equals .05 of the total area under the curve, so for any obtained value of chi-square greater than 12.59, the *p*-value is less than .05 and the null hypothesis can be rejected. In the preceding example, $\chi^2 = 238.17$, a value far to the right of the critical value, so we *reject the null hypothesis*. Based on the evidence presented here, there is a relationship between region and the importance of religion in one's life.

FIGURE 12.2 ■ Chi-square Distribution, *df* = 6

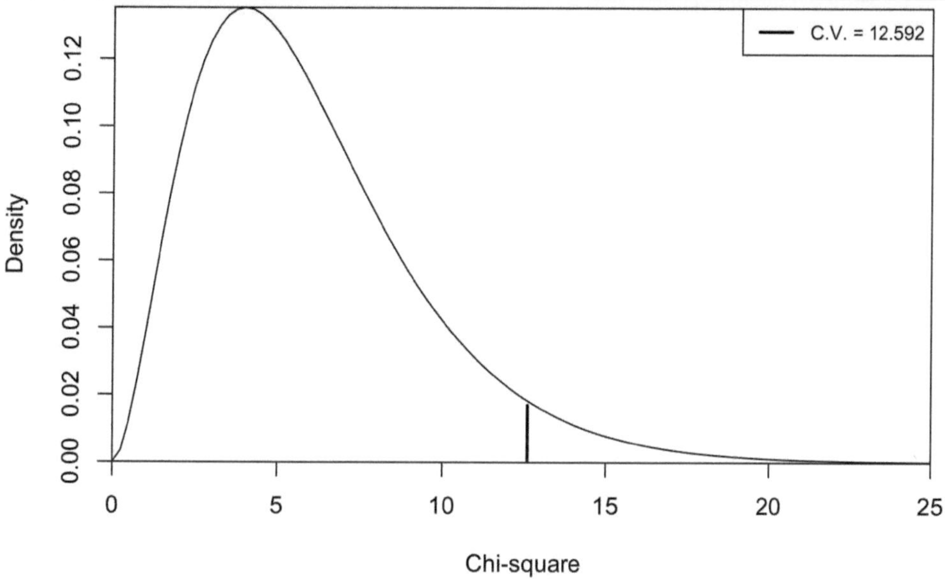

You can get the value of χ^2, along with degrees of freedom and the *p*-value from R, by adding `chisq = T` to the crosstab command. You can also get it from R by using the chisq.test function, as shown next.

```
#Get chi-square statistic from R
chisq.test(anes20$relig_imp, anes20$V203003)

    Pearson's Chi-squared test

data:   anes20$relig_imp and anes20$V203003
X-squared = 238, df = 6, p-value < 2e-16
```

Here, we get confirmation of the value of χ^2 (238) and degrees of freedom (6). We also get a more precise report of the p-value (area on the tail of the distribution, beyond our chi-square value). In this case, we get scientific notation (p < 2e - 16), which tells us that $p <$ .0000000000000002, a number far less than the cutoff point of .05. In cases like this, you can just report the $p < .05$.

Education and Religiosity

Let's take a look at another example, using the same dependent variable but switching to a different independent variable, level of educational attainment. Of course the null and alternative hypotheses are as follows:

- H_0: No relationship. Importance of religion and educational attainment are statistically independent.
- H_1: There is a relationship. Importance of religion and educational attainment are not statistically independent.

The first step is to relabel (shorten) the categories for educational attainment so they fit better in the crosstab.

```
#create new education variable
anes20$educ <- ordered(anes20$V201511x)

#shorten the category labels for education
levels(anes20$educ) <- c("LT HS", "HS", "Some Coll", "4yr degr", "Grad
   degr")
```

Let's take a look at the crosstab and mosaic plot for this pair of variables.

```
crosstab(anes20$relig_imp, anes20$educ,
        prop.c = T,
        chisq = T,
        xlab = "Educational Attainment",
        ylab = "Importance of religion")
```

```
   Cell Contents
|-------------------------|
|                   Count |
|          Column Percent |
|-------------------------|

=======================================================================
```

```
                    anes20$educ
anes20$relig_imp      LT HS        HS   Some Coll   4yr degr   Grad degr  Total
--------------------------------------------------------------------------------
Low                      96       365         860        789         650   2760
                      25.5%     27.4%       30.9%      38.4%       41.0%
--------------------------------------------------------------------------------
Moderate                 93       284         535        389         275   1576
                      24.7%     21.3%       19.2%      18.9%       17.4%
--------------------------------------------------------------------------------
High                    187       684        1387        875         660   3793
                      49.7%     51.3%       49.9%      42.6%       41.6%
--------------------------------------------------------------------------------
Total                   376      1333        2782       2053        1585   8129
                       4.6%     16.4%       34.2%      25.3%       19.5%
================================================================================

Statistics for All Table Factors

Pearson's Chi-squared test
----------------------------------------------------------------
Chi^2 = 108.2       d.f. = 8       p < 2e-16

        Minimum expected frequency: 72.9
```

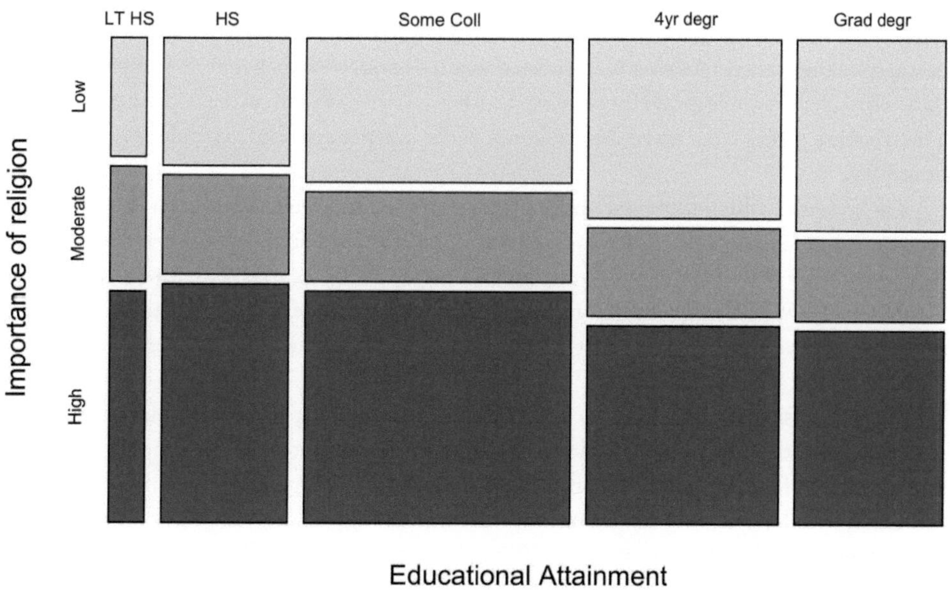

Once again, the first step is to focus on any pattern you see in the column percentages. The crosstab shows that the percentage in the "Low" row increases somewhat steadily from 25.5% to 41% as you move from the lowest to highest education groups. However, the changes in column percentages are not as large or as consistent in the "Moderate" and "High" religious importance rows. This pattern is confirmed and perhaps clearer in the mosaic plot, where the most noticeable changes in the height of the column boxes occur in the "Low" row. In comparison to the analysis of regional differences, the pattern here does not seem as strong. Nevertheless, based on the χ^2 (108.23) and associated p-value ($< .05$), the pattern in this table is different enough from what you would expect if there were no relationship between these variables that we can confidently reject the null hypothesis.

DIRECTIONAL PATTERNS IN CROSSTABS

We have two different types of independent variables in the two examples. In the first crosstab, the independent variable was a nominal variable with each category representing a different region of the country, whereas in the second example the independent variable was ordered, with categories for educational attainment ranging from less than high school to a post-graduate degree. In both cases, the dependent variable is ordered from low to high levels of importance attached to religion. The difference in level of measurement for the independent variables has implications for the types of statements we can make about the patterns in the tables. In both cases, we can discuss the fact that the column percentages vary from one column to the next, indicating that outcomes on the independent variable seem to affect outcomes on the dependent variable. We can go farther and also address the *direction* of the relationship between education and religious importance because the measure of educational attainment has ordered categories. Generally, with ordered data, we are interested not just in whether the column percentages change but also in whether the relationship is positive or negative. Are high values on the independent variable generally associated with low or high values on the dependent variable? This question makes sense when the categories of the independent and dependent variables can be ordered.

The extent to which there is a positive or negative relationship depends on the pattern of the percentages in the table. Table-level statistics that are based on crosstabs treat *the cell in the upper-left corner of the table as the lowest-ranked category for both variables, regardless of how the categories are labeled.* This is why it is usually a good idea to recode and relabel ordered variables so the substantive meaning of the first category can be thought of as "low" or "negative" in value on whatever the scale is.

When assessing directionality, we want to know how moving across the columns from low to high outcomes of the independent variable changes the likelihood of outcomes being in low or high values of the dependent variable. If low values of the independent variable are generally associated with low values of the dependent variable and high values with high values, then we are talking about a positive relationship. When low values of one variable are associated with high values of another, then the relationship is negative. Check out Table 12.2, which provides generic illustrations of what positive and negative relationships might look like in a crosstab.

TABLE 12.2 ■ Demonstration of Positive and Negative Patterns in Crosstabs

	Positive Relationship			Negative Relationship		
	Low (1)	Medium (2)	High (2)	Low (1)	Medium (2)	High (2)
Low (1)	50%	33%	20%	20%	33%	50%
Medium (2)	30%	34%	30%	30%	34%	30%
High (2)	20%	33%	50%	50%	33%	20%

On the left side of Table 12.2 (positive relationships), the column percentages in the "Low" row drop as you move from the low to high columns and increase in the "High" row as you move from the low to high column. So, low outcomes tend to be associated with low values of the independent variable, and high outcomes with high values of the independent variable. On the right side (negative relationships), low outcomes tend to be associated with high values of the independent variable and high outcomes with low values of the independent variable.

Let's reflect back on the crosstab between education and religious importance. Does the pattern in the table suggest a positive or negative relationship, or is it hard to tell? Though it is not as clear as in the hypothetical pattern in Table 12.2, there is a negative pattern in the first crosstab: The likelihood of being in the "low religious importance" row increases across columns as you move from the lowest to highest levels of education, and there is a somewhat weaker tendency for the likelihood of being in the "high religious importance" row to decrease across columns as you move from the lowest to highest levels of education. High values on one variable tend to be associated with low values on the other, the hallmark of a negative relationship.

Age and Religious Importance

Let's take a look at another example, one where directionality is more apparent. The crosstab that follows still uses importance of religion as the dependent variable but switches to a five-category variable for age as the independent variable. Both the table percentages and the mosaic plot show a much stronger pattern in the data than in either of the previous examples. The percentage of respondents who assign a low level of importance to religions drops precipitously from 50.4% among the youngest respondents to just 19.4% among the oldest respondents. At the same time, the percentage assigning a high level of importance to religions grows steadily across columns, from 30.7% in the youngest group to 63.1% in the oldest group. As you might expect, given the strength of this pattern, the chi-square statistic is quite large and the *p*-value is very close to zero. We can reject the null hypothesis. There is a relationship between these two variables.

```
#Collapse age into fewer categories
anes20$age5 <- cut2(anes20$V201507x, c(30, 45, 61, 76, 100))
#Assign labels to levels
```

```
levels(anes20$age5) <- c("18-29", "30-44", "45-60", "61-75", "76 + ")
#Crosstab with mosaic plot and chi-square
crosstab(anes20$relig_imp, anes20$age5, prop.c = T,
         plot = T, chisq = T,
         xlab = "Age",
         ylab = "Religious Importance")
```

```
    Cell Contents
|-------------------------|
|                   Count |
|          Column Percent |
|-------------------------|

==================================================================
                     PRE: SUMMARY: Respondent age
anes20$relig_imp    18-29   30-44    45-60   61-75     76+   Total
------------------------------------------------------------------
Low                   505     877      666     522     135    2705
                    50.4%   43.7%    32.3%   24.4%   19.4%
------------------------------------------------------------------
Moderate              189     357      396     470     122    1534
                    18.9%   17.8%    19.2%   22.0%   17.5%
------------------------------------------------------------------
High                  308     774     1003    1149     440    3674
                    30.7%   38.5%    48.6%   53.7%   63.1%
------------------------------------------------------------------
Total                1002    2008     2065    2141     697    7913
                    12.7%   25.4%    26.1%   27.1%    8.8%
==================================================================

Statistics for All Table Factors

Pearson's Chi-squared test
------------------------------------------------------------
Chi^2 = 396.7      d.f. = 8       p < 2e-16

        Minimum expected frequency: 135.1
```

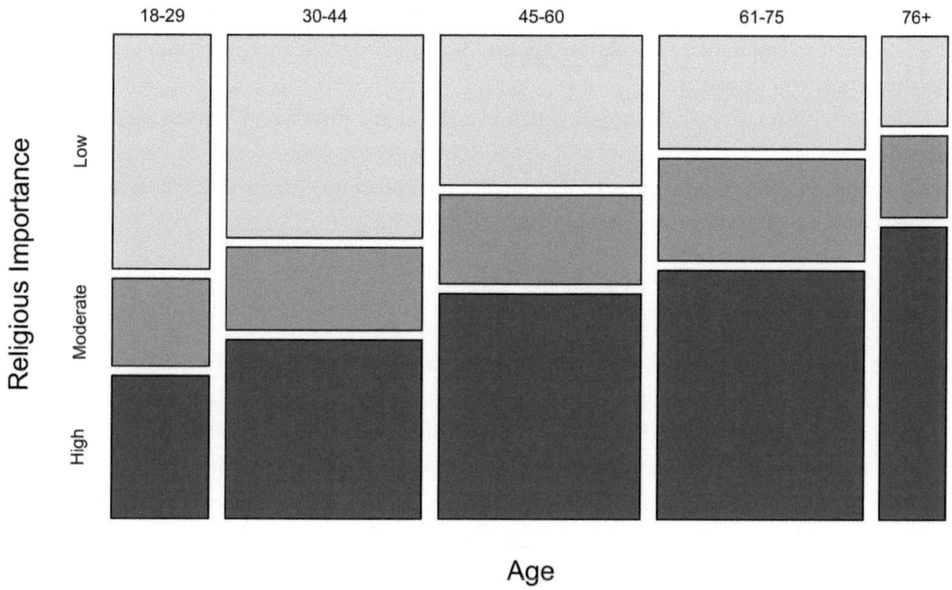

The pattern is also clear in the mosaic plot, in which the height of the light gray boxes (representing low importance) shrinks steadily as age increases, while the height of the dark boxes (representing high importance) grows steadily as age increases.

We should recognize that both variables are measured on ordinal scales, so it is perfectly acceptable to evaluate the table contents for directionality. So what do you think? Is this a positive or negative relationship? Whether focusing on the percentages in the table or the size of the boxes in the mosaic plot, the pattern of the relationship stands out pretty clearly: The importance respondents assign to religion increases steadily as age increases. There is a positive relationship between age and importance of religion.

LIMITATIONS OF CHI-SQUARE

Like the *t*-score, *z*-score, and *F*-ratio, chi-square is a test statistic that can be used as a measure of statistical significance. Based on the value of chi-square, we can determine if an observed relationship between two variables is different enough from what would be expected under the null hypothesis that we can reject H_0.

Like other tests of statistical significance, chi-square does not tell how strong the relationship is. As you can see from the three crosstabs examined in this chapter, all three relationships are statistically significant, but there seems to be a lot of variation in the strength of the relationships. Of course, the other tests for statistical significance also don't directly address the strength of the relationship between two variables. That is not their purpose, so it is not really a "problem" with measures of significance. The problem, though, is the tendency to treat statistically significant relationships as important without paying close attention to effect size. It does not hurt to emphasize an important point: Statistical significance does not equal substantive importance or impact.

One other limitation of chi-square is that even though a directional hypothesis for the relationship between age (or education) and religious importance makes sense, chi-square does not test for directionality; it just tests to see if the observed outcomes are sufficiently different from the expected outcomes that we can reject the null hypothesis.

NEXT STEPS

As you might have guessed from the last few paragraphs, we will move on from here to explore ways in which we can use *measures of association* to assess the strength and direction of relationships in crosstabs. Following this, the rest of the book focuses on estimating the strength and statistical significance of relationships between numeric variables. In some ways, everything in the first 13 chapters of this book establishes the foundation for exploring relationships between numeric variables. I think you will find these next few chapters particularly interesting and surprisingly easy to follow. Of course, that's because you've laid a strong base with the work you've done throughout the rest of the book.

EXERCISES

Concepts and Calculations

1. The following table shows the relationship between age (18 to 49, 50 plus) and support for building a wall on the U.S. border with Mexico.
   - State the null and alternative hypotheses for this table.
   - There are three cells missing the column percentages (Oppose/18–49, Neither/50+, and Favor/18–49). Calculate the missing column percentages.
   - After filling in the missing percentages, describe the relationship in the table. Does the relationship appear to be different from what you would expect if the null hypothesis were true? Explain yourself with specific references to the data.

```
================================================
Anes20$wall            SUMMARY: Respondent age
                       18 to 49       50+   Total
------------------------------------------------
Oppose                     1858      1803    3661
                         ------     41.6%
------------------------------------------------
Neither                     735       674    1409
                          20.6%     -----
------------------------------------------------
```

```
Favor                      980      1859     2839
                        ------     42.9%
--------------------------------------------------
Total                     3573      4336     7909
                         45.2%     54.8%
==================================================
```

2. The same two variables from Problem 1 are shown again in the following table. In this case, the cells show the observed (top) and expected (bottom) frequencies, with the exception of three cells.
   - Estimate the expected frequencies for the three cells in which they are missing.
   - By definition, what are expected frequencies? Note the formula, the substantive meaning of "expected."
   - Estimate the degrees of freedom and critical value of χ^2 for this table.

```
==================================================
Anes20$wall         SUMMARY: Respondent age
                    18 to 49        50+    Total
--------------------------------------------------
Oppose                   1858      1803     3661
                       ------    2007.1
--------------------------------------------------
Neither                   735       674     1409
                        636.5     -----
--------------------------------------------------
Favor                     980      1859     2839
                       ------    1556.4
--------------------------------------------------
Total                    3573      4336     7909
==================================================
```

3. Using the observed and expected values from the results in Problem 2, calculate the value of χ^2 for the table, compare it to the critical value, and decide if you should reject the null hypothesis.

4. The following mosaic plot illustrates the relationship between respondent race/ethnicity and support for building a wall at the U.S. border with Mexico. What does the mosaic plot tell you about both the distribution of the independent variable and the relationship between racial and ethnic identity and attitudes toward the border wall? Is this a directional or non-directional relationship?

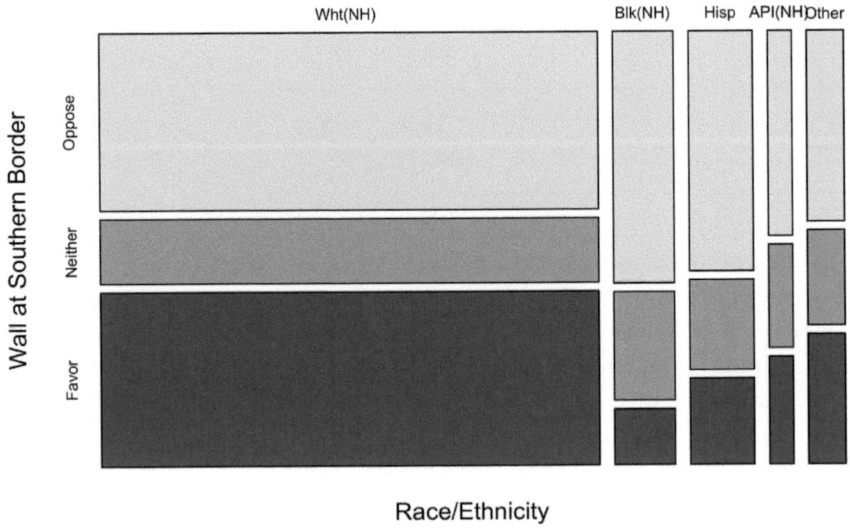

5. The following table illustrates the relationship between marital status and support for a wall at the southern U.S. border. Describe the relationship, paying special attention to statistical significance, strength of relationship, and (if appropriate) direction of relationship. Make sure to use column percentages to bolster your case.

```
   Cell Contents
|-------------------------|
|                   Count |
|          Column Percent |
|-------------------------|


=====================================================
               anes20$marital
anes20$wall      Married  Never Married   Other  Total
-----------------------------------------------------
Oppose              1795           1128     844   3767
                   41.6%          58.0%   43.5%
-----------------------------------------------------
Neither              734            375     355   1464
                   17.0%          19.3%   18.3%
-----------------------------------------------------
Favor               1781            443     742   2966
                   41.3%          22.8%   38.2%
-----------------------------------------------------
```

```
Total                4310            1946    1941    8197
                     52.6%           23.7%   23.7%
=======================================================


Statistics for All Table Factors


Pearson's Chi-squared test
-------------------------------------------------------------
Chi^2 = 215.6      d.f. = 4      p < 2e-16


        Minimum expected frequency: 346.7
```

R Problems

1. I want to look at a crosstab that shows the relationship between religious importance (`relig_imp`) and whether people support or oppose adoption by gay and lesbian couples (`V201415`). I've done something wrong and can't get the command to work. What am I doing wrong? Correct the code and show the output. Hint: there are three errors.

```
crosstab(anes20$V201415 ~ Anes20$relig_imp,
        prop.c = T,
        chisqu = T,
        plot = F)
```

For the remaining problem, use the following code, borrowed from Chapter 4, to create a three-category variable based on responses to the Democratic and Republican feeling thermometer ratings.

```
#Create net party feeling thermometer rating
anes20$netpty_ft <- (anes20$dempty_ft-anes20$reppty_ft)

#Collapse "netpty_ft" and assign the result to "anes20$netpty_ft.3"
anes20$netpty_ft.3 <- ordered(cut2(anes20$netpty_ft, c(0, 1)))

#Assign meaningful level names
levels(anes20$netpty_ft.3) <- c("Favor Reps", "Same", "Favor Dems")
```

2. Repeat the analysis from Problem 1 but using `anes20$age5` as the independent variable. Describe the relationship between these two variables, paying attention to statistical significance and the strength and direction of the relationship. Do you think the relationship in this table is stronger, weaker, or about the same as the relationship in the table from Question 1? Explain.

3. Create a crosstab that shows the relationship between religious importance (`anes20$relig_imp`) created earlier in this chapter and the newly created net feeling thermometer rating, `anes20$netpty_ft.3`. In this case, religious importance is the independent variable and the net feeling thermometer rating is the dependent variable. Be sure to include column percentages, the chi-square statistic, and the mosaic plot. Using this information, describe the relationship between these two variables, paying attention to statistical significance and the strength and direction of the relationship. Comment on whether you find it easier to use the crosstab or the mosaic plot to evaluate the relationship. Explain.

13 MEASURES OF ASSOCIATION

GET READY

This chapter examines multiple ways to measure the strength of relationships between two variables in a contingency table. To follow along in R, you should load the `anes20` dataset, as well as the libraries for the `DescTools`, `descr`, and `dplyr` packages. You might also want to keep a calculator handy or be ready to use R as a calculator.

GOING BEYOND CHI-SQUARED

As discussed at the end of the previous chapter, chi-squared is a good measure of statistical significance, but it is not appropriate to use it to judge the strength or direction of a relationship. By "strength of a relationship" in crosstabs, we generally mean the extent to which values of the dependent variable are conditioned by (or depend upon) values of the independent variable. As a general rule, when chi-squared is not significant, the outcomes of the dependent variable vary only slightly and randomly around their expected (null hypothesis) outcomes. If chi-squared is statistically significant, we know there is a relationship between the two variables, but we don't know the degree to which the outcomes on the dependent variable are conditioned by the values of the independent variables—only that they differ enough from the expected outcomes under the null hypothesis that we can conclude there is a real relationship between the two variables.

Of course, we can use the column percentages to try to get a general sense of how strong the relationship is, but it can be difficult to do so with any precision. Consider the following descriptions of column percentages from the three crosstabs we examined in Chapter 12:

- *Region and religiosity*: The percentage assigning a low level of importance to religion ranges from 26.4% for respondents in the South, to 32.6% for Midwesterners, to 39.4% for Northeastern respondents, to 43.5% for respondents from the West. The opposite pattern is found in the "High" religious importance row, with Midwestern (48%) and Southern (55.7%) respondents having the highest column percentages, and Northeastern (37.7%) and Western (37.4%) respondents having the lowest. There isn't much of a regional pattern in the "Moderate" row.
- *Education and religiosity*: The percentage assigning low importance to religion increases steadily from 25.5% among those with less than a high school degree to 41% among those with a graduate degree, an increase of 15.5 percentage points. The percentage

assigning a moderate level of importance to religion declines by about 7 percentage points for the lowest and highest level of education, and there is a similar drop off in the percentage assigning high importance across levels of education.

- *Age and religiosity*: The percentage of respondents who assign a low level of importance to religion drops from 50.4% among the youngest respondents to just 19.4% among the oldest respondents. At the same time, the percentage assigning a high level of importance to religion grows steadily across columns, from 30.7% in the youngest group to 63.1% in the oldest group. The percentage assigning moderate importance to religion does not change much across age groups.

Describing the differences in column percentages like this is an important part of telling the story of the relationship between two variables, so we should always give them a close look and be comfortable discussing them. However, we should not rely on them alone, as they do not provide a clear, consistent, and standard basis for judging the strength of the relationship in the table. How strong are the relationships that we just described? We might have a sense that the region and age are more strongly related to religiosity than education is, based on the percentage differences, but we still can't tell how much stronger or even how strong any of the relationships are on their own. What we need is to be able to complement the discussion of column percentages with statistics that take into account the outcomes across the entire table and weight those outcomes according to their share of the overall sample.

MEASURES OF ASSOCIATION FOR CROSSTABS

Measures of association for crosstabs are statistics that summarize the strength of the relationship between two variables. They provide the same type of information we got from the measures of effect size examined in previous chapters (Cohen's *D* and eta^2), summarizing the extent to which differences in outcomes on the dependent variable are related to differences in outcomes on the independent variable.

Cramer's *V*

One useful measure of association for crosstabs, especially if one of the variables is a nominal-level variable, is Cramer's *V*. We begin with this measure because it is derived from chi-square. One of the many virtues of chi-square is that it is based on information from every cell of the table and uses cell frequencies rather than percentages, so each cell is weighted appropriately. Recall, though, that the "problem" with chi-square is that its size reflects not just the strength of the relationship but also the sample size and the size of the table. Cramer's *V* discounts the value of chi-squared for both the sample size and the size of the table by incorporating them in the denominator of the formula:

$$\text{Cramer's } V = \sqrt{\frac{\chi^2}{N * min(r-1,\, c-1)}}$$

Here we take the square root of chi-square divided by the sample size times either the number of rows minus 1 or the number of columns minus 1, whichever is smaller.[1]

Interpreting Cramer's V is relatively straightforward, as long as you understand that it is bounded by 0 and 1, where a value of zero means there is absolutely no relationship between the two variables, and a value of one means there is a perfect relationship between the two variables. Figure 13.1 illustrates these two extremes.

FIGURE 13.1 ■ Hypothetical Outcomes for Cramer's *V*

No Relationship

	Low	High
Low	33.%	33.%
High	67%	67%

Cramer's V = 0

Perfect Relationship

	Low	High
Low	100%	0%
High	0%	100%

Cramer's V = 1.0

In the table on the left, there is no change in the column percentages as you move across columns. We know from the discussion of chi-square in the last chapter that this is exactly what statistical independence looks like—the row outcome does not depend upon column outcomes. In the table on the right, you can perfectly predict the outcome of the dependent variable based on levels of the independent variable.

Table 13.1 provides the χ^2 value, number of rows and columns, and the sample size for each of the three crosstabs reported in Chapter 12 and described earlier.

TABLE 13.1 ■ Summary Statistics From Crosstabs in Chapter 12

Independent Variable	χ^2	Rows × Columns	Sample Size
Region	238.2	3 × 4	8249
Education	108.2	3 × 5	8129
Age	396.7	3 × 4	7913

Let's use this information to calculate the Cramer's *V* statistic for these relationships.

Region:

$$V=\sqrt{\frac{238.2}{8249*2}}=.12$$

Education:

$$V = \sqrt{\frac{108.2}{8129 * 2}} = .08$$

Age:

$$V = \sqrt{\frac{396.7}{7913 * 2}} = .16$$

The Cramer's *V* values reported show that age has the greatest impact on religiosity, followed by region and then education, and that all three relationships are relatively weak.

Are you surprised by the value of Cramer's *V* (.16) for the impact of age on religiosity? The differences in the column percentages reported earlier were fairly large, so we might have expected that Cramer's *V* would come in at a somewhat higher value. This highlights an inherent limitation of focusing on just a few differences in column percentages rather than taking in information from the entire table. In the specific case of this relationship, by focusing on the large differences in the low and high religious importance rows, we ignored the tiny differences in the moderate row. More to the point, it is difficult to process the outcomes in all twelve cells of the table and assign them appropriate weight by just "eyeballing' the table. Again, column percentages are important—they tell us what is happening between the two variables—but focusing on them exclusively invariably provides an incomplete accounting of the relationship. Good measures of association such as Cramer's *V* take into account information from all cells in a table and provide a more complete assessment of the relationship.

Getting Cramer's *V* in R

The `CramerV` function is found in the `DescTools` package and can be used to get the values of Cramer's *V* in R.

```
#If needed, use code from Chapter 12 to create anes20$relig_imp
#V for region and religious importance:
CramerV(anes20$relig_imp, anes20$V203003)
[1] 0.12015

#V for education and religious importance:
CramerV(anes20$relig_imp, anes20$educ)
[1] 0.081592

#V for age and religious importance:
CramerV(anes20$relig_imp, anes20$age5)
[1] 0.15831
```

These results confirm the earlier calculations.

Lambda

Another sometimes useful statistic for judging the strength of a relationship is lambda (λ). Lambda is referred to as a proportional reduction in error (PRE) statistic because it summarizes how much we can reduce the level of error from guessing or predicting the outcome of the dependent variable by using information from an independent variable. The concept of proportional reduction in error plays an important role in many of the topics included in the last several chapters of this book.

The formula for lambda is

$$\text{Lambda}(\lambda) = \frac{E_1 - E_2}{E_1},$$

where

- E_1 = error by guessing with no information on the independent variable, and
- E_2 = error by guessing with information about an independent variable.

Okay, so what does this mean? Let's think about it in terms of the relationship between region and religious importance. The following table below includes just the raw frequencies for these two variables.

```
#Cell frequencies for religious importance by region
crosstab(anes20$relig_imp, anes20$V203003,plot = F)

   Cell Contents
|-------------------------|
|                    Count|
|-------------------------|

=======================================================================
                      SAMPLE: Census region
anes20$relig_imp  1. Northeast   2. Midwest   3. South   4. West   Total
-----------------------------------------------------------------------
Low                        549          649        811       782    2791
-----------------------------------------------------------------------
Moderate                   320          385        548       343    1596
-----------------------------------------------------------------------
High                       525          956       1710       671    3862
-----------------------------------------------------------------------
Total                     1394         1990       3069      1796    8249
=======================================================================
```

Now, suppose that you had to "guess" or "predict" the value of the dependent variable for each of the 8249 respondents from this table. What would be your best guess? By "best guess," I mean which category would give the least overall error in guessing? As a rule, with ordinal or nominal dependent variables, it is best to guess the modal outcome if you want to minimize error in guessing (you may recall this from Chapter 4). In this case, that outcome is the "High" row, which has 3862 respondents. If you guess this, you will be correct 3862 times and wrong 4387 times. This seems like a lot of error, but no other guess can give you an error rate this low (go ahead, try).

From this, we get $E_1 = 4387$.

E_2 is the error we get when we are able to guess the value of the dependent variable based on the value of independent variable. Here's how we get E_2: We look within each column of the independent variable and choose the category of the dependent variable that will give us the least overall error for that column (the modal outcome within each column). For instance, for "Northeast" we would guess "Low," and we would be correct 549 times and wrong 845 times (make sure you understand how I got these numbers); for "Midwest," we would guess "High" and would be wrong 1034 times; for "South," we would guess "High" and be wrong 1359 times; and for "West," we could guess "Low" and be wrong 1014 times. Now that we have given our best guess within each category of the independent variable, we can estimate E_2, which is the sum of all of the errors from guessing when we have information from the independent variable:

$$E_2 = 845 + 1034 + 1359 + 1014 = 4252$$

Note that E_2 (4252) is less than E_1 (4387). This means that the independent variable reduced the error in guessing the outcome of the dependent variable by 135 ($E_1 - E_2 = 135$). On its own, it is difficult to tell if reducing error by 135 represents a lot or a little reduction in error. Because of this, we express reduction in error as a *proportion of the original error*:

$$\text{Lambda}(\lambda) = \frac{4387 - 4252}{4387} = \frac{135}{4387} = .031$$

What this tells us is that knowing the region of residence for respondents in this table reduces the error in predicting the dependent variable by 3.1%.

Lambda ranges from 0 to 1, where 0 means that the independent variable does not account for (reduce) any error in predicting the dependent variable, and 1 means the independent variable accounts for all of the error in predicting the dependent variable. You might notice that this interpretation is very similar to the interpretation of eta^2 in Chapter 11. That's because both statistics are measuring the amount of error (variation) in the dependent variable that is explained by the independent variable. Lambda and eta^2 are both proportional reduction in error (PRE) statistics.

One of the nice things about lambda, compared to Cramer's *V*, is that it has a very intuitive meaning. Think about how we interpret Cramer's *V* for the relationship between region and religious importance (.12). We know that .12 on a scale from 0 to 1 is rather low, but it is hard to ascribe more meaning to it than that. Now, contrast that with lambda (.031), which also indicates a weak relationship but at the same time conveys an additional piece of information, that knowing the outcome of the independent variable leads to a 3.1% decrease in error when predicting the dependent variable.

Getting Lambda From R

The `Lambda` function in R can be used to calculate lambda for a pair of variables. Along with variable names `(dependent, independent)`, you need to specify two additional pieces of information in the Lambda function: `direction = "row"` means that we are predicting row outcomes, and `conf.level = .95` adds a 95% confidence interval around lambda.

```
#For Region and Religious importance (dependent, independent)
Lambda(anes20$relig_imp, anes20$V203003,
       direction = "row", conf.level = .95)

    lambda     lwr.ci     upr.ci
 0.0307727  0.0086624  0.0528831
```

Here, we get confirmation that the calculations for the region/religious importance table were correct (lambda = .03). Also, since the 95% confidence interval does not include 0 (barely!), we can reject the null hypothesis and conclude that there is a statistically significant but quite small reduction in error predicting religious importance with region.

We can also see the values of lambda for the other two tables: Lambda is 0 for education and religious importance and .07 for age and religious importance.

```
#For Education and Religious importance
Lambda(anes20$relig_imp, anes20$educ,
       direction = "row", conf.level = .95)

   lambda    lwr.ci    upr.ci
        0         0         0
#For Age and Religious importance
Lambda(anes20$relig_imp, anes20$age5,
       direction = "row", conf.level = .95)

   lambda    lwr.ci    upr.ci
 0.070771  0.048647  0.092896
```

The finding for the relationship between education and religious importance points to one limitation of lambda: It can equal 0 even if there is a statistically significant relationship in the table. The problem is that the relationship is not strong enough to be a *proportional reduction in error* relationship, even if there is a significant pattern (according to chi-square) in the data. This happens most often when the relationship is weak and the distribution of the dependent variable heavily concentrated in the modal outcome. As a result, having information about the independent variable does not change guesses across columns, and the errors from guessing in E_2 are the same as the error in E_1, resulting in lambda = 0.

Lambda and Cramer's *V* are using different standards to evaluate the strength of relationship. Lambda focuses on whether using the independent variable leads to less error in guessing outcomes of the dependent variable. Cramer's *V*, on the other hand, focuses on how different the pattern in the table is from what you would expect if there was no relationship.

ORDINAL MEASURES OF ASSOCIATION

Cramer's *V* and lambda measure the association between two variables by focusing on deviation from expected outcomes or reduction in error, respectively, two useful pieces of information. One drawback to these measures though is that they do not express information about the strength of a relationship from *a directional perspective*. Consider, for example, the impact of education and age on religious importance, as just discussed. In both cases, there are clear directional expectations based on the ordinal nature of both variables: Religious importance should decrease as levels of education increase (negative relationship), and it should increase as age increases (positive relationship). The findings based on the column percentage reported earlier supported these expectations. One problem, however, is that the measures of association we've looked at so far are not designed to capture the extent to which the pattern in a crosstab follows a particular directional pattern. What we need, then, are measures of association that incorporate the idea of directionality into their calculations.

To demonstrate some of these ordinal measures of association, I use a somewhat simpler 3x3 crosstab that still focuses on religious importance as the dependent variable (the smaller table works better for demonstration). Here, I use `anes20$ideol3`, a three-category version of `anes20$V201200`, to look at how political ideology is related to religious importance. You might question whether political ideology is an ordinal variable. After all, none of the category labels carry the obvious hints, such as "Low" and "High," or "Less than" and "Greater than." Instead, as pointed out in Chapter 1, you can think of the categories of political ideology as growing more conservative (and less liberal) as you move from "Liberal" to "Moderate" to "Conservative".

```
#Create three-category ideology variable
anes20$ideol3 <- anes20$V201200
levels(anes20$ideol3) <- c("Liberal","Liberal","Liberal","Moderate",
                           "Conservative","Conservative",
                           "Conservative")
#Crosstab of religious importance by ideology
crosstab(anes20$relig_imp, anes20$ideol3,
         prop.c = T, chisq = T,
         plot = F)
```

```
   Cell Contents
|-------------------------|
|                   Count |
|          Column Percent |
|-------------------------|

======================================================================
                  PRE: 7pt scale liberal-conservative self-placement
anes20$relig_imp  Liberal       Moderate       Conservative    Total
----------------------------------------------------------------------
Low                   1410           570                489     2469
                      56.6%         31.5%              17.9%
----------------------------------------------------------------------
Moderate               428           427                498     1353
                      17.2%         23.6%              18.2%
----------------------------------------------------------------------
High                   654           815               1750     3219
                      26.2%         45.0%              63.9%
----------------------------------------------------------------------
Total                 2492          1812               2737     7041
                      35.4%         25.7%              38.9%
======================================================================

Statistics for All Table Factors

Pearson's Chi-squared test
---------------------------------------------------------------------
Chi^2 = 997.01     d.f. = 4      p < 2e-16

      Minimum expected frequency: 348.19
```

```
CramerV(anes20$relig_imp, anes20$ideol3, conf.level = .95)
```

```
Cramer V   lwr.ci   upr.ci
 0.26608  0.24915  0.28221
```

```
Lambda(anes20$relig_imp, anes20$ideol3, direction = "row", conf.level = .95)
```

```
lambda   lwr.ci   upr.ci
0.19780 0.17694 0.21867
```

Here's what I would say about this relationship based on what we've learned thus far. There is a positive relationship between political ideology and the importance people attach to religion. Looking from the liberal column to the conservative column, there is a steady decrease in the percentage who attach low importance to religion and a steady increase in the percentage who attach high importance to religion. Almost 57% of liberals are in the low row, compared to only about 18% among conservatives; and only 26% of liberals are in the high row, compared to about 64% of conservatives. The *p*-value for chi-square is near 0, so we can reject the null hypothesis; the value of Cramer's *V* (.27) suggests a weak to moderate relationship, and the value of lambda shows that ideology reduces the error in guessing religious importance by about 20%. Both of these measures of association indicate that ideology is more strongly related to religious importance than any of the other independent variables we've looked at so far (region, education, age).

There is nothing wrong with using lambda and Cramer's *V* for this table, as long as you understand that they are not addressing the directionality in the data. For that, we need to use an *ordinal measure of association*, one that assesses the degree to which high values of the independent variable are related to high or low values of the dependent variable. In essence, these statistics focus on the ranking of categories and summarize how well we can predict the ordinal ranking of the dependent variable based on values of the independent variable. Ordinal-level measures of association *range in value from -1 to +1*, with 0 indicating no relationship, -1 indicating a perfect negative relationship, and + 1 indicating a perfect positive relationship.

Gamma

One example of an ordinal measure of association is gamma (γ). Very generally, what gamma does is calculate how much of the table follows a positive pattern and how much it follows a negative pattern, but using different terminology. More formally, gamma is based on the number of similarly ranked and differently ranked pairs of observations in the contingency table. Similarly ranked pairs can be thought of as pairs or observations that follow a positive pattern, and differently ranked pairs are those that follow a negative pattern:

$$\text{Gamma} = \frac{N_{\text{similar}} - N_{\text{different}}}{N_{\text{similar}} + N_{\text{different}}}$$

If you look at the parts of this equation, it is easy to see that gamma is a measure of the degree to which positive or negative pairs dominate the table. When positive (similarly ranked) pairs dominate, gamma will be positive; when negative (differently ranked) pairs dominate, gamma will be negative; and when there is no clear trend, gamma will be zero or near zero.

But how do we find these pairs? Let's look at similarly ranked pairs in a generic 3x3 table first.

For each cell in the table in Figure 13.2, we multiply that cell's number of observations times the sum of the observations that are found in all cells ranked higher in value on both the independent and dependent variables (below and to the right). We refer to these cells as "similarly ranked" because they have the same ranking on both the independent and dependent variables, relative to the cell of interest. These pairs of cells are highlighted in Figure 13.2.

For differently ranked pairs, we need to match each cell with cells that are inconsistently ranked on the independent and dependent variables (higher in value on one and lower in value on the other). Because of this inconsistent ranking, we call these differently ranked pairs. These cells are below and to the left of the reference cell, as highlighted in Figure 13.3.

FIGURE 13.2 ■ Similarly Ranked Pairs

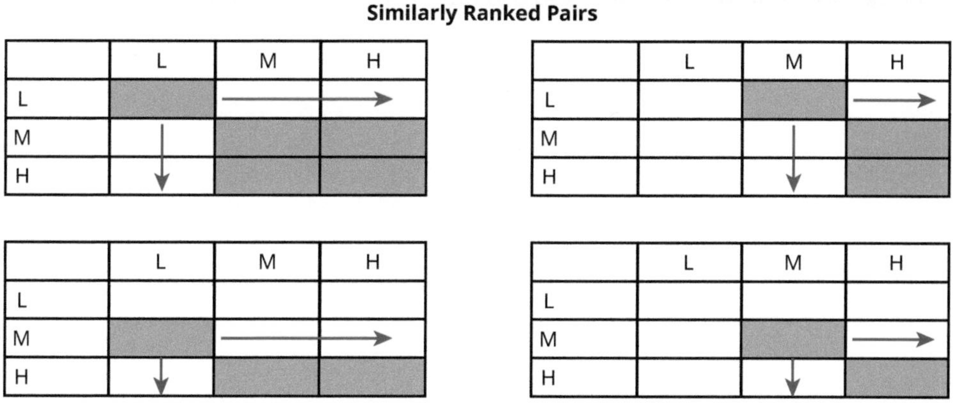

FIGURE 13.3 ■ Differently Ranked Pairs

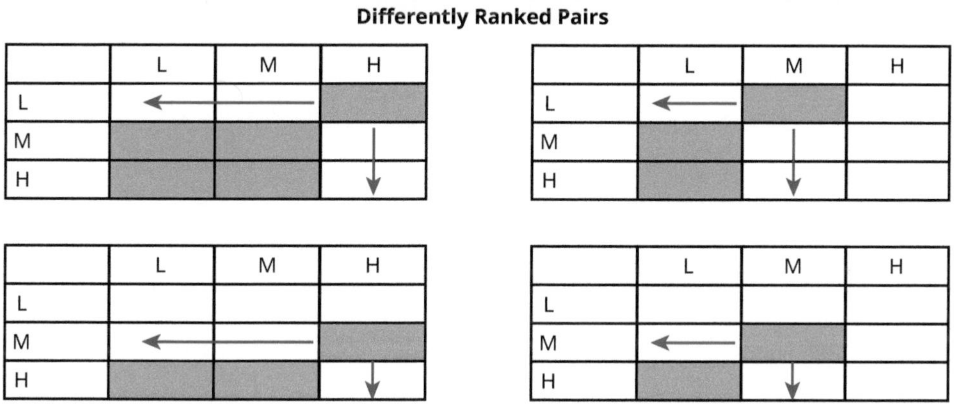

Using this information to calculate gamma may seem a bit cumbersome—and it is—but it becomes clearer after working through it a few times. We'll work through this once (I promise, just once) using the crosstab for ideology and religious importance. The raw frequencies for this crosstab are reproduced in the next section.

Let's start with the similarly ranked pairs. In the following calculation, I begin in the upper-left corner (liberal/low) and pair those 1410 respondents with all respondents in cells below and to the right (1410 x (427 + 498 + 815 + 1750)). Note that the four cells below and to the right are all higher in value on both the independent and dependent variables than the reference cell. Then I go to the liberal/moderate cell and match those 428 respondents with the two cells below and to the right (428 x (815 + 1750)). Next, I go to the moderate/low cell and match those 570 respondents with the two cells below and to the right (570 x (498 + 1750)), and finally, I go to the moderate/moderate cell and match those 427 respondents with respondents in the one cell below and to the right (427 x 1750).

```
#Calculate similarly ordered pairs
similar <- 1410*(427 + 498 + 815 + 1750) + 428*(815 + 1750) +
  570*(498 + 1750) + (427*1750)
similar
[1] 8047330
```

This gives us 8,047,330 similarly ranked pairs. That seems like a big number, and it is, but we are really interested in how big it is relative to the number of differently ranked pairs.

```
Cell Contents
|-------------------------|
|                   Count |
|-------------------------|


=================================================================================
                  PRE: 7pt scale liberal-conservative self-placement
anes20$relig_imp    Liberal    Moderate     Conservative    Total
--------------------------------------------------------------------------------
Low                    1410         570              489     2469
--------------------------------------------------------------------------------
Moderate                428         427              498     1353
--------------------------------------------------------------------------------
High                    654         815             1750     3219
--------------------------------------------------------------------------------
Total                  2492        1812             2737     7041
================================================================================
```

To calculate differently ranked pairs, I begin in the upper-right corner (conservative/low) and pair those 489 respondents with respondents in all cells below and to the left (489 x (427 + 428 + 654 + 815)). Note that the four cells below and to the left are all lower in value on the independent variable and higher in value on the dependent variable than the reference cell. Then, I match the 570 respondents in the moderate/low cell with all respondents in cells below and to the left (570 x (428 + 654)). Next, I match the 498 respondents in the conservative/moderate cell with all respondents in cells below and to the left (498 x (654 + 815)). And, finally, I match the 427 respondents in the moderate/moderate cell with all respondents in the one cell below and to the left (427 x 654).

```
#Calculate differently ordered pairs
different = 489*(427 + 428 + 654 + 815) + 570*(428 + 654) +
  498*(654 + 815) + 427*654
different
[1] 2763996
```

This gives us 2,763,996 differently ranked pairs. Again, this is a large number, but we now have a basis of comparison, the 8,047,330 similarly ranked pairs. Since there are more similarly ranked pairs, we know that there is a positive relationship in this table. Let's plug these values into the gamma formula to get a more definitive sense of this relationship.

For this table,

$$\text{gamma} = \frac{8047330 - 2763996}{8047330 + 2763996} = \frac{5283334}{10811326} = .4887$$

Now, let's check this with the R function for gamma.

```
#Get Gamma from R
GoodmanKruskalGamma(anes20$relig_imp, anes20$ideol3, conf.level = .95)

gamma    lwr.ci  upr.ci
0.48869 0.46254 0.51483
```

There are two things here: The value of gamma is the same as what we calculated on our own, and the 95% confidence interval around the point estimate for gamma does not include 0, so we can reject the null hypothesis (H_0:$gamma = 0$). Further, we can interpret the value of gamma (.49) as meaning that we there is a moderate-to-strong, positive relationship between ideology and religious importance.

A Problem With Gamma

Gamma is a useful statistic from an instructional perspective in part because the logic of taking the balance of similar and different rankings to determine directionality is so intuitively clear. However, in practice, gamma does come with a bit of baggage. The main problem with gamma is that it tends to overstate the strength of relationships. For instance, in the example of the relationship between ideology and religious importance, the value of gamma far outstrips the values of Cramer's V (.266) and lambda (.198). We don't expect to get the same results, but when two measures of association suggest a somewhat modest relationship and another suggests a much stronger relationship, it is a bit hard to reconcile the difference. The reason gamma tends to produce stronger effects is because it focuses only on the diagonal pairs. In other words, gamma is calculated only on the basis of pairs of observations that follow clearly positive (similarly ranked) or negative (differently ranked) patterns. But we know that not all pairs of observations in a table follow these directional patterns. Many of the pairs of observations in the table are "tied" in value on one variable but have different values on the other variable, so they don't follow a positive or negative pattern. If tied pairs were also taken into account, the denominator would be much larger and gamma would be somewhat smaller.

The two tables in Figure 13.4 illustrate examples of tied pairs. In the first table, the shaded area represents the cells that are tied in their value of Y (low) but have different values of X (low, medium, and high). In the second table, the highlighted cells are tied in their value of X (low) but differ in their value of Y (low, medium, high). And there are many more tied pairs throughout the tables. None of these pairs are included in the calculation of gamma, but they can constitute a substantial part of the table, especially if the directional pattern is weak. To the extent there are a lot of tied pairs in a given table, gamma is likely to significantly overstate the magnitude of the relationship between *X* and *Y* because the denominator for gamma is smaller than it should be if it is intended to represent the entire table.

FIGURE 13.4 ■ Tied Pairs in a Crosstab

Tied on Y = Low

	Low	Medium	High
Low			
Medium			
High			

Tied on X = Low

	Low	Medium	High
Low			
Medium			
High			

Tau-*b* and Tau-*c*

There are many useful alternatives to gamma. The ones I like to use are tau-*b* and tau-*c*, both of which maintain the focus on the number of similar and differently ranked pairs in the numerator but also take into account the number of tied pairs in the denominator, albeit in somewhat different ways.[2] In practice, the primary difference between the two is that tau-*b* is appropriate for square tables (e.g., 3x3 or 4x4), while tau-*c* is appropriate for rectangular tables (e.g., 2x3 or 3x4). Like gamma, tau-*b* and tau-*c* are bounded by -1 (perfect negative relationship) and +1 (perfect positive relationship), with 0 indicating no relationship.

The R functions for tau-*b* and tau-*c* are similar to the function for gamma. Although tau-*b* is the most appropriate measure of association for the relationship between ideology and religion (a square table), both statistics are reported next in order to demonstrate the commands.

```
#Get tau-b and tau-c for religious importance by ideology
KendallTauB(anes20$relig_imp, anes20$ideol3, conf.level = .95)

  tau_b   lwr.ci  upr.ci
0.33091  0.31151 0.35030

StuartTauC(anes20$relig_imp, anes20$ideol3, conf.level = 0.95)

   tauc   lwr.ci  upr.ci
0.31971  0.30104 0.33839
```

The results for tau-*b* and tau-*c* are very close in magnitude (they usually are), both pointing to a moderate, positive relationship between ideology and religion. You should also note that the confidence interval does not include 0, so we can reject the null hypothesis (H_0: tau-*b*/tau-*c* = 0). You probably noticed that these values are somewhat stronger than those for Cramer's *V* and lambda but somewhat weaker than the result obtained from gamma. In my experience, this is usually the case.

With the exception of gamma, because it tends to overstate the magnitude of relationships, you can use the Table 13.2 as a rough guide to how the measures of association discussed here are connected to judgments of effect size. Using this information in conjunction with the contents of a crosstab should enable you to provide a fair and substantive assessment of the relationship in the table. Always remember that the column percentages tell the story of how the variables are connected, but you still need a measure of association to summarize the strength and direction of the relationship. At the same time, the measure of association is not very useful on its own without referencing the contents of the table.

TABLE 13.2 ■ Measures of Association and Effect Size Interpretations

Absolute Value	Effect Size
.05	
.15	Weak
.25	
.35	Moderate
.45	
.55	Strong
.65	
.75	Very Strong
.85	

Let's circle back to the other two ordinal independent variables we looked at earlier while focusing on Cramer's *V* and lambda: education and age. These were both 3x5 crosstabs, so we will look at values for gamma and tau-*c*. Cramer's *V* is also included here for the purpose of comparison.

```
#Get stats for religious importance by education
CramerV(anes20$relig_imp, anes20$educ, conf.level = 0.95)

Cramer V   lwr.ci    upr.ci
0.081592 0.063227 0.094511

StuartTauC(anes20$relig_imp, anes20$educ, conf.level = 0.95)
     tauc     lwr.ci     upr.ci
-0.088665 -0.107838 -0.069491
```

```
GoodmanKruskalGamma(anes20$relig_imp, anes20$educ, conf.level = 0.95)

   gamma    lwr.ci    upr.ci
-0.12453 -0.15138 -0.09769

#Get stats for religious importance by age

CramerV(anes20$relig_imp, anes20$age5, conf.level = 0.95)

Cramer V   lwr.ci    upr.ci
 0.15831  0.14126  0.17256

StuartTauC(anes20$relig_imp, anes20$age5, conf.level = 0.95)

   tauc     lwr.ci    upr.ci
0.19269   0.17361   0.21177

GoodmanKruskalGamma(anes20$relig_imp, anes20$age5, conf.level = 0.95)

  gamma     lwr.ci    upr.ci
0.26282   0.23718   0.28846
```

As you can see, the tau-*c* coefficients are close in value to the Cramer's *V* coefficients, except now tau-*c* indicates the direction of the relationship. Also, as expected, the gamma coefficients are larger in value than the others, reflecting its focus on just the directional pairs in the tables.

REVISITING THE GENDER GAP IN ABORTION ATTITUDES

Back in Chapter 10, we looked at the gender gap in abortion attitudes, using *t*-tests for the sex-based differences in two different outcomes, the proportion who thought abortion should always be illegal and the proportion who thought abortion should be available as a matter of choice. Those tests found no significant gender-based difference for preferring that abortion be illegal and a significant but small difference in the proportion preferring that abortion be available as a matter of choice. With crosstabs, we can do better than just testing these two polar opposite positions by utilizing a wider range of preferences, as shown next.

```
#Create abortion attitude variable
anes20$abortion <- factor(anes20$V201336)

#Change levels to create four-category variable
levels(anes20$abortion) <- c("Illegal", "Rape/Incest/Life",
                 "Additional Conditions", "Choice",
                 "Additional Conditions")

#Create numeric indicator for sex
anes20$Rsex <- factor(anes20$V201600)
```

```
levels(anes20$Rsex) <- c("Male", "Female")

crosstab(anes20$abortion, anes20$Rsex, prop.c = T,
        chisq = T, plot = F)

Cell Contents
|-------------------------|
|                   Count |
|          Column Percent |
|-------------------------|

=========================================================================
                        anes20$Rsex
anes20$abortion              Male          Female          Total
-------------------------------------------------------------------------
Illegal                       388             475            863
                            10.4%           10.7%
-------------------------------------------------------------------------
Rape/Incest/Life              929             978           1907
                            24.9%           22.1%
-------------------------------------------------------------------------
Additional Conditions         688             706           1394
                            18.5%           16.0%
-------------------------------------------------------------------------
Choice                       1723            2263           3986
                            46.2%           51.2%
-------------------------------------------------------------------------
Total                        3728            4422           8150
                            45.7%           54.3%
=========================================================================

Statistics for All Table Factors

Pearson's Chi-squared test
------------------------------------------------------------------------------
Chi^2 = 24.499    d.f. = 3    p = 0.0000196
        Minimum expected frequency: 394.76
```

Here, the dependent variable ranges from wanting abortion to be illegal in all circumstances to allowing it only in cases of rape, incest, or threat to the life of the mother, to allowing it in additional circumstances, to allowing it generally as a matter of choice. These categories can be seen as ranging from most to least restrictive views on abortion. As expected, based on the analysis in Chapter 10, the differences between men and women on this issue are not very great. In fact, the greatest difference in column percentages is in the "Choice" row, where women are about five percentage points more likely than men to favor this position. This weak effect is also reflected in the measures of association reported next.

```
#Get Cramer's and Tau-c
CramerV(anes20$abortion, anes20$Rsex)
[1] 0.054827

StuartTauC(anes20$abortion, anes20$Rsex, conf.level = 0.95)

    tauc     lwr.ci   upr.ci

0.041292 0.018106 0.064479
```

This relationship provides another demonstration of the need to go beyond just reporting on statistical significance. Had we simply reported that chi-square is statistically significant, with a *p*-value of .0000196, we might have concluded that sex plays an important role in shaping attitudes on abortion. However, by focusing some attention on the column percentages and measures of association, we can more accurately report that sex is related to how people answered this question but its impact is quite small.

When to Use Which Measure

A number of measures of association are available for describing the relationship between two variables in a contingency table. Not all measures of association are appropriate for all tables, however. To determine the appropriate measure of association, you need to know the level of measurement for your table.

For tables in which at least one variable is nominal, such as the table featuring region and religious importance, you must use non-directional measures of association. For these tables, Cramer's *V* and lambda serve as excellent measures of association. You can certainly use both, or you may decide that you are more comfortable with one over the other and use it on its own. Just make sure that if you are using just one of these two statistics, you are not doing so because it makes the relationship look stronger.

For tables in which both variables are ordinal, you can choose between gamma, tau-*b*, and tau-*c*. My own preference is to rely on tau-*b* and tau-*c*, since they incorporate more information from the table than gamma does and, hence, provide a more realistic impression of the relationship in the table. If you use gamma, you should remember and acknowledge that it tends to overstate the magnitude of relationships because it focuses just on the positive and negative patterns in the table. You should also remember that tau-*b* is appropriate for square tables, as tau-*c* is for rectangular tables.

NEXT STEPS

The next several chapters build on this and earlier chapters and turn to assessing relationships between numeric variables. In Chapter 14, we examine how to use scatter plots and correlations coefficients to assess relationships between numeric variables. The correlation coefficient is an interval-ratio version of the same type of measures of association you learned about in this chapter. Like other directional measures of association, it measures the extent to which the relationship between the independent and dependent variable follows a positive versus negative pattern and provides a sense of both the direction and strength of the relationship. The remaining chapters focus on regression analysis, first examining how single variables influence a dependent variable, then how we can assess the impact of multiple independent variables on a single dependent variable. If you've been able to follow along so far, the remaining chapters won't pose a problem for you.

EXERCISES

Concepts and Calculations

1. Identify the appropriate measures of association for each of the pairs of variables listed in the following table.

	Dependent Variable	Independent Variable
A.	Presidential approval (five-categories, strongly disapprove to strongly approve)	Party identification (Democrat, Independent, Republican)
B.	Presidential approval	Marital status (married, widowed, divorced, never married, other)
C.	Party identification	Region
D.	Evaluation of the economy (negative, mixed, positive)	Party identification (Democrat, Independent, Republican)
E.	Republican primary preference (Trump, Haley, Christie, DeSantis, other)	Importance of religion (low, moderate, high)

2. The following table provides the raw frequencies for the relationship between respondent sex and political ideology in the 2020 ANES survey.
   - What percentage of female respondents identify as liberal?
   - What percentage of male respondents identify as liberal?
   - What about conservatives? What percentage of male and female respondents identify as conservative?
   - Do the differences between male and female respondents seem substantial?

	Male	Female	Total
Liberal	1043	1443	2486
Moderate	812	984	1796
Conservative	1443	1283	2726
Total	3298	3710	7008

Chi-square = 66.23

3. Calculate and interpret Cramer's *V* for this table. Show your work.
4. Calculate and interpret lambda for this table. Show your work.
5. Calculate and interpret gamma for this table. Show your work.
6. If you were using tau-*b* or tau-*c* for this table, which would be most appropriate? Why?

R Problems

1. In the wake of the *Dobbs v. Jackson* Supreme Court decision, which overturned the longstanding precedent from *Roe v. Wade*, it is interesting to consider who is most likely to be upset by the court's decision. You should examine how three different independent variables influence `anes20$upset_court`, a slightly transformed version of `anes20$V201340`. This variable measures responses to a question that asked respondents if they would be pleased or upset if the Supreme Court reduced abortion rights. The three independent variables you should use are `anes20$V201600` (sex of respondent), `anes20$ptyID3` (party identification), and `anes20$age5` (age of the respondent).

Copy and run the two code chunks that follow to create `anes20$upset_court` and `anes20$age5`.

```
#Create anes20$upset_court
anes20$upset_court <- anes20$V201340
levels(anes20$upset_court) <- c("Pleased", "Upset", "Neither")
anes20$upset_court <- ordered(anes20$upset_court,
                              levels = c("Upset", "Neither", "Pleased"))

#Collapse age into fewer categories
anes20$age5 <- cut2(anes20$V201507x, c(30, 45, 61, 76))

#Assign label to levels
levels(anes20$age5) <- c("18-29", "30-44", "45-60", "61-75", "76 + ")
```

2. Create a crosstab that shows the relationship between sex of respondent (`anes20$V201600`) and reaction to potential changes in abortion rights (`anes20$upset_court`). Do not include the mosaic plots.
   - Discuss the contents of the table focusing on strength and direction of the relationship (if direction is appropriate). This discussion should rely primarily on the

column percentages. Think about this as telling whatever story there is to tell about how the variables are related.

- Decide which measure(s) of association is most appropriate for summarizing the relationship in the table, run the commands for the measure(s) of association, and discuss the results. What do the results from the measure(s) of association tell you about the strength, direction, and statistical significance of the relationship?

3. Create a crosstab that shows the relationship between respondent party identification (`anes20$ptyID3`) and reaction to potential changes in abortion rights (`anes20$upset_court`). Do not include the mosaic plots.
   - Discuss the contents of the table focusing on strength and direction of the relationship (if direction is appropriate). This discussion should rely primarily on the column percentages. Think about this as telling whatever story there is to tell about how the variables are related.
   - Decide which measure(s) of association is most appropriate for summarizing the relationship in the table, run the commands for the measure(s) of association, and discuss the results. What do the results from the measure(s) of association tell you about the strength, direction, and statistical significance of the relationship?

4. Create a crosstab that shows the relationship between respondent age (`anes20$age5`) and reaction to potential changes in abortion rights (`anes20$upset_court`). Do not include the mosaic plots.
   - Discuss the contents of the table focusing on strength and direction of the relationship (if direction is appropriate). This discussion should rely primarily on the column percentages. Think about this as telling whatever story there is to tell about how the variables are related.
   - Decide which measure(s) of association is most appropriate for summarizing the relationship in the table, run the commands for the measure(s) of association, and discuss the results. What do the results from the measure(s) of association tell you about the strength, direction, and statistical significance of the relationship?

5. Thinking of the results from all three crosstabs, which independent variable seems to have the greatest impact on the dependent variable? What evidence leads you to this conclusion? Were you surprised by how weak or strong any of the relationships were? Explain.

14 CORRELATION AND SCATTERPLOTS

GET READY

This chapter expands the discussion of measures of association to include methods commonly used to measure the strength and direction of relationships between numeric variables. To follow along in R, you should load the `countries2` dataset and attach the libraries for the following packages: `descr`, `DescTools`, `Hmisc`, `ppcor`, and `readxl`.

RELATIONSHIPS BETWEEN NUMERIC VARIABLES

Crosstabs, chi-square, and measures of association are valuable and important techniques when using factor variables, but they do not provide a sufficient means of assessing the strength and direction of relationships when the independent and dependent variables are numeric. For instance, I have data from roughly 190 different countries, and I am interested in explaining cross-national differences in life expectancy (measured in years). One potential explanatory variable is the country-level fertility rate (births per woman of childbearing age), which I expect to be negatively related to life expectancy. These are both ratio-level variables, and a crosstab for these two variables would have more than 100 columns and more than 100 rows since very few countries will have exactly the same values on these variables. This would not be a useful crosstab.

Before getting into the specifics of how to evaluate this type of relationship, let's look at some descriptive information for these two variables.

```
#Display histograms side-by-side (one row, two columns)
par(mfrow = c(1,2))
hist(countries2$lifexp, xlab = "Life Expectancy",
     main = "",
     cex.lab = .7, #Reduce the size of the labels and
     cex.axis = .7) # axis values to fit

hist(countries2$fert1520, xlab = "Fertility Rate",
     main = "",
     cex.lab = .7, #Reduce the size of the labels and
     cex.axis = .7) # axis values to fit
```

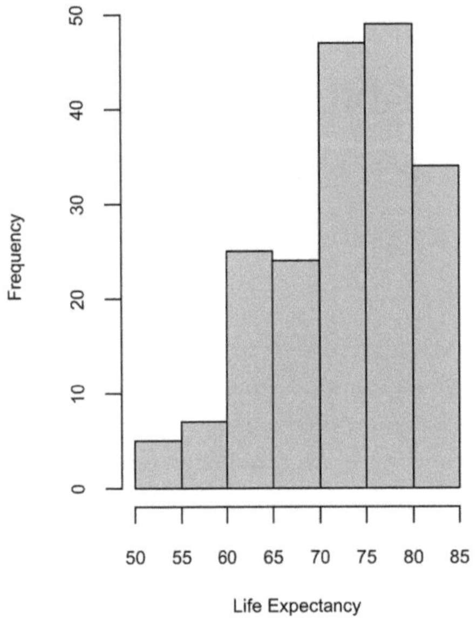

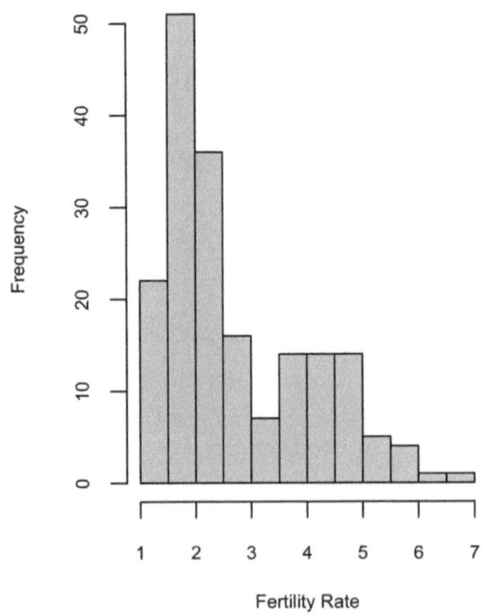


```
#Return to default display setting (one row, one column)
par(mfrow = c(1,1))
```

On the left, life expectancy presents with a bit of negative skew (mean = 72.6, median = 73.9, skewness = −.5), the modal group is in the 75 to 80 years-old category, and there is quite a wide range of outcomes, spanning from around 53 to 85 years. When you think about the gravity of this variable—how long people tend to live—this range is consequential. In the graph on the right, the pattern for fertility rate is a mirror image of life expectancy—very clear right skew (mean = 2.74, median = 2.28, skewness = .93), modal outcomes toward the low end of the scale, and the range is from 1 to 2 births per woman in many countries to more than 5 for several countries.[1]

Before looking at the relationship between these two variables, let's think about what we expect to find. I anticipate that countries with low fertility rates have higher life expectancy than those with high fertility rates (negative relationship), for a couple of reasons. First, in many places around the world, pregnancy and childbirth are significant causes of death in women of childbearing age. While maternal mortality has declined in much of the "developed" world, it is still a serious problem in many of the poorest regions. Second, higher birth rates are also associated with high levels of infant and child mortality. Together, these outcomes associated with fertility rates auger for a negative relationship between fertility rate and life expectancy.

SCATTERPLOTS

Let's look at this relationship with a *scatterplot*. Scatterplots are like crosstabs in that they display joint outcomes on both variables, but they look a lot different due to the nature of the data.

```
#Scatterplot(Independent variable, Dependent variable)
plot(countries2$fert1520, countries2$lifexp,
```

```
xlab = "Fertility Rate",
ylab = "Life Expectancy")
```

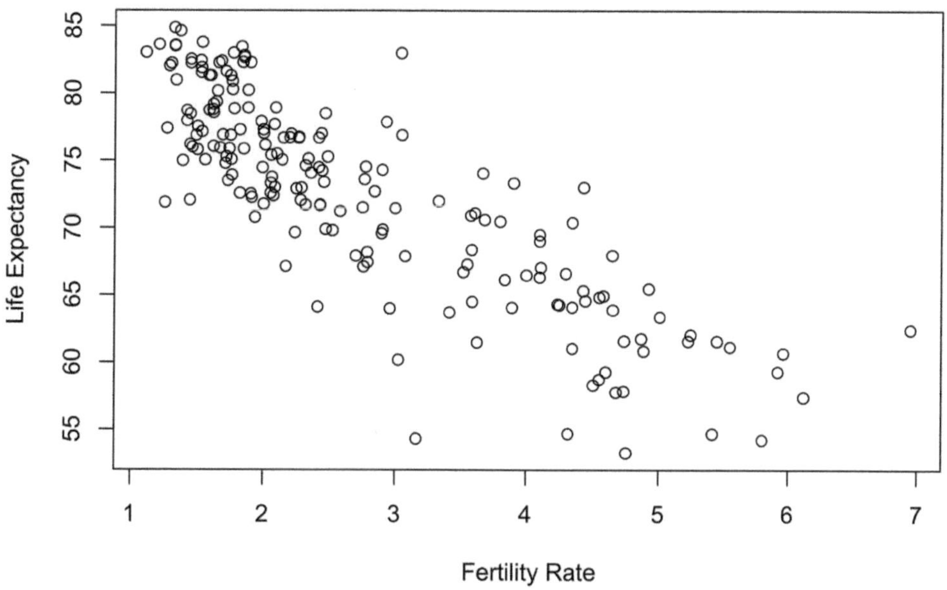

In the preceding scatterplot, the values for the dependent variable span the vertical axis, values of the independent variable span the horizontal axis, and each circle represents the value of both variables for a single country. There appears to be a strong pattern in the data: Countries with low fertility rates tend to have high life expectancy, and countries with high fertility rates tend to have low life expectancy. In other words, as the fertility rate increases, life expectancy declines. You might recognize this from the discussion of directional relationships in crosstabs as the description of a negative relationship. But you might also notice that the pattern looks different (sloping down and to the right) from a negative pattern in a crosstab. This is because the low values for both the independent and dependent variables are in the lower left-corner of the scatterplot while they are in the upper-left corner of the crosstabs.

When looking at scatterplots, it is sometimes useful to imagine something like a crosstab overlay, especially since you've just learned about crosstabs. If there are relatively empty corners with the markers clearly following an upward or downward diagonal, the relationship is probably strong, like a crosstab in which observations are clustered in diagonal cells. We can overlay a horizontal line at the mean of the dependent variable and a vertical line at the mean of the independent variable to help us think of this in terms similar to a crosstab.

```
plot(countries2$fert1520, countries2$lifexp, xlab = "Fertility Rate",
     ylab = "Life Expectancy")

#Add horizontal and vertical lines at the variable means
abline(v = mean(countries2$fert1520, na.rm = T),
```

```
       h=mean(countries2$lifexp, na.rm = T), lty = c(1, 2))

#Add legend
legend("topright", legend = c("Mean Y", "Mean X"), lty = c(1, 2),
   cex = .9)
```

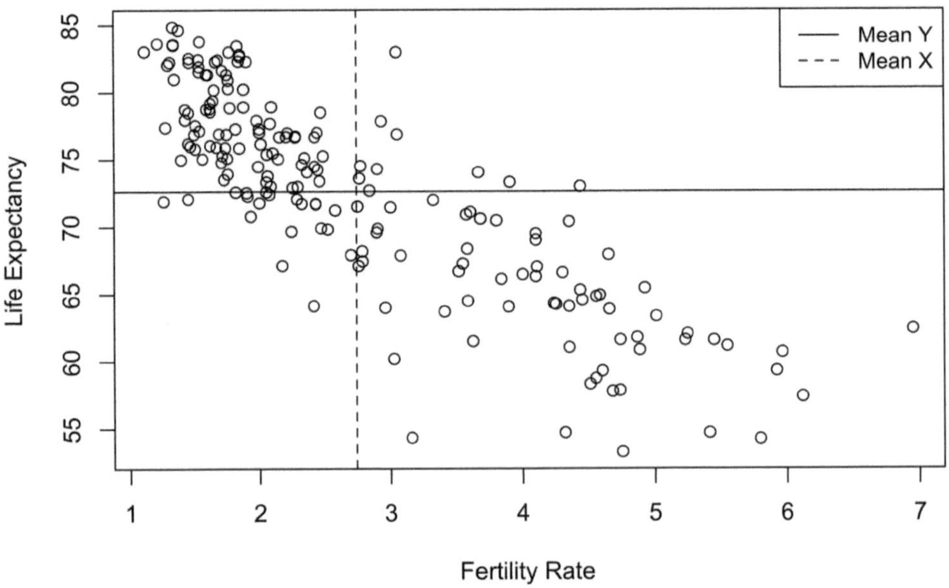

Using this framework helps illustrate the extent to which high or low values of the independent variable are associated with high or low values of the dependent variable. The vast majority of cases that are below average on fertility are above average on life expectancy (upper-left corner), and most of the countries that are above average on fertility are below average on life expectancy (lower-right corner). There are only a few countries that don't fit this pattern, found in the upper-right and lower-left corners. Even without the horizontal and vertical lines for the means, it is clear from looking at the pattern in the data that the typical outcome for the dependent variable declines as the value of the independent variable increases. This is what a strong negative relationship looks like.

Just to push the crosstab analogy a bit farther, we can convert these data into high and low categories (relative to the means) on both variables, create a crosstab, and check out some of measures of associations used in the Chapter 13.

```
#Collapsing "fert1520" at its mean value into two categories
countries2$Fertility.2 <- cut2(countries2$fert1520,
                       cut = c(mean(countries2$fert1520, na.rm = T)))

#Assign labels to levels
levels(countries2$Fertility.2) <-  c("Below Average", "Above Average")

#Collapsing "lifexp" at its mean value into two categories
```

```
countries2$Life_Exp.2 = cut2(countries2$lifexp,
                   cut = c(mean(countries2$lifexp, na.rm = T)))

#Assign labels to levels
levels(countries2$Life_Exp.2) <- c("Below Average", "Above Average")

crosstab(countries2$Life_Exp.2, countries2$Fertility.2,
         prop.c = T,
         plot = F)

   Cell Contents
|-------------------------|
|                   Count |
|          Column Percent |
|-------------------------|

=======================================================================
                         countries2$Fertility.2
countries2$Life_Exp.2    Below Average          Above Average     Total
-----------------------------------------------------------------------
Below Average                       20                     63        83
                                 17.9%                  86.3%
-----------------------------------------------------------------------
Above Average                       92                     10       102
                                 82.1%                  13.7%
-----------------------------------------------------------------------
Total                              112                     73       185
                                 60.5%                  39.5%
=======================================================================
```

This crosstab reinforces the impression that this is a strong negative relationship: 86.3% of countries with above average fertility rates have below average life expectancy, and 82.1% of countries with below average fertility rates have above average life expectancy. The measures of association listed next confirm this. According to lambda, information from the fertility rate variable reduces error in predicting life expectancy by 64%. In addition, the values of Cramer's *V* (.67) and tau-*b* (−.67) confirm the strength and direction of the relationship. Note that tau-*b* also confirms the negative relationship.

```
#Get measures of association
Lambda(countries2$Life_Exp.2, countries2$Fertility.2, direction = c("row"),
       conf.level = .05)

lambda    lwr.ci    upr.ci
0.63855 0.63467 0.64244
```

```
CramerV(countries2$Life_Exp.2, countries2$Fertility.2, conf.level = .05)

Cramer V  lwr.ci   upr.ci
0.67262   0.66801  0.67723

KendallTauB(countries2$Life_Exp.2, countries2$Fertility.2, conf.
   level = .05)

tau_b      lwr.ci   upr.ci
-0.67262 -0.67608 -0.66915
```

The crosstab and measures of association together point to a strong negative relationship. However, a lot of information is lost by just focusing on a two-by-two table that divides both variables at their means. There is substantial variation in outcomes on both life expectancy and fertility *within* each of the crosstab cells, variation that can be exploited to get a more complete sense of the relationship. For example, countries with above average levels of fertility range from about 2.8 to 7 births per woman of childbearing age, and countries with above average life expectancy have outcomes ranging from about 73 to 85 years of life expectancy. But the crosstab treats all countries in each cell as if they have the same outcome—high or low on the independent and dependent variables. Even if we expanded this to a 4x4 or 5x5 table, we would still be losing information by collapsing values of both the independent and dependent variables. What we need is a statistic that utilizes all the variation in both variables, as represented in the scatterplot, and summarizes the extent to which the relationship follows a positive or negative pattern.

PEARSON'S *R*

Similar to ordinal measures of association, a good interval/ratio measure of association is positive when values of x that are relatively high tend to be associated with values of y that are relatively high and values of x that are relatively low are associated with values of y that are relatively low. And, of course, a good measure of association should be negative when values of x that are relatively high tend to be associated with values of y that are relatively low and values of x that are relatively low are associated with values of y that are relatively high.

One way to capture these positive and negative patterns is to express the value of all observations as deviations from the means of both the independent variable $(x_i - \bar{x})$ and the dependent variable $(y_i - \bar{y})$. Then, we can multiply the deviation from $\bar{x}$ times the deviation from $\bar{y}$ to see if the observation follows a positive or negative pattern.

$$(x_i - \bar{x})(y_i - \bar{y})$$

If the product is negative, the observation fits a negative pattern (think of it like a dissimilar pair from crosstabs); if the product is positive, the observation fits a positive pattern (think similar pair). We can sum these products across all observations to get a sense of whether the relationship, on balance, is positive or negative, in much the same way we did by subtracting dissimilar pairs from similar pairs for gamma in Chapter 13.

Pearson's correlation coefficient (r) does this and can be used to summarize the strength and direction of a relationship between two numeric variables:

$$r = \frac{\sum(x_i - \bar{x})(y_i - \bar{y})}{\sqrt{\sum(x_i - \bar{x})^2 \sum(y_i - \bar{y})^2}}$$

Although this formula may look a bit dense, it is quite intuitive. The numerator is exactly what we just discussed, a summary of the positive or negative pattern in the data. The denominator standardizes the numerator by the overall levels of variation in X and Y and provides -1 to $+1$ boundaries for Pearson's r. As was the case with the ordinal measures of association, values near 0 indicate a weak relationship, and the strength of the relationship grows stronger moving from 0 to -1 or $+1$.

Calculating Pearson's *r*

Let's work our way through these calculations just once for the relationship between fertility rate and life expectancy, using a *random sample of only ten countries* (Afghanistan, Bhutan, Costa Rica, Finland, Indonesia, Lesotho, Montenegro, Peru, Singapore, and Togo). The following scatterplot shows that even with just ten countries, a pattern emerges that is similar to that found in the scatterplot from the full set of countries—as fertility rates increase, life expectancy decreases.

```
#import "10countries" Excel file. Check file path if you get an error
X10countries <- read_excel("10countries.xlsx")

#Plot relationship for 10-country sample
plot(X10countries$fert1520, X10countries$lifexp,
     xlab = "Fertility Rate",
     ylab = "Life Expectancy")
```

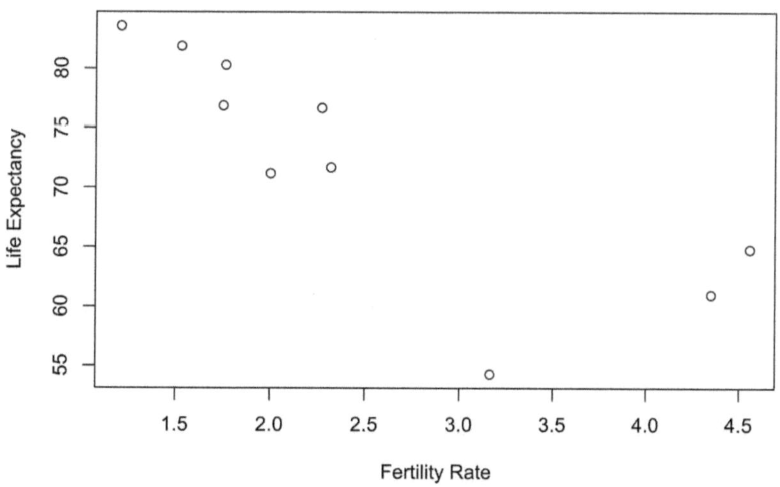

We can use data on the dependent and independent variables for these ten countries to generate the components necessary for calculating Pearson's *r*.

```
#Deviation of y from its mean
X10countries$ydev <- X10countries$lifexp-mean(X10countries$lifexp)

#Deviation of x from its mean
X10countries$xdev <- X10countries$fert1520-mean(X10countries$fert1520)

#Cross-product of x and y deviations from their means
X10countries$ydev_xdev <- X10countries$ydev*X10countries$xdev

#Squared deviation of y from its mean
X10countries$ydevsq <- X10countries$ydev^2

#Squared deviation of x from its mean
X10countries$xdevsq <- X10countries$xdev^2
```

Table 14.1 includes the country names, the values of the independent and dependent variables, and all the pieces needed to calculate Pearson's *r* for the sample of ten countries shown in the scatterplot. It is worth taking the time to work through the calculations to see where the final figure comes from.

TABLE 14.1 ■ Data for Calculation of Pearson's *r* Using a Sample of Ten Countries

Country	lifexp	fert1520	ydev	xdev	ydev_xdev	ydevsq	xdevsq
Afghanistan	64.8	4.555	−7.44	2.064	−15.355	55.354	4.259
Bhutan	71.2	2.000	−1.04	−0.491	0.511	1.082	0.241
Costa Rica	80.3	1.764	8.06	−0.727	−5.863	64.964	0.529
Finland	81.9	1.530	9.66	−0.961	−9.287	93.316	0.924
Indonesia	71.7	2.320	−0.54	−0.172	0.093	0.292	0.030
Lesotho	54.3	3.164	−17.94	0.673	−12.069	321.844	0.453
Montenegro	76.9	1.751	4.66	−0.741	−3.452	21.716	0.549
Peru	76.7	2.270	4.46	−0.221	−0.987	19.892	0.049
Singapore	83.6	1.209	11.36	−1.282	−14.568	129.050	1.644
Togo	61	4.352	−11.24	1.860	−20.908	126.338	3.460

Here, the first two columns of numbers are the outcomes for the dependent and independent variables, respectively, and the next two columns express these values as deviations from their means. The fifth column of data represents the cross-products of the mean deviations for each observation, and the sum of that column is the numerator for the equation. Since 8 of the

10 of the cross-products are negative, the numerator is negative (−81.885), as shown next. We can think of this numerator as measuring the *covariation* between *x* and *y*.

```
#Multiply and sum the y and x deviations from their means
numerator = sum(X10countries$ydev_xdev)

numerator
[1] -81.885
```

The last two columns measure squared deviations from the mean for both *y* and *x*, and their column totals (summed next) capture the total variation in *y* and *x*, respectively. The square root of the product of the variation in *y* and *x* (100.61) constitutes the denominator in the equation, giving us the following.

```
var_y = sum(X10countries$ydevsq)
var_x = sum(X10countries$xdevsq)
denominator = sqrt(var_x*var_y)

denominator
[1] 100.61
```

Now, let's plug in the values for the numerator and denominator and calculate Pearson's *r*.

$$r = \frac{\sum(x_i - \bar{x})(y_i - \bar{y})}{\sqrt{\sum(x_i - \bar{x})^2 \sum(y_i - \bar{y})^2}} = \frac{-81.885}{100.61} = -.81391$$

This correlation confirms a strong negative relationship between fertility rate and life expectancy in the sample of ten countries. Before getting too far ahead of ourselves, we need to be mindful of the fact that we are testing a hypothesis regarding these two variables. In this case, the null hypothesis is that there is no relationship between these variables:

$$H_0: \rho = 0$$

And the alternative hypothesis is that we expect a negative correlation:

$$\mathrm{H}_1: \rho < 0$$

A *t*-test can be used to evaluate the null hypothesis that Pearson's *r* equals zero. We can get the results of the *t*-test and check the calculation of the correlation coefficient for this small sample using the `cor.test` function in R.

```
#cor.test(variable1, variable2)
cor.test(X10countries$lifexp, X10countries$fert1520)


    Pearson's product-moment correlation


data: X10countries$lifexp and X10countries$fert1520
t = -3.96, df = 8, p-value = 0.0042
alternative hypothesis: true correlation is not equal to 0
```

```
95 percent confidence interval:
  -0.95443 -0.37798
sample estimates:
     cor
-0.81391
```

Here, we find the correlation coefficient (−.81), a t-score (−3.96), a p-value (.004), and a confidence interval for r (−.95, −.38). R confirms the earlier calculations and that the correlation is statistically significant (reject H_0), even though the sample is so small.

We are really interested in the correlation between fertility rate and life expectancy among the *full set of countries*, so let's see what this looks like for the original dataset.

```
#Result from the full data set
cor.test(countries2$lifexp, countries2$fert1520)

    Pearson's product-moment correlation

data: countries2$lifexp and countries2$fert1520
t = -21.2, df = 183, p-value < 2e-16
alternative hypothesis: true correlation is not equal to 0
95 percent confidence interval:
  -0.88042 -0.79581
sample estimates:
     cor
-0.84326
```

For the full 185 countries shown in the first scatterplot, the correlation between fertility rate and life expectancy is −.84. There are a few things to note about this. First, this is a very strong relationship. Second, the negative sign confirms that the predicted level of life expectancy decreases as the fertility rate increases. There is a strong tendency for countries with above average levels of fertility rate to have below average levels of life expectancy. Finally, it is interesting to note that the correlation from the small sample of ten countries is similar to that from the full set of countries in both strength and direction of the relationship. Three cheers for random sampling!

Other Independent Variables

Of course, we don't expect that differences in fertility rates across countries are the only thing that explains differences in life expectancy. In fact, there are probably several other variables that play a role in shaping country-level differences in life expectancy. The mean level of education is one such factor. The following scatterplot shows the relationship between the mean years of education and life expectancy. Here, you see a pattern opposite of that shown in the first figure: as education increases so does life expectancy.

```
#Scatterplot for "lifexp" by "mnschool"
plot(countries2$mnschool, countries2$lifexp,
     xlab = "Mean Years of School",
     ylab = "Life Expectancy")
```

Again, we find a couple of corners in this graph relatively empty, signaling a fairly strong relationship. In this case, the upper-left (low on *x*, high on *y*) and lower-right (high on *x*, low

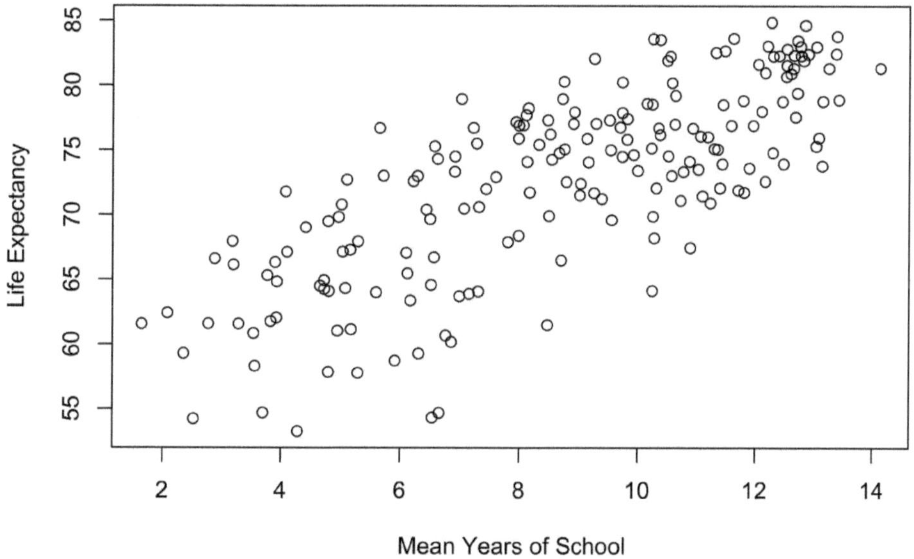

on *y*) corners are empty, which is what we expect with a strong, positive relationship. Similar to the relationship between fertility and life expectancy, there is a clear pattern in the data that doesn't require much effort to see. This is the first sign of a strong relationship. If you have to look closely to try to determine if there is a directional pattern to the data, then the relationship probably isn't a very strong one. While there is a clear positive trend in the data, the observations are not quite as tightly clustered as in the first scatterplot, so the relationship might not be quite as strong. Eyeballing the data like this is not very definitive, so we should have R produce the correlation coefficient to get a more precise sense of the strength and direction of the relationship.

```
cor.test(countries2$lifexp, countries2$mnschool)

    Pearson's product-moment correlation

data: countries2$lifexp and countries2$mnschool
t = 16.4, df = 187, p-value < 2e-16
alternative hypothesis: true correlation is not equal to 0
```

```
95 percent confidence interval:
 0.70301 0.82125
sample estimates:
     cor
0.76862
```

This confirms a strong (r = .77), statistically significant, positive relationship between the level of education and life expectancy. Though not quite as strong as the correlation between fertility and life expectancy (r = −.84), the two correlations are similar enough in value that I consider them to be of roughly the same magnitude.

The first two independent variables illustrate what strong positive and strong negative relationships look like. In reality, many research ideas don't pan out and we end up with correlations near zero. If two variables are unrelated, there should be no clear pattern in the data points; that is, the data points appear to be randomly dispersed in the scatterplot. The following figure shows the relationship between the logged values of the population size[2] and country-level outcomes on life expectancy.

```
#Take the log of population size (in millions)
countries2$lg10pop <- log10(countries2$pop19_M)

#Scatterplot of "lifexp" by "lg10pop"
plot(countries2$lg10pop, countries2$lifexp,
      xlab = "Logged Value of Country Population",
      ylab = "Life Expectancy")
```

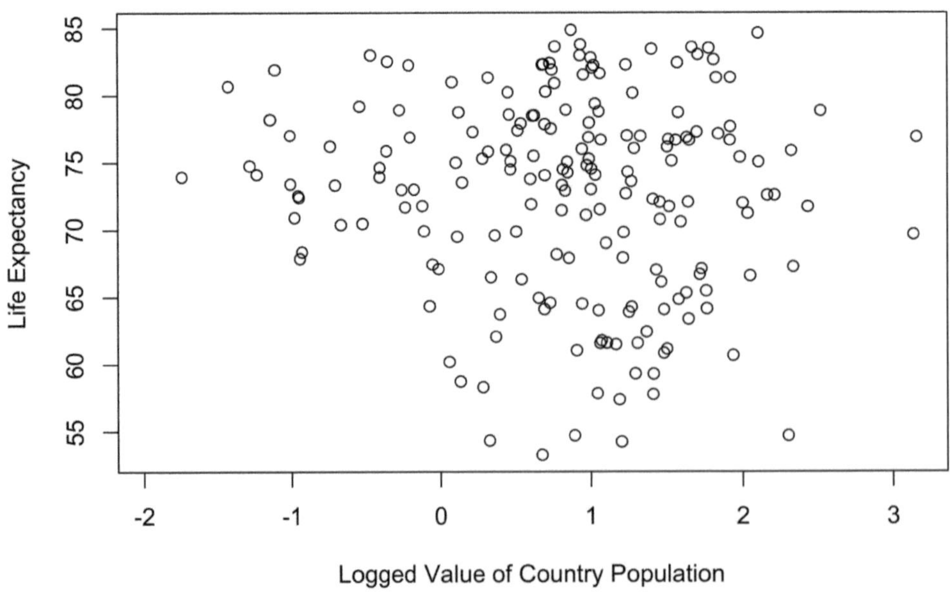

As you can see, this is a seemingly random pattern (there is no clear pattern), and correlation for this relationship (below) is −.08 and not statistically significant. This is one example of what a scatterplot looks like when there is no discernible relationship between two variables.

```
cor.test(countries2$lifexp, countries2$lg10pop)

    Pearson's product-moment correlation

data: countries2$lifexp and countries2$lg10pop
t = -1.17, df = 189, p-value = 0.24
alternative hypothesis: true correlation is not equal to 0
95 percent confidence interval:
 -0.223909 0.058057
sample estimates:
     cor
-0.08462
```

VARIATION IN STRENGTH OF RELATIONSHIPS

The examples presented so far have shown two very strong relationships and one very weak relationship. Given the polarity of these examples, it is worth examining a wider range of patterns you are likely to find when using scatterplots and correlations coefficients. The following graphs shown in Figure 14.1 illustrate positive and negative relationships ranging from relatively weak (.25, −.25), to moderate (.50, −.50), to strong (.75, −.75) patterns.

FIGURE 14.1 ■ Generic Positive and Negative Correlations and Scatterplot Patterns

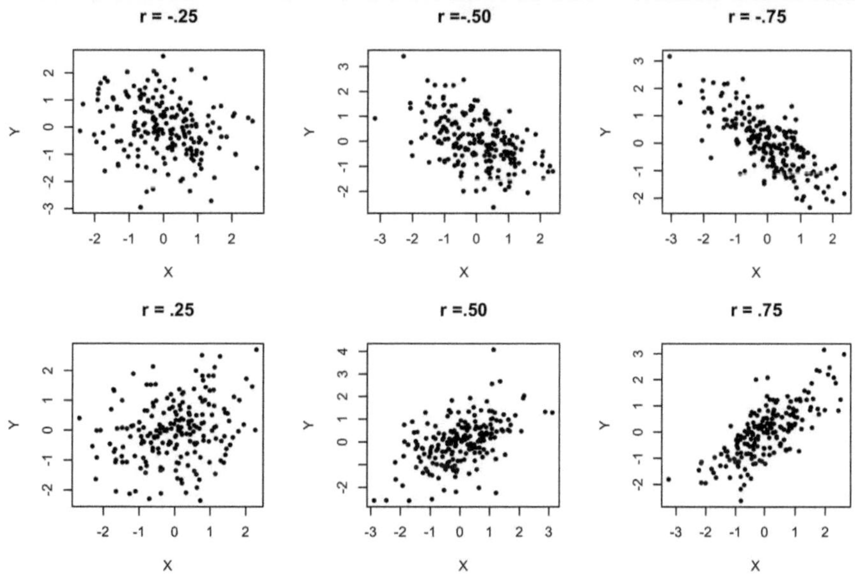

The key takeaway from this figure is that as you move from left to right, it is easier to spot the negative (top row) or positive (bottom row) patterns in the data, and the increasingly clear pattern in the graphs is reflected in the increasingly large correlation coefficients. It can be difficult to spot positive or negative patterns when the relationship is weak, as evidenced by the two plots in the left column. In the top left graph, $r = -.25$, and in the bottom left graph, $r = .25$, but if you had seen these plots without the correlation coefficients attached to them, you might have assumed that there was no relationship between X and Y in either of them. This is the value of the correlation coefficient; it complements the scatterplots using information from all observations and provides a standard basis for judging the strength and direction of the relationship.

PROPORTIONAL REDUCTION IN ERROR

One of the nice things about Pearson's r is that it can be used as a proportional reduction in error (PRE) statistic. Actually, it is the square of the correlation coefficient that acts as a PRE statistic and tells us how much of the variation (error) in the dependent variable is explained by variance in the independent variable. This statistic (r^2) can be interpreted roughly the same way as lambda and eta^2, except that it does not have lambda's limitations. The values of (r^2) for fertility rate, mean level of education, and population are .71, .59, and .007, respectively. This means that using fertility rate as an independent variable reduces error in predicting life expectancy by 71%, while using education level as an independent variable reduces error by 59%, and using population size has no real impact of prediction error. We will return to this discussion in much greater detail in Chapter 15.

CORRELATION AND SCATTERPLOT MATRICES

One thing to consider is that the independent variables are not just related to the dependent variable but also to each other. This is an important consideration that we explore here and will take up again in Chapter 16. For now, let's explore some handy tools for looking at inter-item correlations—the correlation matrix and scatterplot matrix. For both, we need to copy the subset of variables that interest us into a new object and then execute the correlation and scatterplot commands using that object.

```
#Copy the dependent and independent variables to a new object
lifexp_corr <- na.omit(countries2[, c("lifexp", "fert1520",
                                      "mnschool", "lg10pop")])

#Use "cor" function to get a correlation matrix from the new object
cor(lifexp_corr)

                 lifexp      fert1520      mnschool      lg10pop
lifexp         1.000000     -0.840604      0.771580    -0.038338
```

fert1520	-0.840604	1.000000	-0.765023	0.069029
mnschool	0.771580	-0.765023	1.000000	-0.094533
lg10pop	-0.038338	0.069029	-0.094533	1.000000

Each entry in this matrix represents the correlation of the intersection column and row variables. For instance, if you read down the first column, you see how each of the independent variables is correlated with the dependent variable (the r = 1.00 entries on the diagonal are the intersections of each variable with itself). If you look at the intersection of the `fert1520` column and the `mnschool` row, you can see that the correlation between these two independent variables is a rather robust −.77, whereas scanning across the bottom row, we see that population is not related to fertility rate or to level of education (r = .07 and −.09, respectively). You might have noticed that the correlations with the dependent variable are very slightly different from the separate correlations reported earlier. This is because I used `na.omit` when taking a subset of variables to delete all missing cases from the matrix. When considering the effects of multiple variables at the same time, an observation that is missing on one of the variables is missing on all of them.

We can also look at a scatterplot matrix to gain a visual understanding of what these relationships look like.

```
#Use the copied subset of variables for scatterplot matrix
plot(lifexp_corr, cex = .6) #reduce size of markers with "cex"
```

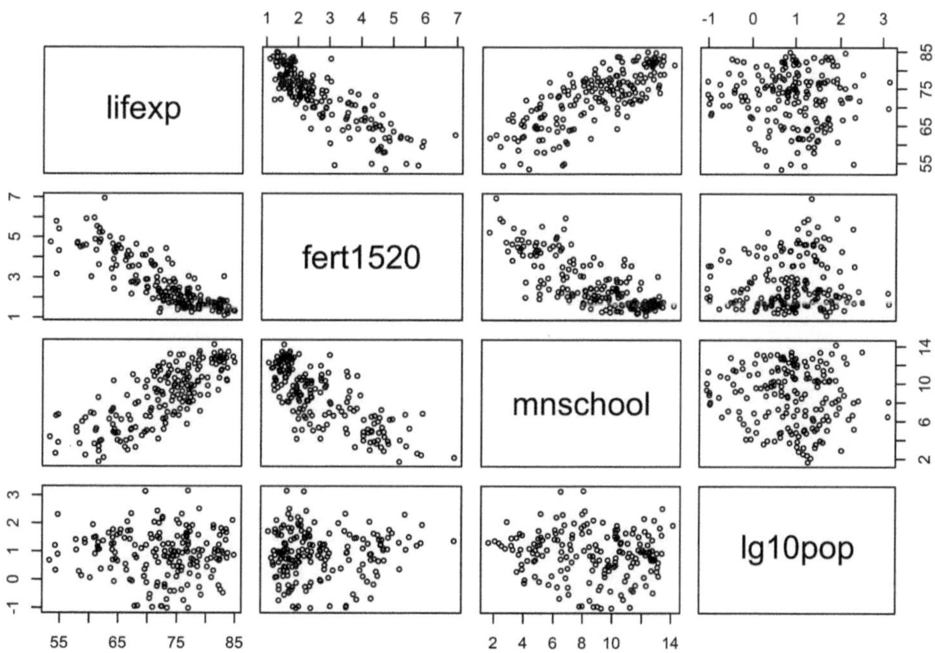

A scatterplot matrix is fairly easy to read. The row variable is treated as the dependent variable, and the plot at the intersection of each column shows how that variable responds to the column variable. The top row shows the scatterplots for the dependent variable and all independent variables, the second row shows how the fertility rate is related to the other variables, and so on. The plots are condensed a bit, so they are not as clear as the individual scatterplots we looked at previously, but the matrix does provide a quick and accessible way to look at all the relationships at the same time.

OVERLAPPING EXPLANATIONS

With these results in mind, we might conclude that fertility rate and education both are strongly related to life expectancy. However, the simple bivariate (also called zero-order) findings rarely summarize the true impact of one variable on another. This issue was raised back in Chapter 1, in the discussion of the need to control for the influence of potentially confounding variables in order to have confidence in the independent effect of any given variable. The primary reason for this is that often some other variable is related to both the dependent and independent variables and could be "producing" the observed bivariate relationships. In the current example, we need not look any farther than the two variables in question as potentially confounding variables for each other; the level of education and the fertility rate are not just related to life expectancy but are also strongly related to each other (see the correlation matrix above). Therefore, we might be overestimating the impact of both variables by considering them in isolation.

This idea is easier to understand if we think about explaining variation in the dependent variable. As reported earlier, fertility rate and educational attainment explain 71% and 59% of the variance (error) in the dependent variable, respectively. However, these estimates are likely overstatements of the **independent impact** of both variables because this would mean that together they explain 130% of the variance in the dependent variable, which is not possible (you can't explain more than 100%). What's happening here is that the two independent variables co-vary with each other AND with the dependent variable, so there is some double-counting of variance explained.

Consider the Venn diagram in Figure 14.2,[3] which illustrates how a dependent variable (Y) is related to two independent variables (X_1 and X_2) and how those independent variables can be related to each other.

The large circles represent the total variation in Y, X_1, and X_2, and the overlapping areas represent the correlations between the three variable. Areas D and G reflect the correlation between Y and X_1; Areas E and G reflect the correlation between Y and X_2; and Areas G and F reflect the correlation between X_1 and X_2. Both X_1 and X_2 explain significant portions of variation in Y but also account for significant portions of each other. Area G, where all three circles overlap, is the area where X_1 and X_2 share variance with each other and with Y. If we count Area G as the portion of Y that is explained by X_1 and as the portion of Y that is explained by X_2, then we are double-counting this portion of the variance in Y and overestimating the impact of both X_1 and X_2.

FIGURE 14.2 ■ Overlapping Explanations of a Dependent Variable

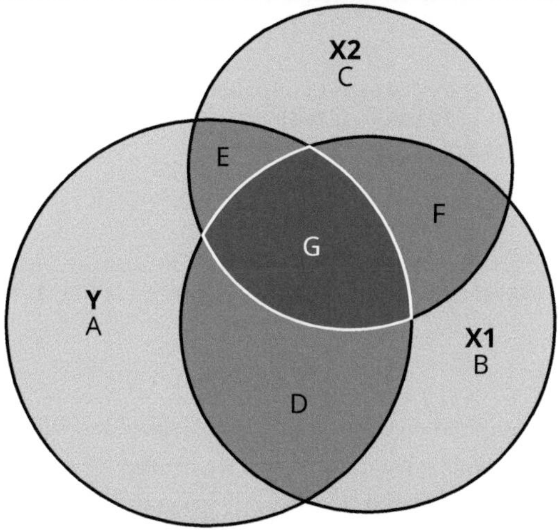

Note: The code for creating this Venn diagram was adapted from examples in *Exploring* R^2 *and regression variance with Euler/Venn Diagrams* by Andrew Heiss, 2021. https://www.andrewheiss.com/blog/2021/08/21/r2-euler/#relationship-between-three-variables). CC BY 4.0 (https://creativecommons.org/licenses/by/4.0/.

The more overlap between the Y, X_1 and X_2 circles, the greater the level of shared explanation of the variation in the Y. If we treat the Y, X_1, and X_2 circles as representing life expectancy, fertility rate, and level of education, respectively, then this all begs the question, "What are the independent effects of fertility rate and level of education on life expectancy, after controlling for the overlap between the two independent variables?"

One important technique for addressing this issue is the **partial correlation coefficient**, which considers not just how an independent variable is related to a dependent variable but also how it is related to other independent variables and how those other variables are related to the dependent variable. The partial correlation coefficient for a model with two independent variables is usually identified as $r_{yx_1 \cdot x_2}$ (correlation between x_1 and y, controlling for x_2) and is calculated using the following formula:

$$r_{yx_1 \cdot x_2} = \frac{r_{x_1y} - (r_{yx_2})(r_{x_1x_2})}{\sqrt{1 - r^2_{yx_2}}\,\sqrt{1 - r^2_{x_1x_2}}}$$

The key to understanding the partial correlation coefficient lies in the numerator. Here we see that the original bivariate correlation between x_1 and y is discounted by the extent to which a third variable (x_2) is related to both x_1 and y. If there is a weak relationship between x_1 and x_2 or between y and x_2, then the partial correlation coefficient for x_1 will be close in value to the zero-order correlation. If there is a strong relationship between x_1 and x_2 or between y and x_2, then the partial correlation coefficient for x_1 will be significantly lower in value than the zero-order correlation.

Let's calculate the partial correlations separately for fertility rate and mean level of education for women, using information from the correlation matrix.

```
# partial correlation for fertility, controlling for education levels:
pcorr_fert <- (-.8406-(.77158*-0.76502))/
   (sqrt(1-.77158^2)*sqrt(1- 0.76502^2))
pcorr_fert
[1] -0.61104

# partial correlation for education levels, controlling for fertility:
pcorr_educ <- (.77158-(-.8406*-0.76502))/(sqrt(1-.8406^2)*sqrt(1-.76502^2))
pcorr_educ
[1] 0.36839
```

Here, we see that the original correlation between fertility and life expectancy (−.84) overstated the impact of this variable quite a bit. When we control for mean years of education, the relationship is reduced to −.61, which is significantly lower but still suggests a somewhat strong relationship. We see a more dramatic reduction in the influence of level of education. When controlling for fertility rate, the relationship between level of education and life expectancy drops from .77 to .37, a fairly steep decline. These findings clearly illustrate the importance of thinking in terms of *multiple, overlapping explanations* of variation in the dependent variable.

You can get the same results using the `pcor` function in the `ppcor` package. The format for this command is `pcor.test(y, x, z)`. Note again that when copying the subset of variables to the new object, I used `na.omit` to drop any observations that had missing data on any of the three variables.

```
#Copy subset of variables, dropping missing data
partial <- na.omit(countries2[, c("lifexp", "fert1520", "mnschool")])

#Partial correlation for fert1520, controlling for mnschool
pcor.test(partial$lifexp, partial$fert1520, partial$mnschool)

  estimate        p.value   statistic    n  gp   Method
1 -0.61105     5.1799e-20     -10.356  183   1  pearson

#Partial correlation for mnschool, controlling for fert1520
pcor.test(partial$lifexp, partial$mnschool, partial$fert1520)

  estimate        p.value   statistic    n  gp   Method
1  0.36838  0.00000031119      5.3162  183   1  pearson
```

The calculations were right on target!

NEXT STEPS

What you've learned about Pearson's r in this chapter forms the basis for the rest of the chapters in this book. Using this as a jumping-off point, we investigate a number of more advanced methods that can be used to describe relationships between numeric variables, all falling under the general rubric of "regression analysis." In the next chapter, we look at how we can expand upon Pearson's r to predict specific outcomes of the dependent variable using a linear equation that summarizes the pattern seen in the scatterplots. In fact, for the remaining chapters, most scatterplots will also include a trend line that fits the linear pattern in the data. Following that, we will explore how to include several independent variables in a single linear model to predict outcomes of the dependent variable. In doing so, we will also address more fully the issue highlighted in the correlation and scatterplot matrices, highly correlated independent variables. In the end, you will be able to use multiple independent variables to provide a more comprehensive and interesting statistical explanation of the dependent variable.

EXERCISES

Concepts and calculations

1. Use the information in the table that follows to determine whether there is a positive or negative relationship between X and Y. As a first step, you should note whether the values of X and Y for each observation are above or below their respective means. Then, use this information to explain why you think there is a positive or negative relationship.

	Y	Above or Below Mean of *Y*?	*X*	Above or Below Mean of *X*?
	10		17	
	5		10	
	18		26	
	11		21	
	16		17	
	7		13	
Mean	11.2		17.3	

2. Match each of the following correlations to the corresponding scatterplot shown next.
   - **A.** $r = .80$
   - **B.** $r = -.50$
   - **C.** $r = .40$
   - **D.** $r = -.10$

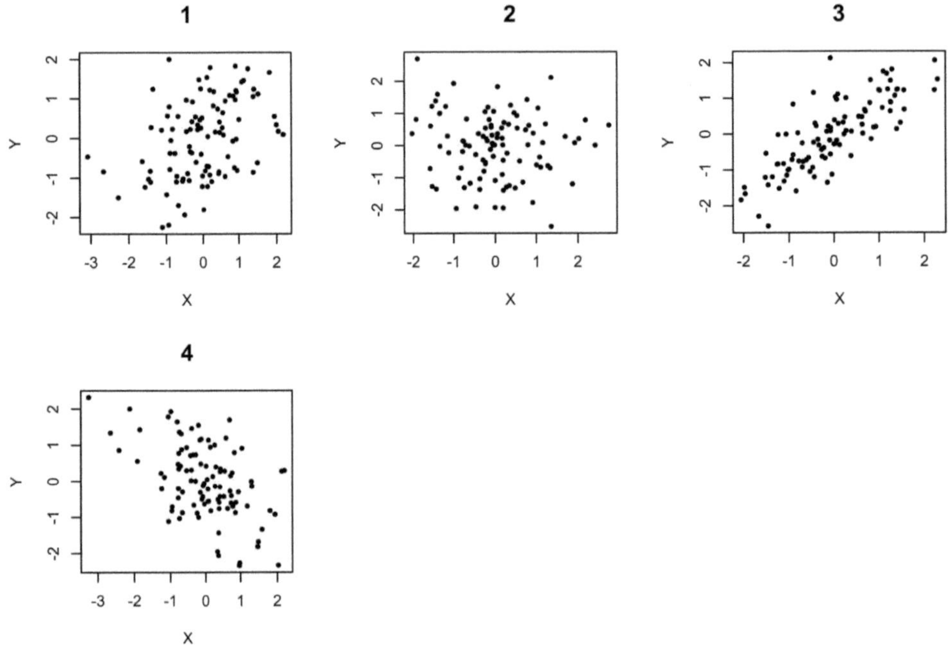

3. What is the proportional reduction in error for each of the correlations in Question 2?
4. A researcher is interested in differences in health outcomes across counties. They are particularly interested in whether obesity rates affect diabetes rates in counties, expecting that diabetes rates increase as obesity rates increase. Using a sample of 500 counties, they produce the scatterplot that follows to support their hypothesis.

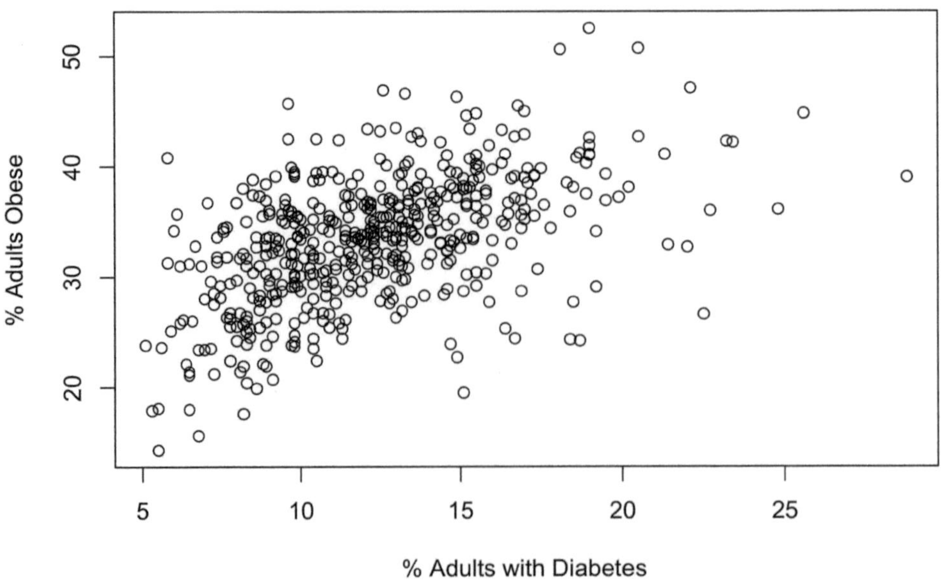

- Based on the discussion, what is the dependent variable?
- Is the pattern in the figure positive or negative? Explain your answer.
- Is the pattern in the figure strong or weak? Explain your answer.
- Based just on eyeballing the data, what's your best guess for the correlation between these two variables?
- This graph contains an important flaw. Can you spot it?

5. In which of the following cases are the partial correlations between y and x_1 likely to be substantially lower than the zero-order correlations, and in which cases is there likely to be very little difference. Explain. (No calculations necessary.)

	r_{yx_1}	r_{yx_2}	$r_{x_1x_2}$
A.	.55	.75	.11
B.	−.68	.09	.55
C.	.59	.55	−.68
D.	−.63	.19	.25
E.	.49	.64	.73

R Problems

For these exercises, you will use the `states20` dataset to identify potential explanations for state-to-state differences in infant mortality (`states20$infant_mort`). Show all graphs and statistical output and make sure you use appropriate labels on all graphs.

1. Generate a scatterplot showing the relationship between per capita income (`states20$PCincome2020`) and infant mortality in the states. Describe contents of the scatterplot, focusing on direction and strength of the relationship. Offer more than just "weak and positive" or something similar. What about the scatterplot makes it look strong or weak, positive or negative? Then get the correlation coefficient for the relationship between per capita income and infant mortality. Interpret the coefficient and comment on how well it matches your initial impression of the scatterplot.

2. Repeat the same analysis from Question 1, except now use `lowbirthwt` (% of births that are technically low birth weight) as the independent variable. Again, show all graphs and statistical results. Use words to describe and interpret the results.

3. Repeat the same analysis from Question 1, except now use `teenbirth` (the % of births to teen mothers) as the independent variable. Again, show all graphs and statistical results. Use words to describe and interpret the results.

4. Repeat the same analysis from Question 1, except now use an independent variable of your choosing from the `states20` dataset that you expect to be related to infant mortality in the states. Justify the variable you choose. Again, show all graphs and statistical results. Use words to describe and interpret the results.

15 SIMPLE REGRESSION

GET READY

This chapter builds on the correlation and scatterplot techniques discussed in Chapter 14 and begins a four-chapter overview of regression analysis. It is a good idea to follow along in R, not just to learn the relevant commands but also because we will use R to perform some important calculations in a couple of different instances. You should load the `anes20`, `county20large`, `states20`, `votes1620`, and `countries2` datasets as well as the libraries for the `DescTools`, `descr`, and `stargazer` packages.

LINEAR RELATIONSHIPS

In Chapter 14, we began to explore techniques that can be used to analyze relationships between numeric variables, focusing on scatterplots as a descriptive tool and the correlation coefficient as a measure of association. Both of these techniques are essential tools for social scientists and should be used when investigating numeric relationships, especially in the early stages of a research project. As useful as scatterplots and correlations are, they still have some limitations. For instance, suppose we wanted to predict values of *y* based on values of *x*.[1]

Think about the example in the last chapter where we looked at factors that influence average life expectancy across countries. The correlation between life expectancy and fertility rate was -.84, and there was a very strong negative pattern in the scatterplot. Based on these findings, we could "predict" that generally life expectancy tends to be lower when fertility rate is high and higher when the fertility is low. However, we can't really use this information to make specific predictions of life expectancy outcomes based on specific values of fertility rate. Is life expectancy simply .84 of the fertility rate, or -.84 times the fertility rate? No, it doesn't work that way. Instead, with scatterplots and correlations, we are limited to saying how strongly *x* is related to *y* and in what direction. It is important and useful to be able to say this, but it would be nice to be able to say exactly how much *y* increases or decreases for a given change in *x*.

Even when $r = 1.0$ (when there is a perfect relationship between *x* and *y*), we still can't predict *y* based on *x* using the correlation coefficient itself. Is *y* equal to $1 * x$ when the correlation is 1.0? No, not unless *x* and *y* are the same variable.

Consider the following two hypothetical variables.

```
#Create a data set named "perf" with two variables and six cases
x <- c(4, 12, 2, 8, 6, 10)
y <- c(13, 29, 9, 21, 17, 25)
perf <- data.frame(x, y)

#list the values of the data set
perf

   x  y
1  4 13
2 12 29
3  2  9
4  8 21
5  6 17
6 10 25
```

It's a little hard to tell just by looking at the values of these variables, but they are perfectly correlated ($r = 1.0$). Check out the correlation coefficient in R.

```
#Get the correlation between x and y
cor.test(perf$x, perf$y)

      Pearson's product-moment correlation

data:  perf$x and perf$y
t = Inf, df = 4, p-value < 2e-16
alternative hypothesis: true correlation is not equal to 0
95 percent confidence interval:
 1 1
sample estimates:
cor
  1
```

Still, despite the perfect correlation, notice that the values of *x* are different from those of *y* for all six observations, so it is unclear how to predict values of *y* from values of *x* even when $r = 1.0$.

A scatterplot of this relationship provides a hint to how we can use *x* to predict *y*. Notice that all the data points fall along a straight line, which means that *y* is a perfect linear function of *x*. To predict *y* based on values of *x*, we need to figure out what the linear pattern is.

```
#Scatterplot of x and y
plot(perf$x, perf$y,
     main = "An Illustration of a Perfect Positive Relationship",
     xlab = "X",
     ylab = "Y")
```

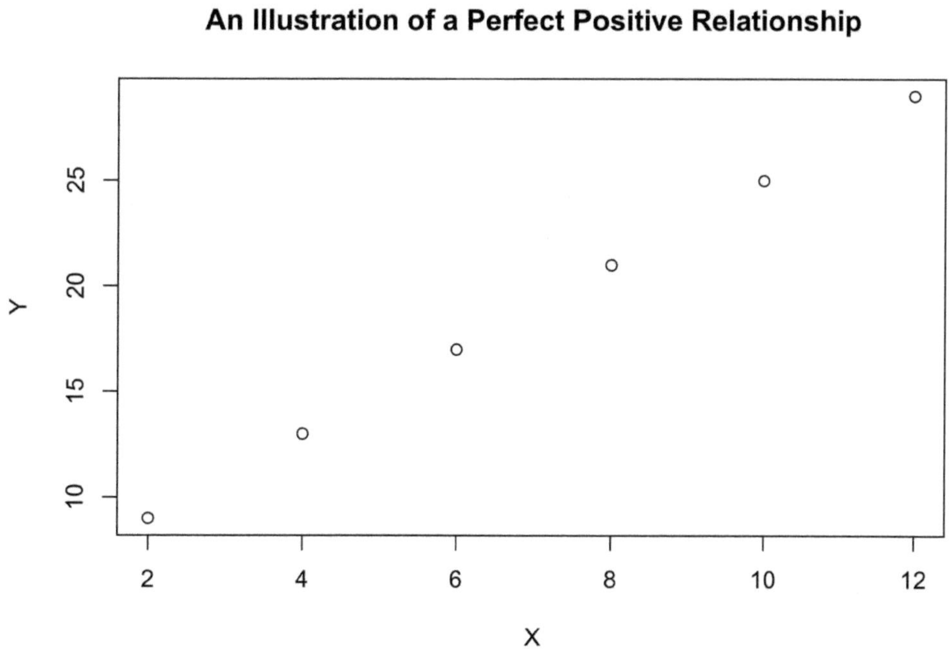

Do you see a pattern in the data points that allows you to predict *y* outcomes based on values of *x*? You might have noticed in the listing of data values that each value of *x* divides into *y at least* two times, but that several of them do not divide into *y* three times. So, we might start out with *y* = 2*x* for each data point and then see what's left to explain. To do this, we create a new variable (`predictedy`) where we predict *y* = 2*x* and then use that to calculate another variable (`leftover`) that measures how much is left over.

```
#Create prediction Y = 2x
perf$predictedy <- 2*perf$x

#subtract "predictedy" from "y"
perf$leftover <- perf$y-perf$predictedy

#List all data
Perf

   x  y predictedy leftover
1  4 13          8        5
2 12 29         24        5
3  2  9          4        5
4  8 21         16        5
5  6 17         12        5
6 10 25         20        5
```

Now we see that all the predictions based on $y = 2x$ under-predicted y by the same amount, 5. In fact, from this we can see that for each value of x, y is exactly equal to $2x + 5$. Plug one of the values of x into this equation so you can see that it perfectly predicts the value of y.

As you may already know from one of your math classes, the equation for a straight line is

$$y = mx + b$$

where m is the slope of the line (how much y changes for a unit change in x) and b is a constant. In the example, $m = 2$ and $b = 5$, and x and y are the independent and dependent variables, respectively:

$$y = 2x + 5$$

In the real-world of data analysis, perfect relationships between x and y do not usually exist, especially in the social sciences. As you know from the scatterplot examples used in Chapter 14, even with very strong relationships, most data points do not come close to lining up on a single straight line.

ORDINARY LEAST SQUARES REGRESSION

For relationships that are not perfectly predicted, the goal is to determine a linear prediction that fits the data points as well as it can, allowing for but minimizing the amount of prediction error. What we need is a methodology that estimates and fits the best possible "prediction" line to the relationship between two variables. That methodology is called **ordinary least squares (OLS) regression.** OLS regression provides a means of predicting y based on values of x that minimizes prediction error.

The **population** regression model is

$$Y_i = \alpha + \beta x_i + \epsilon_i$$

where

Y_i = the outcome on the dependent variable,

α = the constant (a.k.a. the intercept),

β = the slope (expected change in y for a unit change in x),

X_i = the outcome on the independent variable, and

ϵ_i = random error component.

But, of course, we usually cannot know the population values of these parameters, so we estimate an equation based on sample data:

$$y_i = a + bx_i + e_i$$

where

y_i = the outcome on the dependent variable,

a = the sample constant (a.k.a. the intercept),

b = the sample slope (change in y for a unit change in x),

x_i = the outcome on the independent variable, and

e_i = error term; actual values of y minus predicted values of y $(y_i - \hat{y}_i)$.

The sample regression model can also be written as

$$\hat{y}_i = a + bx_i$$

where the caret above the y signals that it is the predicted value and implies e_i.

To estimate the predicted value of the dependent variable ($\hat{y}_i$) for any given observation, we need know the values of a, b, and the outcome of the independent variable for that case (x_i). We can calculate the values of a and b with the following formulas:

$$b = \frac{\sum(x_i - \bar{x})(y_i - \bar{y})}{\sum(x_i - \bar{x})^2}$$

Note that the numerator is the same as the numerator for Pearson's r and, similarly, summarizes the direction of the relationship, while the denominator reflects the variation in x. This focus on the variation in x in the denominator is necessary because we are interested in the expected (positive or negative) change in y for a *unit change* in x. Unlike Pearson's r, the magnitude of b is not bounded by -1 and +1. The size of b is a function of the scale of the independent and dependent variables and the strength of relationship. This is an important point, one we will revisit in Chapter 17.

Once we have obtained the value of the slope, b, we can plug it into the following formula to obtain the constant, a.

$$a = \bar{y} - b\bar{x}$$

Hypothesis Testing and Regression Analysis

The estimates of the slope and constant in regression analysis can be used to evaluate hypotheses about those values in the population. The logic is exactly the same as in earlier chapters: You obtain a sample estimate (b) and then estimate how likely is it that you could get a sample estimate of that magnitude from a population in which the parameter (β) is equal to some other value. In the case of regression analysis, the null and alternative hypotheses for the slope are the following:

H_0: $\beta = 0$ (There is no relationship)

H_1: $\beta \neq 0$ (There is a relationship—two-tailed)

Or, we use can use a one-tailed hypothesis:

H_1: $\beta > 0$ (There is a positive relationship)

H_1: $\beta < 0$ (There is a negative relationship)

As you will see later, the R output reports p-values for the slope estimates, based on the size of the coefficient divided by the standard error of the coefficient. As in previous chapters, the general rule of thumb is that if $p < .05$ (less than a 5% chance of getting b from a population in which H_0 is true), you can reject the null hypothesis.

Calculation Example: Presidential Vote in 2016 and 2020

Let's work through calculating these quantities for a small dataset using data on the 2016 and 2020 U.S. presidential election outcomes from seven states. Presumably there is a close relationship between how states voted in 2016 and how they voted in 2020. We can use regression analysis to analyze that relationship and to predict values in 2020 based on values in 2016 for the seven states listed next.

```
#Assign votes1620 data.frame to "d" for convenience
d <- votes1620
d

  state vote20 vote16
1    AL   37.1   35.6
2    FL   48.3   49.4
3    ME   54.7   51.5
4    NH   53.8   50.2
5    RI   60.6   58.3
6    UT   37.6   39.3
7    WI   50.3   49.6
```

In this dataset, vote20 is Biden's percentage of the two-party vote in 2020 and vote16 is Clinton's percentage of the two-party vote in 2016. This dataset includes two very Republican states (AL and UT), one very Democratic state (RI), two states that lean Democratic (ME and NH), and two very competitive states (FL and WI). First, let's take a look at the relationship using a scatterplot and correlation coefficient to get a sense of how closely the 2020 outcomes mirrored those from 2016.

```
#Plot "vote20" by "vote16"
plot(d$vote16, d$vote20, xlab = "Dem two-party % 2016",
     ylab = "Dem two-party % 2020")
```

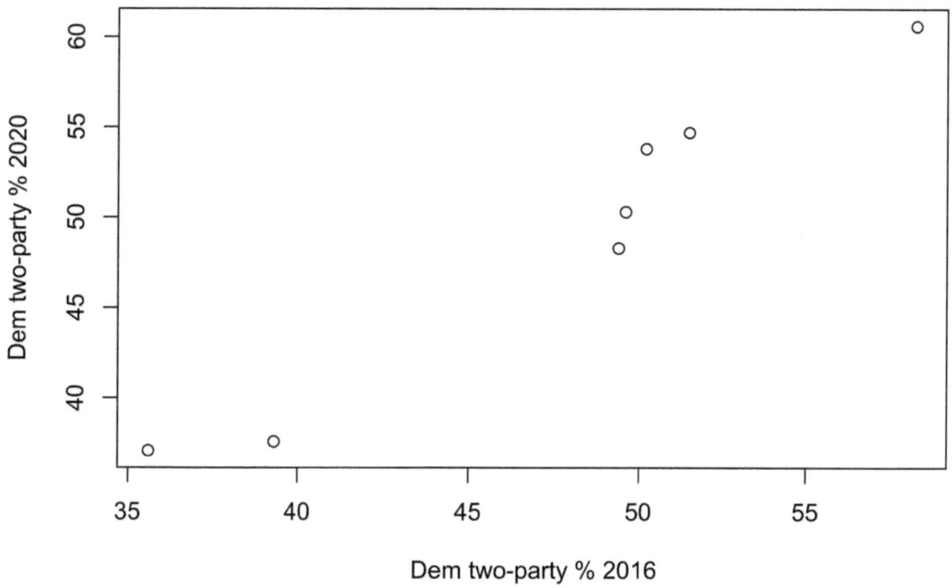


```
#Get correlation between "vote20" and "vote16"
cor.test(d$vote16, d$vote20)


    Pearson's product-moment correlation

data:  d$vote16 and d$vote20
t = 10.5, df = 5, p-value = 0.00014
alternative hypothesis: true correlation is not equal to 0
95 percent confidence interval:
 0.85308 0.99686
sample estimates:
    cor
0.97791
```

Even though the scatterplot and correlation coefficient (.978) together reveal an almost perfect correlation between these two variables, we still need a mechanism for predicting values of y based on values of x. For that, we need to calculate the regression equation.

The following code chunk creates everything necessary to calculate the components of the regression equation, and these components are listed in Table 15.1. Make sure you take a close look at this so you really understand where the regression numbers are coming from.

```
#calculate deviations of y and x from their means
d$Ydev <- d$vote20-mean(d$vote20)
d$Xdev <- d$vote16-mean(d$vote16)

#(deviation for y)*(deviations for x)
d$YdevXdev <- d$Ydev*d$Xdev

#squared deviations for x from its mean
d$Xdevsq <- d$Xdev^2
```

TABLE 15.1 ■ Components for Calculating Regression Slope

state	vote20	vote16	ydev	Xdev	ydev	YdevXdev	Xdevsq
AL	37.1	35.6	-11.814	-12.1	-11.814	142.953	146.410
FL	48.3	49.4	-0.614	1.7	-0.614	-1.044	2.890
ME	54.7	51.5	5.786	3.8	5.786	21.986	14.440
NH	53.8	50.2	4.886	2.5	4.886	12.214	6.250
RI	60.6	58.3	11.686	10.6	11.686	123.869	112.360
UT	37.6	39.3	-11.314	-8.4	-11.314	95.040	70.560
WI	50.3	49.6	1.386	1.9	1.386	2.633	3.610

Now we have all of the components we need to estimate the regression equation. Note that in six of the seven states, the value of $(x_i - \overline{x})(y_i - \overline{y})$ is positive, as is consistent with a strong positive relationship. The sum of this column, headed by `YdevXdev`, is the numerator for *b*, and the sum of the column headed by `Xdevsq` is the denominator.

```
#Create the numerator for slope formula
numerator <- sum(d$YdevXdev)

#Display value of the numerator
numerator
[1] 397.65

#Create the denominator for slope formula
denominator <- sum(d$Xdevsq)

#Display value of the denominator
denominator
[1] 356.52
```

```
#Calculate slope
b <- numerator/denominator

#Display slope
b
[1] 1.1154
```

$$b = \frac{\sum(x_i - \overline{x})(y_i - \overline{y})}{\sum(x_i - \overline{x})^2} = \frac{397.65}{356.52} = 1.115$$

Now that we have a value of b, we can plug that into the formula for the constant.

```
#Get x and y means to calculate constant
mean(d$vote20)
[1] 48.914
mean(d$vote16)
[1] 47.7

#Calculate Constant
a <- 48.914 - (47.7*b)

#Display constant
a
[1] -4.2889
```

$$a = \overline{y} - b\overline{x} = -4.289$$

Which leaves us with the following linear equation:

$$\hat{y} = -4.289 + 1.115x$$

We can now predict values of y based on values of x. Predicted vote in 2020 is equal to -4.289 plus 1.115 times the value of the vote in 2016. We can also say that for every one-unit increase in the value of x (vote in 2016), the expected value of y (vote in 2020) increases by 1.115 percentage points.

We've used the term "predict" quite a bit in this and previous chapters. Let's take a look at what it means in the context of regression analysis. To illustrate, let's predict the value of the Democratic share of the two-party vote in 2020 for two states, Alabama and Wisconsin. The values of the independent variable (Democratic vote share in 2016) are 35.6 for Alabama and 49.6 for Wisconsin. To predict the 2020 outcomes, all we have to do is plug the values of the independent variable into the regression equation and calculate the outcomes.

TABLE 15.2 ■ Predicting 2020 Outcomes Using a Simple Regression Model

State	Predicted Outcome ($\hat{y}$)	Actual Outcome (y)	Error ($y-\hat{y}$)
Alabama	$-4.289 + 1.115(35.6) = 35.41$	37.1	1.70
Wisconsin	$-4.289 + 1.115(49.6) = 51.02$	50.3	-.72.

The regression equation under-estimated the outcome for Alabama by 1.7 percentage points and over-shot Wisconsin by .72 points. These are near-misses and suggest a close fit between predicted and actual outcomes. This ability to predict outcomes of numeric variables based on outcomes of independent variable is something we get from regression analysis that we do not get from other methods we have studied. The error reported in the last column ($y - \hat{y}$) can be calculated across all observations and used to evaluate how well the model explains variation in the dependent variable.

HOW WELL DOES THE MODEL FIT THE DATA?

Figure 15.1 plots the regression line from the previously estimated equation through the seven data points. The regression line represents the predicted outcome of the dependent variable for any value of the independent variable. For instance, the value of the independent variable (2016 vote) for Utah is 39.3, where the upward point arrow intersects the x-axis. The point where the arrowhead meets the regression line is the predicted value of y (2020 vote) when $x = 39.3$. This value is found by following the left-pointing arrow from the regression line to where it intersects the y-axis, 39.5. The closer the scatterplot markers are to the regression line, the less prediction error there is. The linear predictions will always fit some cases better than others. The key thing to remember is that the predictions generated by OLS regression are the best predictions you can get; that is, the estimates that give you the least overall error in prediction for these two variables.

In Figure 15.2, the vertical lines between the markers and the regression line represent the error in the predictions ($y - \hat{y}$). These errors are referred to as *residuals*. As you can see, most data points are very close to the regression line, indicating that the model fits the data well. You can see this by "eyeballing" the regression plot, but it would be nice to have a more precise estimate of the overall strength of the model. An important criterion for judging the model is how well it predicts the dependent variable, overall, based on the independent variable. Conceptually, you can get a sense of this by thinking about the total amount of error in the predictions; that is, the sum of $y - \hat{y}$ across all observations. The problem with this calculation is that you will end up with a value of zero (except when rounding error gives us something very close to zero) because the positive errors perfectly offset the negative errors. We've encountered this type of issue before (Chapter 6) when trying to calculate the average deviation from the mean. As we did then, we can use squared prediction errors to evaluate the amount of error in the model.

FIGURE 15.1 ■ The Regression Line Superimposed in the Scatterplot

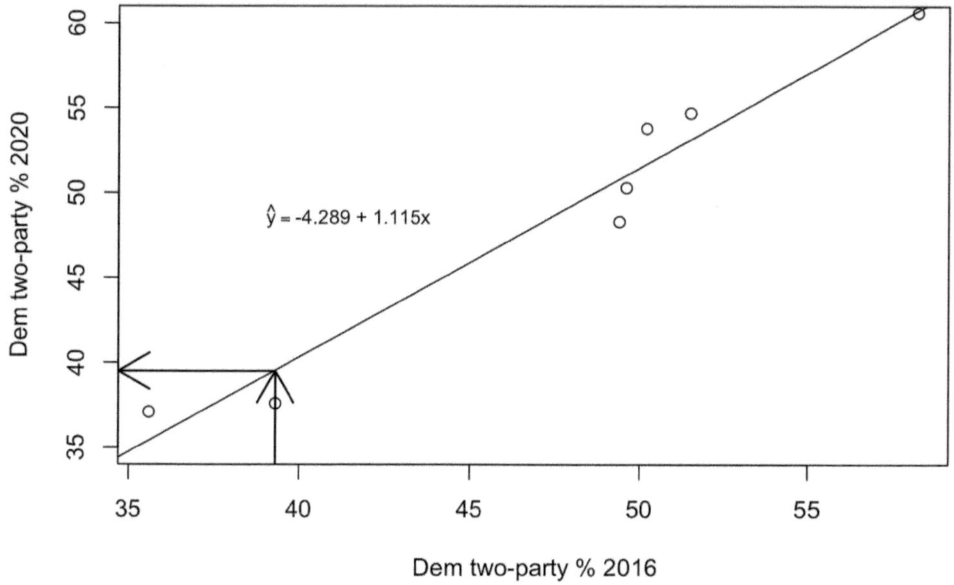

FIGURE 15.2 ■ Prediction Errors (Residuals) in the Regression Model

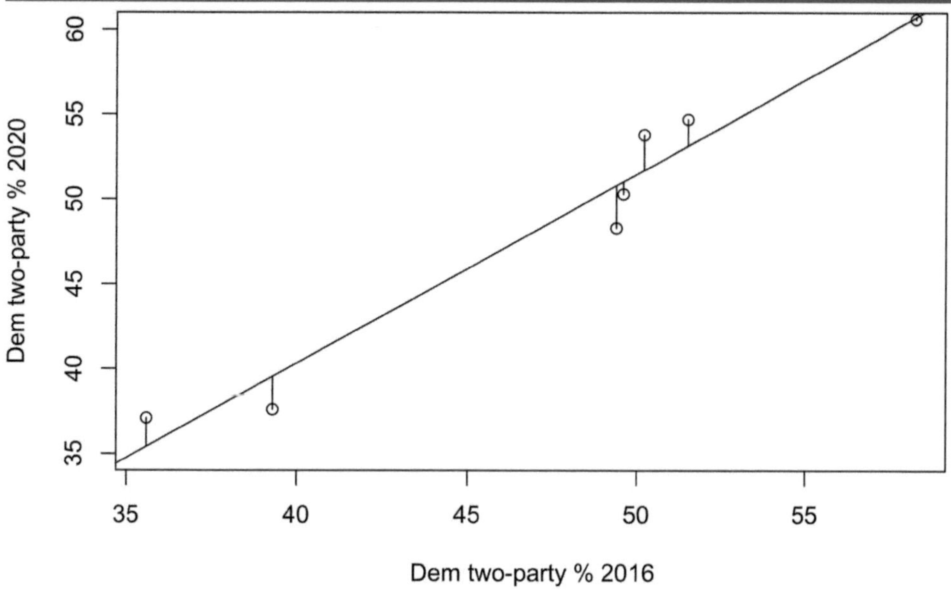

Table 15.3 provides the actual and predicted values of y, the deviation of each predicted outcome from the actual value of y, and the squared deviations (fourth column). The sum of the squared prediction errors is 20.259. This, we refer to as *total squared error* (or variation) in the residuals $(y - \hat{y})$. The "squares" part of ordinary least squares regression refers to the fact that it produces the lowest squared error possible, given the set of variables.

TABLE 15.3 ■ Total Squared Error in the Regression Model

State	y	$\hat{y}$	$y - \hat{y}$	$(y - \hat{y})^2$
AL	37.1	35.418	1.682	2.829
FL	48.3	50.810	-2.510	6.300
ME	54.7	53.153	1.547	2.393
NH	53.8	51.703	2.097	4.397
RI	60.6	60.737	-0.137	0.019
UT	37.6	39.545	-1.945	3.783
WI	50.3	51.033	-0.733	0.537
			Sum	**20.259**

PROPORTIONAL REDUCTION IN ERROR

So, is 20.259 a lot of error? A little error? About average? It's hard to tell without some objective standard for comparison. One way to get a grip on the "Compared to what?" type of question is to think of this issue in terms of improvement in prediction over a model without the independent variable. In the absence of a regression equation or any information about an independent variable, the single best guess we can make for outcomes of interval and ratio-level data—the guess that gives us the least amount of error—is the mean of the dependent variable. One way to think about this is to visualize what the prediction error would look like if we predicted outcomes with a regression line that was perfectly horizontal at the mean value of the dependent variable. Figure 15.3 shows how well the mean of the dependent variable predicts the individual outcomes.

The horizontal line in this figure represents the mean of y (48.914), and the vertical lines between the mean and the data points represent the error in prediction if we used the mean of y as the prediction. As you can tell by comparing this to the previous figure, there is a lot more prediction error when using the mean to predict outcomes than when using the regression model. Just as we measured the total squared prediction error from using a regression model (Table 15.3), we can also measure the level of prediction error without a regression model. Table 15.4 shows the error in prediction from the mean $(y - \bar{y})$ and its squared value. The total squared error when predicting with the mean is 463.789, compared to 20.259 with predictions from the regression model.

FIGURE 15.3 ■ Prediction Error Without a Regression Model

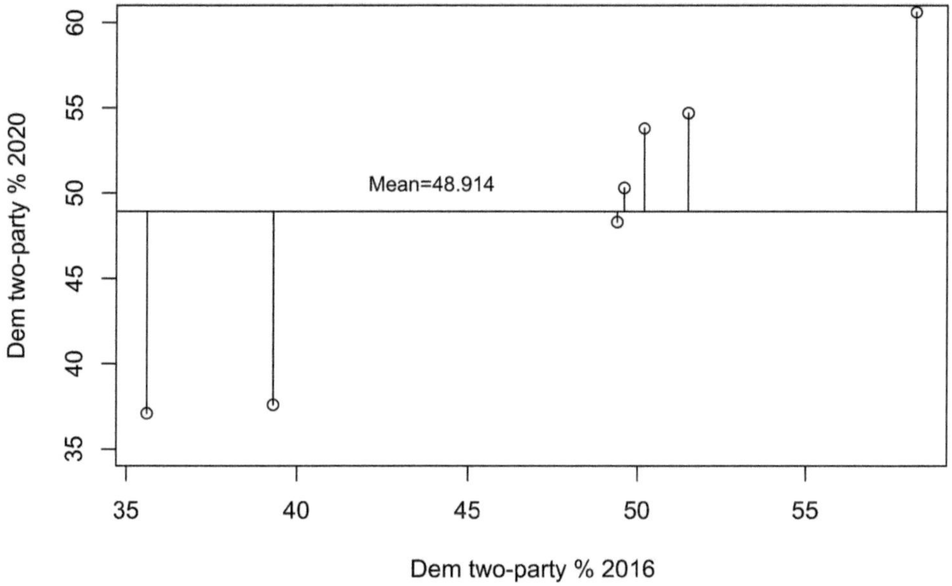

TABLE 15.4 ■ Total Error (Variation) Around the Mean

State	y	$y-\bar{y}$	$(y-\bar{y})^2$
AL	37.1	-11.814	139.577
FL	48.3	-0.614	0.377
ME	54.7	5.786	33.474
NH	53.8	4.886	23.870
RI	60.6	11.686	136.556
UT	37.6	-11.314	128.013
WI	50.3	1.386	1.920
		Sum	**463.789**

We can now use this information to calculate the proportional reduction in error (PRE) that we get from the regression equation. In the discussion of measures of association in Chapter 13 we calculated another PRE statistic, lambda, based on the difference between predicting with no independent variable ($Error_1$) and predicting on the basis of an independent variable ($Error_2$):

$$\text{Lambda}(\lambda) = \frac{E_1 - E_2}{E_1}$$

The resulting statistic, lambda, expresses reduction in error as a proportion of the original error.

We can calculate the same type of statistic for regression analysis, except we will use $\sum(y-\bar{y})^2$ (total error from predicting with the mean of the dependent variable) and $\sum(y-\hat{y})^2$ (total error from predicting with the regression model) in place of E_1 and E_2, respectively. This gives us a statistic known as the coefficient of determination, or r^2 (sound familiar?).

$$r^2 = \frac{\sum(y_i-\bar{y})^2 - \sum(y_i-\hat{y})^2}{\sum(y_i-\bar{y})^2} = \frac{463.789 - 20.259}{463.789} = \frac{443.53}{463.789} = .9563$$

Okay, so let's break this down. The original sum of squared error in the model—the error from predicting with the mean—is 463.789, while the sum of squared residuals (prediction error) based on the model with an independent variable is 20.259. The difference between the two (443.53) is difficult to interpret on its own, but when we express it as a proportion of the original error, we get .9563. This means that we see about a 95.6% reduction in error predicting the outcome of the dependent variable by using information from the independent variable, compared to using just the mean of the dependent variable. Another way to interpret this is to say that we have explained 95.6% of the error (or variation) in y by using x as the predictor.

Earlier in this chapter, we found that the correlation between 2016 and 2020 votes in the states was .9779. If you square this, you get .9563, the value of r^2. This is why it was noted in Chapter 14 that one of the virtues of Pearson's r is that it can be used as a measure of proportional reduction in error. In this case, both r and r^2 tell us that this is a very strong relationship. In fact, this is very close to a perfect positive relationship.

GETTING REGRESSION RESULTS IN R

Let's confirm our calculations using R. First, we use the linear model function to tell R which variables we want to use `lm(dependent ~ independent)` and store the results in a new object named `fit`.

```
#Get linear model and store results in new object "fit"
fit <- lm(d$vote20 ~ d$vote16)
```

Then we have R summarize the contents of the new object.

```
#View regression results stored in 'fit'
summary(fit)


Call:
lm(formula = d$vote20 ~ d$vote16)
```

```
Residuals:
    1      2     3     4      5      6      7
1.682 -2.510 1.547 2.097 -0.137 -1.945 -0.733

Coefficients:
                Estimate Std. Error t value Pr(>|t|)
(Intercept)       -4.289      5.142   -0.83  0.44229
d$vote16           1.115      0.107   10.46  0.00014 ***
---
Signif. codes:  0 '***' 0.001 '**' 0.01 '*' 0.05 '.' 0.1 ' ' 1

Residual standard error: 2.01 on 5 degrees of freedom
Multiple R-squared: 0.956, Adjusted R-squared: 0.948
F-statistic: 109 on 1 and 5 DF, p-value: 0.000138
```

Wow. There is a lot of information here. Figure 15.4 provides an annotated guide.

FIGURE 15.4 ■ Annotated Regression Output From R

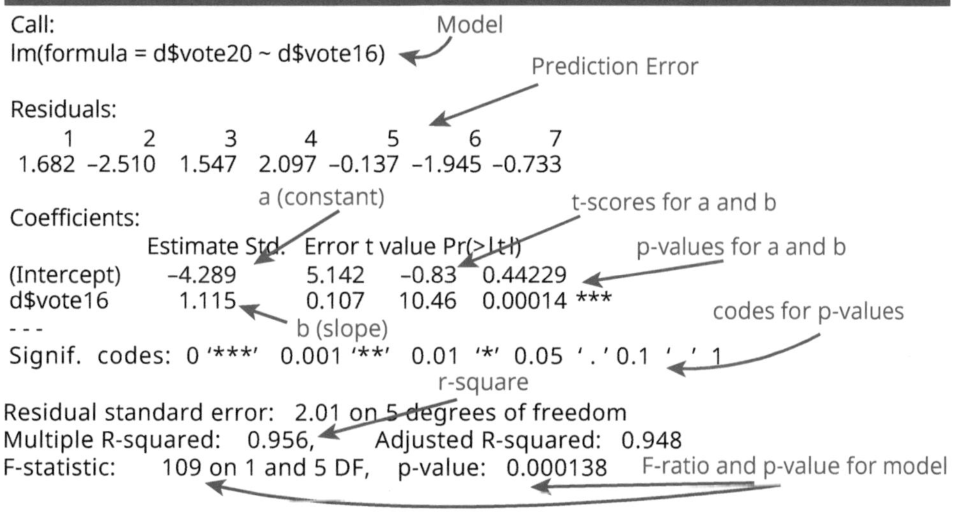

Here we see that our calculations for the constant (4.289), the slope (1.115), and the value of r^2 (.9563) are the same as those produced in the regression output. You will also note here that the output includes values for *t*-scores and *p*-values that can be used for hypothesis testing. The *t*-score is equal to the slope divided by the standard error of the slope. In this case, the *t*-score is 10.46 and the *p*-value is .00014, so we are safe rejecting the null hypothesis. Note also that the *t*-score and *p*-values for the slope are identical to the *t*-score and *p*-values from the correlation coefficient shown earlier. That is because in simple (bi-variate) regression, the results are a strict function of the correlation coefficient.

All Fifty States

We've probably pushed the seven-observation data example about as far as we legitimately can. Before moving on, let's look at the relationship between 2016 and 2020 votes in *all fifty states*, using the `states20` dataset. First, let's take a look at the scatterplot.

```
#Plot 2020 results against 2016 results, all states
plot(states20$d2pty16, states20$d2pty20, xlab = "Clinton % 2016",
     ylab = "Biden % 2020",
     main = "Democratic % Two-party Vote")
```

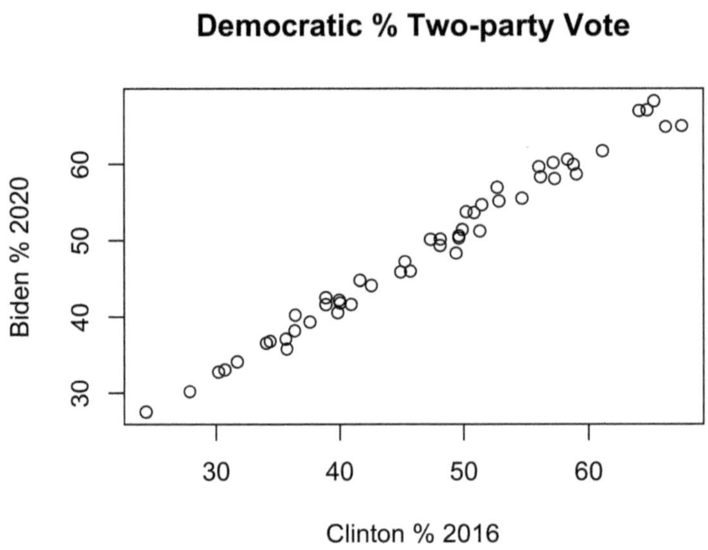

This pattern is similar to that found in the smaller sample of seven states: a strong, positive relationship. This impression is confirmed in the regression results.

```
#Get linear model and store results in new object 'fit50'
fit50 <- lm(states20$d2pty20 ~ states20$d2pty16)
#View regression results stored in 'fit50'
summary(fit50)

Call:
lm(formula = states20$d2pty20 ~ states20$d2pty16)

Residuals:
   Min     1Q Median     3Q    Max
-3.477 -0.703  0.009 0.982 2.665
```

```
Coefficients:
                  Estimate Std. Error t value Pr( > |t|)
(Intercept)         3.4684     0.8533    4.06  0.00018 ***
states20$d2pty16    0.9644     0.0177   54.52  < 2e-16 ***
---
Signif. codes: 0 '***' 0.001 '**' 0.01 '*' 0.05 '.' 0.1 ' ' 1

Residual standard error: 1.35 on 48 degrees of freedom
Multiple R-squared: 0.984, Adjusted R-squared: 0.984
F-statistic: 2.97e + 03 on 1 and 48 DF, p-value: < 2e-16
```

Here we see that the model still fits the data very well (r^2= .98) and that there is a significant, positive relationship between the 2016 and 2020 vote. The linear equation is

$$\hat{y} = 3.468 + .964x$$

The value for *b* (.964) means that for every one unit increase in the observed value of *x* (one percentage point in this case), the expected increase in *y* is .964 units. If it makes it easier to understand, we could express this as a linear equation with substantive labels instead of "*y*" and "*x*":

$$\widehat{\text{2020 Vote}} = 3.468 + .964(\text{2016 Vote})$$

To predict the 2020 outcome for any state, just plug in the value of the 2016 outcome for *x*, multiply that time the slope (.964), and add the value of the constant.

UNDERSTANDING THE CONSTANT

Discussions of regression results typically focus on the estimate for the slope (*b*) because it describes how *y* changes in response to changes in the value of *x*. In most cases, the constant (*a*) isn't of much substantive interest and we rarely spend much time talking about it. But the constant is very important for placing the regression line in a way that minimizes the amount of error obtained from the predictions.

Take a look at the scatterplot that follows, with a regression line equal to $\hat{y} = 3.468 + .964x$ plotted through the data points. To add the regression line to the plot, I use the `abline` function, which we've used before to add horizontal or vertical lines, and specify that the function should use information from the linear model to add the prediction line. Since the results of the regression model are stored in `fit50`, we could also write this as `abline(lm(fit50))`.

```
#Plot 2020 results against 2016 results, all states
plot(states20$d2pty16, states20$d2pty20, xlab = "Clinton % 2016",
     ylab = "Biden % 2020")
#Add prediction line to the graph
abline(lm(states20$d2pty20~states20$d2pty16))
#Add a legend without a box (bty = "n")
legend("right", legend = expression(hat(y)=="3.468 + .964x"), bty = "n",
   cex = .8)
```

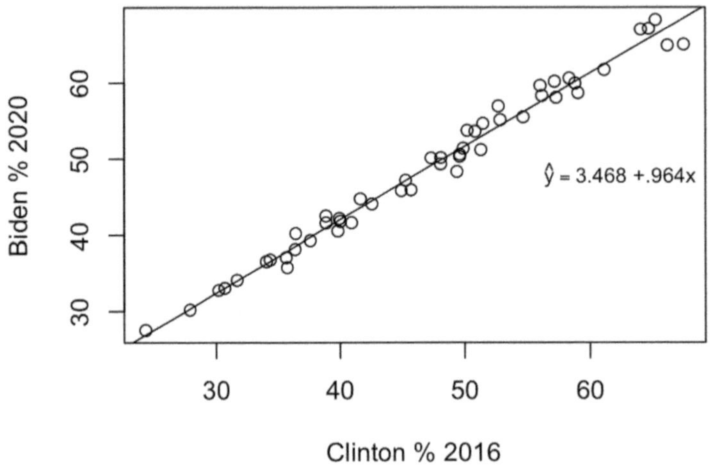

Just to reiterate, this line represents the best fitting line possible for this pair of variables. There is always some error in predicting outcomes in *y*, but no other linear equation will produce less error in prediction than those produced using OLS regression.

The slope (angle) of the line is equal to the value of *b*, and the vertical placement of the line is determined by the constant. Literally, the value of the constant is the predicted value of *y* when *x* equals zero. In many cases, such as this one, zero is not a plausible outcome for *x*, which is part of the reason the constant does not have a really straightforward interpretation. But that doesn't mean it is not important. Look at the scatterplot that follows, in which I keep the value of *b* the same but use the `abline` function to *change the constant to 1.468 instead of 3.468.*

```
#Plot 2020 results against 2016 results, all states
plot(states20$d2pty16, states20$d2pty20, xlab = "Clinton % 2016",
     ylab = "Biden % 2020")
#Add prediction line to the graph, using 1.468 as the constant
abline(a = 1.468, b = .964)
legend("right", legend = expression(hat(y) == "1.468 + .964x"), bty = "n",
   cex = .9)
```

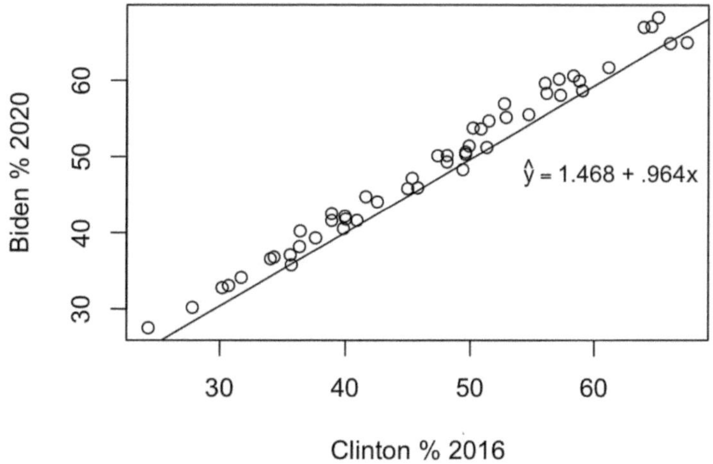

Changing the constant had no impact on the angle of the slope, but the line doesn't fit the data nearly as well. In fact, it looks like the regression predictions underestimate the outcomes in almost every case. The point here is that although we are rarely interested in the substantive meaning of the constant, it plays a very important role in providing the best-fitting regression line.

ORGANIZING THE REGRESSION OUTPUT

Presentation of results is always an important part of the research process. You want to make sure the reader, whether your professor, a client, or your supervisor at work, has an easy time reading and absorbing the findings. Part of this depends upon the descriptions and interpretations you provide, but another important aspect is the physical presentation of the statistical results. You may have noticed the regression output in R is a bit unwieldy and could certainly serve as a barrier to understanding, especially for people who don't work in R on a regular basis.

Fortunately, there is an R package called `stargazer` that can be used to produce publication quality tables that are easy to read and include most of the important information from the regression output. The first step is to run the regression model and assign the results to another object. In the preceding regression example, the fifty-state results were stored in `fit50`, so we will use that. Now, instead of using `summary` to view the results, we use `stargazer` to report the results in the form of a well-organized table. In its simplest form, you just need to tell `stargazer` to produce a "text" table using the information from the object that contains the results of the regression model. Let's give it a try. In the code chunk that follows, I tell R to use stargazer and to take the results stored in `fit50` (the regression results) and generate a table using normal text `type = "text"`. I then use `dep.var.labels =` to assign a label for the dependent variable and `covariate.labels =` to assign a label to the independent variable.[2]

```
#Regression model of Biden vote by Clinton Vote
stargazer(fit50, type = "text",
#Label dependent variable
          dep.var.labels = c("Biden Vote, 2020"),
#Label independent variable
          covariate.labels = c("Clinton Vote, 2016"))

=================================================
                       Dependent variable:
                   ------------------------------
                         Biden Vote, 2020
-------------------------------------------------
Clinton Vote, 2016            0.964***
                              (0.018)


Constant                      3.468***
                              (0.853)


-------------------------------------------------
Observations                    50
R2                             0.984
Adjusted R2                    0.984
Residual Std. Error        1.354 (df = 48)
F Statistic           2,972.400*** (df = 1; 48)
=================================================
Note:                 *p < 0.1; **p < 0.05; ***p < 0.01
```

As you can see, `stargazer` produces a neatly organized, easy to read table, one that stands in stark contrast to the standard regression output from R. Most things in the `stargazer` table are self-explanatory, but there are a couple of things I want to be sure you understand. The values for the slope and the constant are listed in the column headed by a label for the dependent variable, and the standard errors for the slope and constant appear in parentheses just below the their respective values. For instance, the slope for Clinton Vote is .964 and the standard error for the slope is .018. In addition, the significance levels (*p*-value) for the slope and constant are denoted by asterisks, with the corresponding values listed below the table. In this case, the three asterisks denote that the *p*-values for both the slope and the constant are less than .01. These *p*-values are calculated for two-tailed tests, so they are a bit on the conservative side. Because of this, if you are testing a one-tailed hypothesis and the reported *p*-value is "*p < .10", you can reject the null hypothesis since the *p*-value for a one-tailed test is actually less than .05 (half the value of the two-tailed *p*-value).

REVISITING LIFE EXPECTANCY

Before moving on to multiple regression in the next chapter, let's review important aspects of the interpretation of simple regression results by returning to the determinants of life expectancy, the focus of our scatterplot and correlation analysis in Chapter 14. In the *r*-chunk that follows, I have run three different regression models using life expectancy as the dependent variable, one for fertility rate as the independent variable, one for mean years of education, and one for the log of population. One of the upsides to `stargazer` is that we can present all three models in a single table.

```
countries2$lg10pop <- log10(countries2$pop19_M)
#Run life expectancy models
fertility <- lm(countries2$lifexp ~ countries2$fert1520)
education <- lm(countries2$lifexp ~ countries2$mnschool)
pop <- lm(countries2$lifexp ~ countries2$lg10pop)

#Have stargazer use the information in 'fertility', 'education',
#and 'pop' to create a table
stargazer(fertility, education, pop, type = "text",
          dep.var.labels = c("Life Expectancy"),
          covariate.labels = c("Fertility Rate",
                               "Mean Years of Education",
                               "Log10 Population"),
          omit.stat = c("f")) #Drop F-stat to make room for three
             columns
```

```
=========================================================================
                                 Dependent variable:
                    -----------------------------------------------------
                                   Life Expectancy
                           (1)               (2)               (3)
-------------------------------------------------------------------------
Fertility Rate          -4.911***
                         (0.231)

Mean Years of Education                    1.839***
                                           (0.112)

Log10 Population                                             -0.699
                                                             (0.599)

Constant                85.946***         56.663***         73.214***
                         (0.700)           (1.036)           (0.735)
```

```
---------------------------------------------------------------
Observations                185              189              191
R2                         0.711            0.591            0.007
Adjusted R2                0.710            0.589            0.002
Residual Std. Error   4.032 (df = 183) 4.738 (df = 187) 7.423 (df = 189)
===============================================================
Note:                              *p < 0.1; **p < 0.05; ***p < 0.01
```

First, let's examine the impact of fertility rate on life expectancy (Model 1):

- These results can be expressed as a linear equation, where the predicted value of life expectancy is a function of fertility rate: $\hat{y} = 85.946 - 4.911(\mathit{fert}1520)$.
- There is a statistically significant, negative relationship between fertility rate and country-level life expectancy. Generally, countries with relatively high fertility rates tend to have relatively low life expectancy. More specifically, for every one-unit increase in fertility rate, country-level life expectancy decreases by 4.911 years.
- This is a strong relationship. Fertility rates explains approximately 71% of the variation in life expectancy across nations.

Notice that in the summary of this model, I provided a general description of the relationship as well as more specific details and statistics. I think this is good practice, as both the specific and general complement each other well.

Let's look at the impact of level of education on life expectancy (Model 2):

- These results can be expressed as a linear equation, where the predicted value of life expectancy is $\hat{y} = 56.663 + 1.839(\mathit{mnschool})$.
- There is a statistically significant, positive relationship between average years of education and country-level life expectancy. Generally, countries with relatively high levels of education tend to have relatively high levels of life expectancy. More specifically, for every one-unit increase in average years of education, country-level life expectancy increases by 1.839 years.
- This is a fairly strong relationship. Educational attainment explains approximately 59% of the variation in life expectancy across nations.

Finally, let's examine the impact of population size on life expectancy (Model 3):

- Population size has no impact of country-level life expectancy.

The finding for population size provides a perfect opportunity to demonstrate what a null relationship looks like. Note that the constant in the this model (73.214) is very close to the value of the mean of the dependent variable (72.639). This is because the independent variable does not offer a significantly better prediction than we get from simply using $\bar{y}$ to predict y. In fact, in bi-variate regression models in which $b = 0$, the intercept equals the mean of the dependent variable and the "slope" is a perfectly flat horizontal line at $\bar{y}$. Check out the following scatterplot, where the solid line is the prediction based on the regression equation using the log of country population as the independent variable, and the dashed line is the predicted outcome (the mean of Y) if the slope equals exactly 0.

```
#Plot of 'lifexp" by 'lg10(pop)'
plot(countries2$lg10pop, countries2$lifexp, ylab = "Life Expectancy",
     xlab = "Log10 of Population")
#plot regression line from 'pop' model (above)
abline(pop)
#plot a horizontal line at the mean of 'lifexp'
abline(h = mean(countries2$lifexp, na.rm = T), lty = 2)
#Add legend
legend("bottomleft", legend = c(expression(hat(y) == "73.21-.699X"),
       "Mean of Y"), lty = c(1, 2), cex = .7)
```

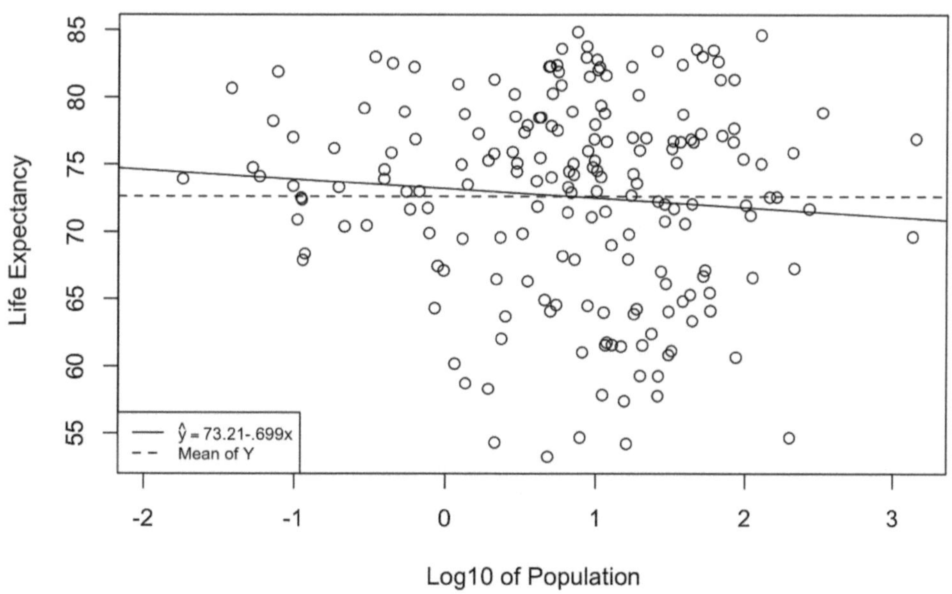

As you can see, the slope of the regression line is barely different from what we would expect if $b = 0$. This is a perfect illustration of support for the null hypothesis. The regression equation offers no significant improvement over predicting y with no regression model; that is, predicting outcomes on y with $\bar{y}$.

IMPORTANT CAVEAT

Ordinary least squares regression is an important and useful tool for political and social research. It provides a way of estimating relationships between variables using a linear model that allows for predicting outcomes of the dependent variable based on values of an independent variable. As you will see in the next two chapters, the regression model is easily extended to incorporate the effects of multiple independent variables at the same time. That said, it is important to mention here that the many virtues of regression analysis depend upon certain **assumptions** we make about the data we are using and the nature of the relationships between the dependent and independent variables. The level of confidence we can have in the results of the regression model depends on the extent to which these assumptions are satisfied. Chapter18 provides a more detailed discussion of regression assumptions, as well as suggestions for what to do if the assumptions are violated.

ADDING REGRESSION INFORMATION TO SCATTERPLOTS

The R assignments for this chapter as well as some portions of the remaining chapters rely on integrating case-specific information from scatterplots with information from regression models to learn more about cases that do and don't fit the model well. In this section, we follow the example of the R problems from the last chapter and use infant mortality in the states (`states20$infant_mort`) as the dependent variable and state per capita income (`states20$PCincome2020`) as the independent variable. First, let's run a regression model and add the regression line to the scatterplot using the `abline` function, as shown earlier in the chapter.

```
#Regression model of "infant_mort" by "PCincome2020"
inf_mort <- lm(states20$infant_mort ~ states20$PCincome2020)
#View results
stargazer(inf_mort, type = "text",
          dep.var.labels = "Infant Mortality",
          covariate.labels = "Per Capita Income")

===============================================
                        Dependent variable:
                    ---------------------------
                         Infant Mortality
-----------------------------------------------
Per Capita Income           -0.0001***
                             (0.00001)


Constant                     10.534***
                             (0.785)


-----------------------------------------------
```

```
Observations                    50
R2                           0.422
Adjusted R2                  0.410
Residual Std. Error     0.880 (df = 48)
F Statistic          35.019*** (df = 1; 48)
==================================================
Note:                 *p < 0.1; **p < 0.05; ***p < 0.01
```

Here, we see a statistically significant and negative relationship between per capita income and infant mortality, and the relationship is fairly strong ($r^2 = .422$).[3]

The scatterplot that follows reinforces both of these conclusions: There is a clear negative pattern to the data. Infant mortality rates tend to be higher in states with low per capita income than in states with high per capita income. The addition of the regression line is useful in many ways, including showing that while most outcomes are close to the predicted outcome, infant mortality is appreciably higher or lower in some states than we would expect based on the per capita income in those states. Note that in order to add the regression line, as illustrated earlier, I just had to reference the store regression output (`inf_mort`) in the `abline` command.

```
#Regression model
plot(states20$PCincome2020, states20$infant_mort,
     xlab = "Per Capita Income",
     ylab = "# Infant Deaths per 1000 live Births")
#Add regression line using stored regression results
abline(inf_mort)
```

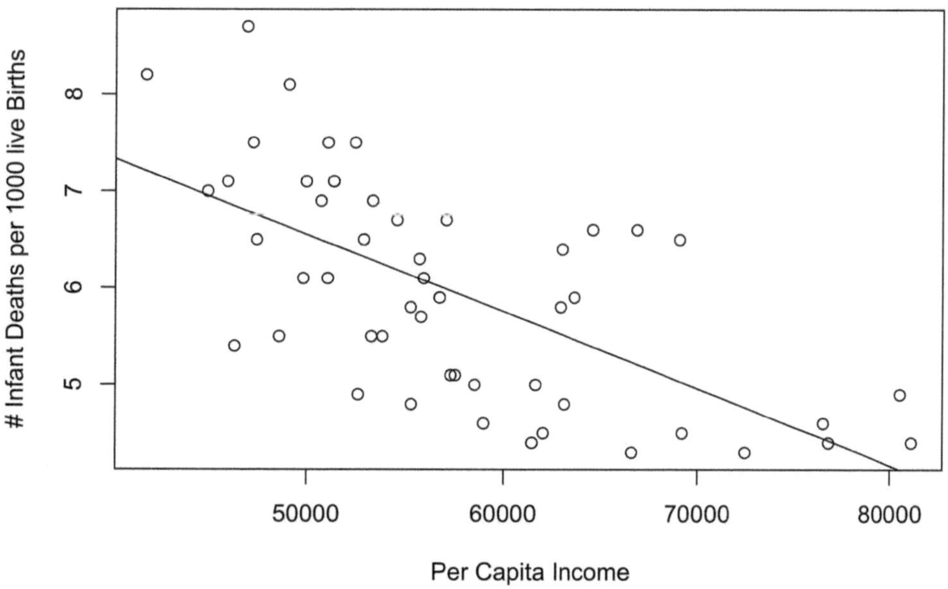

One thing we can do to make the pattern more meaningful is to replace the data points with state abbreviations. Doing this enables us to see which states are high and low on the two variables, as well as which states are relatively close to or far from the regression line. To do this, you need to reduce the size of the original markers (`cex = .001` below) and instruct R to add the state abbreviations using the text function. The format for this is `text(independent_variable, dependent_variable, ID_variable)`, where `ID_variable` is the variable that contains information that identifies the data points (state abbreviations, in this case).

```
#Scatterplot for per capita income and infant mortality
plot(states20$PCincome2020, states20$infant_mort,
     xlab = "Per Capita Income",
     ylab = "# Infant Deaths per 1000 live Births", cex = .001)
#Add regression line
abline(lm(inf_mort))
#Add State abbreviations
text(states20$PCincome2020, states20$infant_mort, states20$stateab,
   cex = .7)
```

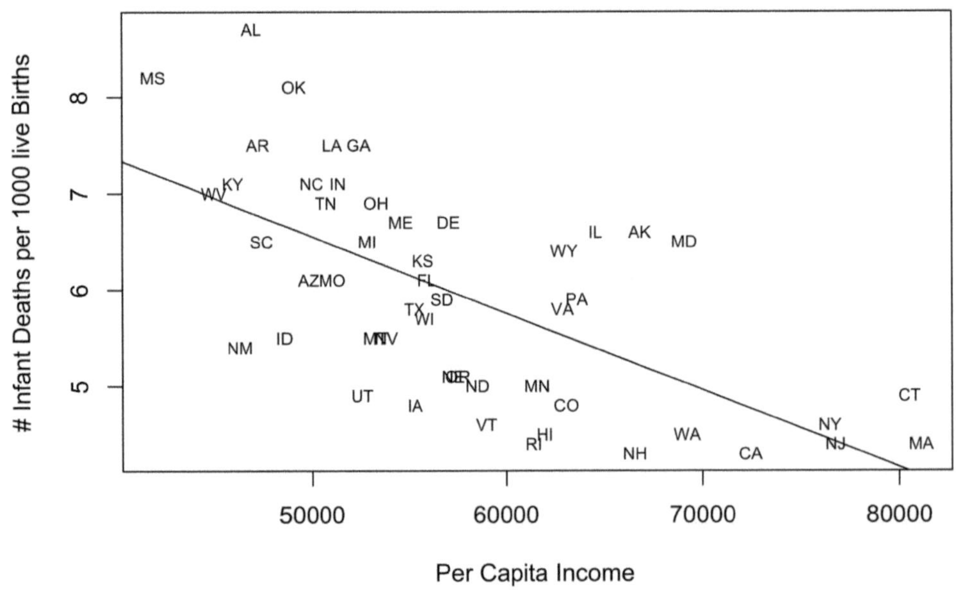

This scatterplot provides much of the same information as the previous one, but now we also get information about the outcomes for individual states. Often, this is useful for connecting with the pattern in the scatterplot, and it is also helpful because now we can see which states are well explained by the regression model and which states are not. States that stand out as having higher than expected infant mortality (farthest above the regression line) given their income level are Alabama, Oklahoma, Alaska, and Maryland. States with lower than expected infant mortality (farthest below the regression line) are New Mexico, Utah, Iowa, Vermont, and Rhode Island. It is not clear if there is a pattern to these outcomes, but there might be a slight tendency for Southern states to have a higher than expected level of infant mortality, once you control for income.

NEXT STEPS

This first chapter on regression provides you with all the tools you need to move on to the next topic: using multiple regression to control for the influence of several independent variables in a single model. The jump from simple regression (covered in this chapter) to multiple regression will not be very difficult, especially because we addressed the idea of multiple influences in the discussion of partial correlations in Chapter 14. Having said that, make sure you have a firm grasp of the material from this chapter to ensure smooth sailing as we move on to multiple regression. Check in with your professor if there are parts of this chapter that you find challenging.

EXERCISES

Concepts and Calculations

1. Identify the parts of this regression equation: $\hat{y} = a + bx$
   - The independent variable is . . .
   - The slope is . . .
   - The constant is . . .
   - The dependent variable is . . .

2. The regression output that follows is from the 2020 ANES and shows the impact of the feeling thermometer rating for conservatives on the number of LGBTQ policies supported (recall this from Chapter 4), which ranges from 0 to 5.

	Dependent variable:
	# LGBTQ Rights Supported
Conservative Feeling Therm	-0.025***
	(0.001)
Constant	4.564***
	(0.032)

```
-------------------------------------------------------------
Observations                          6,830
R2                                    0.210
Adjusted R2                           0.210
Residual Std. Error              1.389 (df = 6828)
F Statistic                 1,816.300*** (df = 1; 6828)
=============================================================
Note:                       *p < 0.1; **p < 0.05; ***p < 0.01
```

- Interpret the coefficient for the feeling thermometer rating. Focus on direction, statistical significance, and how much the dependent variable is expected to change as the values of the independent variable change.
- What are the predicted outcomes (# of LGBTQ policies supported) for three different respondents: one who gave conservatives a 0 on the feeling thermometer, one who gave them a 50, and one who rated them at 100?
- Is the relationship between feelings toward conservatives and number of LGBTQ policies supported strong, weak, or something in between? Explain your answer.

3. The regression model that follows summarizes the relationship between the poverty rate and the percentage (0 to 100, not 0 to 1) of households with access to the internet in a sample of 500 U.S. counties.

```
===============================================
                      Dependent variable:
                  ---------------------------
                       % Internet Access
-----------------------------------------------
% Poverty                 -0.866***
                           (0.043)

Constant                  89.035***
                           (0.760)

-----------------------------------------------
Observations                 500
R2                          0.447
Adjusted R2                 0.446
Residual Std. Error     6.710 f = 498)
F Statistic        402.130*** (df = 1; 498)
=====================================================
Note:              *p < 0.1; **p < 0.05; ***p < 0.01
```

- Write out the results as a linear model (see Question 1 for a generic version of the linear model).
- Interpret the coefficient for the poverty rate. Focus on direction, statistical significance, and how much the dependent variable is expected to change as the values of the independent variable change.
- Interpret the r^2 statistic.
- What is the predicted level of internet access for a county with a 5% poverty rate?
- What is the predicted level of internet access for a county with a 16% poverty rate?
- Is the relationship between poverty rate and internet access strong, weak, or something in between? Explain your answer.

4. Calculate the constant (*a*) for each of the regression models listed in the following table.

	Slope (*b*)	Mean of *X*	Mean of *Y*	Constant (*a*)
Model 1	2.13	6.46	34.38	
Model 2	.105	33.28	8.23	
Model 3	-1.9	5.24	61.88	
Model 4	-.413	22.25	13.38	

R Problems

Use `stargazer` to present the results of all R models.

1. I tried to use stargazer to give me a nice-looking table of my regression results, but I'm having a really hard time getting it to work right. Run the following model and see if you can figure out how to get the results to look the way they should in the stargazer output. (Hint: there are three problems.)

```
prob1 <- lm(states20$PerPupilExp ~ states20$PCincome2020)

stargazer(prob1,
        dep.var.labels = K-12 Per Pupil Expenditures,
        covariate.labels = Per Capita Income)
```

2. Following the example used at the end of this chapter, produce a scatterplot showing the relationship between the percentage of births to teen mothers (`states20$teenbirth`) and infant mortality `states20$infant_mort` in the states (infant mortality is the dependent variable). Make sure you include the regression line and the state abbreviations. Identify any states that you see as outliers—that is, that don't follow the general trend. Is there any pattern to these states? Then, run and interpret a regression model using the same two variables. Discuss the results, focusing on the general direction

and statistical significance of the relationship, as well as the expected change in infant mortality for a unit change in the teen birth rate, and interpret the r^2 statistic.

3. Repeat the same analysis from Question 2, except now use `states20$lowbirthwt` (% of births that are technically low birth weight) as the independent variable. As always, use words. The statistics don't speak for themselves!

4. Run a regression model using the number of LGBTQ protections supported (`anes20$lgbtq_rights`) as the dependent variable and age of the respondent (`anes20$V201507x`) as the independent variable:
   - Describe the relationship between these two variables, focusing on the value and statistical significance of the slope.
   - Is the relationship between these two variables strong, weak, or somewhere in between? How do you judge this?
   - Are you surprised by the strength of this relationship? Explain.
   - Is age more strongly related to support for LGBTQ rights than is the feeling thermometer for conservatives? (See Problem 3 in Concepts and Calculations for the results using the conservative feeling thermometer). Explain your answer.
   - Compare the predicted outcomes for three different hypothetical respondents, one who is 22 years old, one who is 48 years old, and one who is 74 years old.

5. Use the following code to take a sample of 500 counties from `counties20large` and save the sample to a new object named `county500`.

```
set.seed(1234)
#create a sample of 500 rows of data from the county20large data set
county500 <- sample_n(county20large, 500)
```

   - Choose one of the following to use as a dependent variable: `vcrim100k`, `gundeath100k`, `uninsured`, and `cvap_turnout20`.
   - Choose an independent variable that you expect to be related to the dependent variable you selected. Justify your selection and be clear about your expectations. Consult the codebook for the `counties20large` dataset to make sure you understand what these variables are.
   - Run a regression model and report the results, focusing on the slope estimate and the model fit. When answering this part of the question, make sure to provide both a general and specific summary of the results.

16 MULTIPLE REGRESSION

GET READY

This chapter expands the discussion of regression analysis to include the use of multiple independent variables in the same model. To follow along in R, you should load the `countries2` and `states20` datasets and attach the libraries for the `DescTools` and `stargazer` packages.

MULTIPLE REGRESSION

To this point, we have findings from three simple regression models that offer different potential explanations for cross-national variation in life expectancy: Differences in fertility rate and in levels of education are both strongly related to differences in life expectancy, while there is no apparent relationship between size of population and life expectancy. However, we know from the discussion of partial correlations in Chapter 14 that the fertility rate and mean years of education are highly correlated with each other and overlap somewhat in accounting of variation in life expectancy. Because of the degree to which these two variables are correlated with each other, it is highly likely that the results from the simple regression model overestimate their independent effect on the dependent variable.

Just as we were able to extend the discussion from Pearson's *r* to partial correlations to account for overlapping influences in Chapter 14, we can extend the regression model to include several independent variables to accomplish the same goal. With multiple regression analysis, we can use a linear model to consider the impact of each independent variable on the dependent variable, controlling for the influence of all of the other independent variables, as well as how the independent variables are related to each other.

Multiple regression extends the simple model by adding independent variables:

$$y_i = a + b_1x_{1i} + b_2x_{2i} + \ldots + b_kx_{ki} + e_i$$

where

y_i = the predicted value of the dependent variable;

a = the sample constant (a.k.a. the intercept);

x_{1i} = the value of x_1;

b_1 = the partial slope for the impact of x_1, controlling for the impact of other variables;

x_{2i} = the value of x^2;

b_2 = the partial slope for the impact of x^2, controlling for the impact of other variables;

k = the number of independent variables; and

e_i = error term; the difference between the predicted and actual values of y $(y_i - \hat{y}_i)$.

The formulas for estimating the constant and slopes for a model with two independent variables are presented here:

$$b_1 = \left(\frac{s_y}{s_1}\right)\left(\frac{r_{y1} - r_{y2}r_{12}}{1 - r_{12}^2}\right)$$
$$b_2 = \left(\frac{s_y}{s_2}\right)\left(\frac{r_{y2} - r_{y1}r_{12}}{1 - r_{12}^2}\right)$$
$$a = \bar{Y} - b_1\bar{X}_1 - b_2\bar{X}_2$$

We won't calculate the constant and slopes here, but I want to point out that they are based on the relationship between each of the independent variables and the dependent variable and also the interrelationships among the independent variables. You should recognize the right side of the formula for b as very similar to the formula for the partial correlation coefficient. This illustrates that the partial regression coefficient is doing the same thing that the partial correlation does: It provides an estimate of the impact of one independent variable while controlling for how it is related to the other independent variables AND for how the other independent variables are related to the dependent variable. The primary difference is that the partial correlation summarizes the strength and direction of the relationship between x and y, controlling for other specified variables, while the partial regression slope summarizes the expected change in y for a unit change in x, controlling for other specified variables, and can be used to predict outcomes of y. As independent variables are added to the model, the level of complexity for calculating the slopes increases, but the basic principle of isolating the independent effects of multiple independent variables remains the same.

To get multiple regression results from R, you add the independent variables to the linear model function, using the "+" sign to separate them. Also, since you are listing multiple variables from the same dataset, you can save some space and effort by adding `data = DataSetName` to the command and then just listing the variable names, instead of using `DataSetName$Variable` for each variable.[1]

```
#Note the "+" sign before each additional variable.
#See footnote for information about "na.action"

fit <- lm(data = countries2, lifexp ~ fert1520 +
          mnschool + log10(pop19_M), na.action = na.exclude)
summary(fit)
```

```
Call:
lm(formula = lifexp ~ fert1520 + mnschool + log10(pop19_M), data = countries2,
    na.action = na.exclude)

Residuals:
    Min              1Q          Median        3Q      Max
-14.920          -2.297           0.302    2.674    9.632

Coefficients:
                Estimate Std. Error t  value        Pr( > |t|)
(Intercept)       75.468      2.074     36.38       < 2e-16 ***
fert1520          -3.546      0.343    -10.34       < 2e-16 ***
mnschool           0.750      0.140      5.36    0.00000026 ***
log10(pop19_M)     0.296      0.340      0.87          0.38
---
Signif. codes: 0 '***' 0.001 '**' 0.01 '*' 0.05 '.' 0.1 ' ' 1

Residual standard error: 3.77 on 179 degrees of freedom
  (12 observations deleted due to missingness)
Multiple R-squared:    0.748, Adjusted R-squared:    0.743
F-statistic:    177 on 3 and 179 DF,    p-value: < 2e-16
```

The raw R output is shown in the preceding example, so you can see that it changes very little when you add variables. The `stargazer` output is shown next.

```
#Use 'stargazer' to produce a table for the three-variable model
stargazer(fit, type = "text",
          dep.var.labels = c("Life Expectancy"),
          covariate.labels = c("Fertility Rate",
                               "Mean Years of Education",
                               "Log10 Population"))

=========================================================================
                                             Dependent variable:
                                        ---------------------------
                                              Life Expectancy
-------------------------------------------------------------------------
Fertility Rate                                    -3.546***
                                                  (0.343)

Mean Years of Education                            0.750***
                                                  (0.140)

```

```
Log10 Population                              0.296
                                             (0.340)


Constant                                    75.468***
                                             (2.074)

-----------------------------------------------------------------------
Observations                                   183
R2                                               0.748
Adjusted R2                                    0.743
Residual Std. Error                      3.768 (df = 179)
F Statistic                          176.640*** (df = 3; 179)
=======================================================================
Note:                                *p < 0.1; **p < 0.05; ***p < 0.01
```

This output looks a lot like that from the individual models, except now we have the coefficients for each of the independent variables in a single model.

Here's how I would interpret the results:

- Two of the three independent variables are statistically significant: Fertility rate has a negative impact on life expectancy, and educational attainment has a positive impact. The *p*-values for the slopes of these two variables are less than .01. Population size has no impact on life expectancy.
- For every additional unit of fertility rate, life expectancy is expected to decline by 3.546 years, controlling for the influence of other variables. So, for instance, if two countries are the same on all other variables but one country's fertility rate is one unit higher than the other, we would predict life expectancy to be 3.546 years lower in the country with the higher fertility rate than in the other country.
- For every additional year of mean educational attainment, life expectancy is predicted to increase by .75 years, controlling for the influence of other variables. If two countries are the same on all other variables but one country's mean years of education is one unit higher than the other, we would predict life expectancy to be .75 years higher in the country with the higher level of education than in the other country.
- Together, these three variables explain 74.8% of the variation in country-level life expectancy.
- You can also write this out as a linear equation, if that helps you with interpretation:

$$\widehat{\text{lifexp}} = 75.468 - 3.546(\text{fert1520}) + .75(\text{mnschool}) - .296(\text{log10(pop)})$$

It is also important to note that the slopes for fertility and level of education are much smaller than they were in the bivariate (one independent variable) models, reflecting the consequence of controlling for overlapping influences. For fertility rate, the slope changed from −4.911 to −3.546, a 28% decrease; while the slope for education went from 1.839 to .75, a 59% decrease in impact. This is to be expected when two highly correlated independent variables are put in the same multiple regression model.

One thing you might have noticed about this discussion of findings is that it includes both a discussion of general findings and a discussion of specific statistics. I have found that this approach generally works well.

Missing Data

When working with multiple regression or any other statistical technique that involves using multiple variables at the same time, it is important to pay attention to the number of *missing cases*. Missing cases occur on any given variable when cases do not have any valid outcomes. We discussed this a bit earlier in the book in the context of public opinion surveys, where missing data might occur because people refuse to answer questions or do not have an opinion to offer. When working with aggregate cross-national data, as we are here, missing outcomes usually occur because the data are not available for some countries on some variables. For instance, some countries may not report data on some variables to the international organizations (e.g., World Bank, United Nations, etc.) that are collecting data, or perhaps the data-gathering organizations collect certain types of data from certain types of countries but not for others.

This is a more serious problem for multiple regression than simple regression because multiple regression uses "listwise" deletion of missing data, meaning that if a case is missing on one variable it is dropped from the analysis. This is why it is important to pay attention to missing data as you add more variables to the model. It is possible that one or two variables have a lot of missing data, and you could end up making generalizations based on a lot fewer data points than you realize. The output reported in `stargazer` format reports the number of observations (183) but does not report the number of missing cases. The raw R output reports the number of missing as "12 observations deleted due to missingness." There are relatively few missing data points in this model, but it is important to monitor the number of observations as you add independent variables.

Assessing the Substantive Impact

Sometimes, calculating predicted outcomes for different combinations of the independent variables can give you a better appreciation for the substantive importance of the model. We can use the model estimates to predict the value of life expectancy for countries with different combinations of outcomes on these variables. To do this, you need to plug in the values of the independent variables, multiplying them times their respective slopes, and then add the constant, similar to the calculations made in Chapter 15 (Table 15.2). Consider two hypothetical countries that have very different (but realistic) outcomes on fertility and education but are the same on population, as shown in Table 16.1.

TABLE 16.1 ■ Hypothetical Values for Independent variables

Country	Fertility Rate	Mean Education	Logged Population
Country A	1.8	10.3	.75
Country B	3.9	4.9	.75

Country B has a higher fertility rate and lower value on educational attainment than Country A, so we expect it to have an overall lower predicted outcome on life expectancy. Let's plug in the numbers and see what we get.

For Country A

$$\hat{y} = 75.468 - 3.546(1.8) + .75(10.3) + .296(.75) = 77.03$$

For Country B

$$\hat{y} = 75.468 - 3.546(3.9) + .75(4.9) + .296(.75) = 65.54$$

The predicted life expectancy in Country A is *almost 11.5 years higher* than in Country B, based on differences in fertility and level of education. Given the gravity of this dependent variable—how long people are expected to live—this is a very meaningful difference. It is important to understand that the difference in predicted outcomes would be much narrower if the model did not fit the data as well as it does. If fertility and education levels were not as strongly related to life expectancy, then differences in their values would not predict as substantial a difference in life expectancy.

MODEL ACCURACY

Adjusted R^2

The interpretation and importance of the R^2 statistic in multiple regression is much the same as in simple regression, with a couple of minor differences. First, in multiple regression, the R^2 for any given model describes how well *the group* of variables explains variation in the dependent variable, not the explanatory power of any single variable. Second, especially when focusing on differences in R^2 values across different models, it is important to take into account the number of variables in the models. In the current example, the new model R^2 (.748) appears to be an improvement over the strongest of the bivariate models (fertility rate), which had an R^2 of .711. However, with multiple regression analysis, the R^2 value will increase every time a variable is added to the model, even if only by some slight magnitude. Given this, it is important to determine whether there was a *real increase* in the R^2 value of the life expectancy model beyond what could be expected simply due to adding two variables. One way to address this is to "adjust" the R^2 to take into account the number of variables in the model. The *adjusted* R^2 value can then be used to assess the fit of the model.

The formula for the adjusted R^2 is

$$R^2_{adj} = 1 - (1 - R^2)\left(\frac{N-1}{N-k-1}\right)$$

Here, $N - k - 1$ is the degrees of freedom for the regression model, where N = sample size (183), and k = number of variables (3). One thing to note about this formula is that the impact of additional variables on the value of R^2 is greatest when the sample size is relatively small. As N increases, the difference between R^2 and adjusted R^2 for additional variables grows smaller.

For the preceding model, with the R^2 value of .7475 (carried out four places to the right of the decimal),

$$R^2_{adj} = 1 - (.2525)\left(\frac{182}{179}\right) = .743$$

This new estimate is slightly smaller than the *unadjusted* R^2 but still represents a real improvement over the strongest of the bivariate models, in which the adjusted R^2 was .710. Usually, when discussing model fit, people refer to the R^2 of the model rather than the adjusted R^2. This is fine, especially because the two are usually close in value, but it is important to report the adjusted R^2 in the table of results so readers can evaluate it alongside the original R^2.

Root Mean Squared Error

The R^2 (or adjusted R^2) statistic is a nice measure of the explanatory power of the models, and a higher R^2 for a given model generally means less prediction error, within the context of that model. However, the R^2 statistic does not tell us exactly how much error there is in the model estimates. For instance, the preceding results tell us that the model explains about 74% of the variation in life expectancy, but that piece of information does not speak to the typical error in prediction $(y_1 - \hat{y})$. Yes, the model reduces error in prediction by quite a lot, but how much error is there, on average? For that, we can use a different statistic, the root mean squared error (RMSE). The RMSE reflects the typical error in the model; it is the square root of the mean squared error:

$$\text{RMSE} = \sqrt{\frac{\sum (y_i - \hat{y})^2}{N-k-1}}$$

The sum of squared residuals (error) constitutes the numerator, and the model degrees of freedom constitute the denominator. Let's run through these calculations for the regression model.

```
#Calculate squared residuals using the saved residuals from "fit"
residsq <- residuals(fit)^2

#Sum the squared residuals, add "na.rm" because of missing data
sumresidsq <- sum(residsq, na.rm = T)
sumresidsq
[1] 2541.9

#Divide the sum of squared residuals by N-K-1 and take the square root

RMSE = sqrt(sumresidsq/179)
RMSE
[1] 3.7684
```

The resulting RMSE can be taken as the typical prediction error. For the preceding model, RMSE = 3.768. This appears in the regression table as **"Residual Std Error: 3.768 (df = 179)"**.

One drawback to RMSE is that "3.768" has no standard meaning. Whether it is a lot or a little error depends on the scale of the dependent variable. For a dependent variable that ranges from 1 to 15, this could be a substantial amount of error. However, for a variable like life expectancy, which ranges from about 52 to 85, this is not much error. The best use of the RMSE lies in comparison of error across models that use the same dependent variable. For instance, the RMSE values for the fertility, education, and population regression results in Chapter 15 were 4.032, 4.738, and 7.423, respectively. The RMSE for the multiple regression model is smaller than any of the individual models.

Both the R^2 and RMSE are valuable tools for evaluating the level of error in regression models. Best practice is to use each of them for what they can tell you.

PREDICTED OUTCOMES

It is useful to think of the strength of the model in terms of how well its predictions correlate overall with the dependent variable. In the initial discussion of the simple bivariate regression model, I pointed out that the square root of the R^2 is the correlation between the independent and dependent variables. We can think of the multiple regression model in a similar fashion: The square root of R^2 is the correlation (Multiple R) between the dependent variable and the predictions from the regression model. By this, I mean that Multiple R is literally the correlation between the observed values of the dependent variable (y) and the values predicted by the regression model ($\hat{y}$).

In the example shown next, predicted outcomes are generated based on the three-variable model, and a scatterplot of the predicted and actual outcomes is provided to gain an appreciation for how well the model explains variation in the dependent variable. First, let's generate predicted outcomes (`yhat`) for all observations, based on their values of the independent variables, using information stored in `fit`.

```
#Use information stored in "fit" to predict outcomes
countries2$yhat <- (predict(fit))
```

The scatterplot illustrating the relationship between the predicted and actual values of life expectancy is presented next.

```
#Use predicted values as an independent variable in the scatterplot
plot(countries2$yhat, countries2$lifexp,
     xlab = "Predicted  Life Expectancy",
     ylab = "Actual  Life Expectancy")
#Plot the regression line
abline(lm(countries2$lifexp ~ countries2$yhat))
```

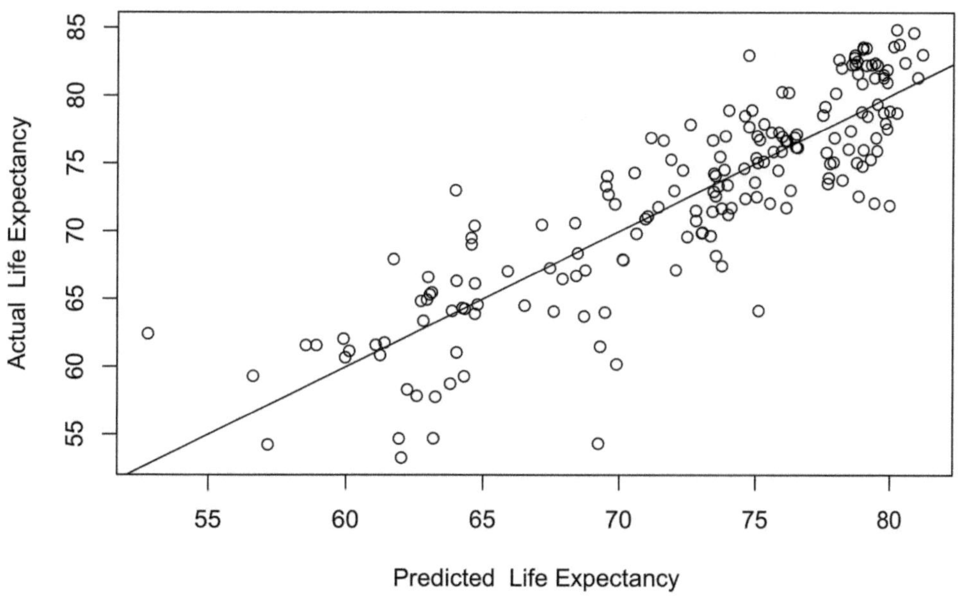

Here we see the predicted values along the horizontal axis, the actual values on the vertical axis, and a fitted (regression) line. The key takeaway from this plot is that the model fits the data fairly well. The correlation (shown next) between the predicted and actual values is .865, which, when squared, equals .748, the R^2 of the model.

```
#get the correlation between y and yhat
cor.test(countries2$lifexp, countries2$yhat)


    Pearson's product-moment correlation


data: countries2$lifexp and countries2$yhat
t = 23.1, df = 181, p-value < 2e-16
alternative hypothesis: true correlation is not equal to 0
95 percent confidence interval:
 0.82270 0.89713
sample estimates:
    cor
0.86458
```

In addition to plotting the predicted and actual data points, you might want to identify some of the observations that are relative outliers (that don't fit the pattern very well). In the preceding scatterplot, there are no extreme outliers, but there are a few observations whose actual life expectancy is quite a bit less than their predicted outcome. These observations are the farthest below the prediction line. Changing the `plot` command so that it lists the country code for each country allows us to identify these and any other outliers.

```
plot(countries2$yhat, countries2$lifexp,
     xlab = "Predicted Life Expectancy",
     ylab = "Actual Life Expectancy",
     cex = .001)
abline(lm(countries2$lifexp ~ countries2$yhat))
text(countries2$yhat, countries2$lifexp, countries2$ccode, cex = .6)
```

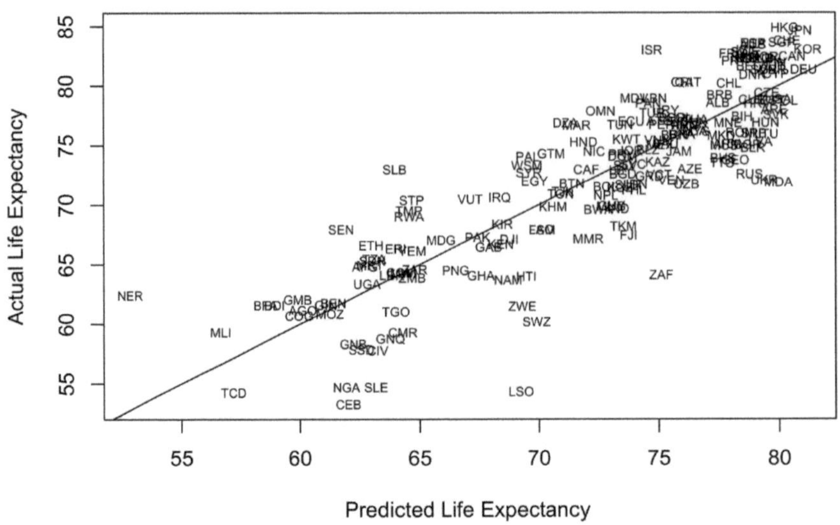

Based on these labels, several countries stand out as having substantially lower than predicted life expectancy: Central African Republic (CEB), Lesotho (LSO), Nigeria (NGA), Sierra Leone (SLE), South Africa (ZAF), and Swaziland (SWZ). This information is particularly useful if it helps you identify patterns that might explain outliers of this sort. For instance, if these countries shared an important characteristic that we had not yet put in the model, we might be able to improve the model by including that characteristic. In this case, the first thing that stands out is that these are all sub-Saharan African countries. At the same time, predictions for most sub-Saharan African countries are closer to the regression line, and one of the countries that stands out for having higher than expected life expectancy is Niger (NER), another sub-Saharan African country. While it is important to be able to identify the data points and think about what the outliers might represent, it is also important to focus on broad, theoretically plausible explanations that might contribute to the model in general and could also help explain the outliers. One thing that is missing in this model that might help explain differences in life expectancy is access to health care, which we will take up in the next chapter.

REVISITING PRESIDENTIAL VOTES IN THE STATES

In this section, I use the presidential vote data from Chapter 15 to further review how to use and interpret the results of a multiple regression model. The dependent variable is Joe Biden's percentage of the two-party vote in the states (`states20$d2pty20`), and the independent variables are the percentage of the state population who are white males (`states20$whitemale`), the percentage of the non-agricultural workforce who are unionized (`states20$union`), and a dichotomous variable identifying whether the state is in the South (1) or non-South (0) (`states20$south`). The expectations are that Biden support is negatively related to the percentage of white males in the population and the Southern region indicator, and positively related to the percentage of the workforce who are unionized.

Interpreting Dichotomous Independent Variables

You might have noticed that one of the independent variables, `south`, is not quite like the others. Both `whitemale` and `union` are continuous numeric variables, whereas `south` is dichotomous variable scored 0 and 1 to indicate the presence of a qualitative characteristic, the Southern region of the country. In Chapter 4 (measures of central tendency), we discussed treating dichotomous indicators of qualitative outcomes as numeric variables that indicate the presence or absence of some characteristic; in this case, that characteristic is being a Southern state.

In order to appreciate what kind of information dichotomous variables provide, let's take a look at the simple mean difference in Biden's percentage of the two-party vote between 13 Southern and 27 non-Southern states, using the `compmeans` command. Remember, non-Southern states are scored 0, and Southern states are scored 1 on this variable.

```
#Get group means
compmeans(states20$d2pty20, states20$south, plot = F)

Warning in compmeans(states20$d2pty20, states20$south, plot = F): Warning:
"states20$south" was converted into factor!

Mean value of "states20$d2pty20" according to "states20$south"
          Mean    N Std.   Dev.
0         50.969 37      10.8849
1         42.643 13       7.0943
Total     48.804 50      10.6293
```

Here, we see that the mean Democratic percentage in Southern states in 2020 was 42.64%, compared to 50.97% in non-Southern states, for a difference of 8.33 points. Of course, we can add a bit more to this analysis by using the *t*-test for the difference.

```
#Get t-test or region-based difference in "d2pty20"
t.test(states20$d2pty20 ~ states20$south, var.equal = T)

   Two Sample t-test

data: states20$d2pty20 by states20$south
t = 2.56, df = 48, p-value = 0.014
alternative hypothesis: true difference in means between group 0 and
    group 1 is not equal to 0
95 percent confidence interval:
  1.797 14.855
sample estimates:
mean in group 0 mean in group 1
         50.969          42.643
```

The *t*-score for the difference is 2.56, and the *p*-value is .014 (< .05), so we conclude that there is a statistically significant difference between the two groups.

Now, let's take a look at a simple regression model to see that we can get exactly the same kind of information through regression analysis.

```
#Run regression model
south_demo <- lm(data = states20, d2pty20 ~ south)
#View regression results
stargazer(south_demo, type = "text",
          dep.var.labels = "Biden % of Two-party Vote",
          covariate.labels = "Southern State (0,1)")

=======================================================================
```

	Dependent variable:
	Biden % of Two-party Vote
Southern State (0,1)	-8.326**
	(3.247)
Constant	50.969***
	(1.656)
Observations	50
R2	0.120
Adjusted R2	0.102
Residual Std. Error	10.072 (df = 48)
F Statistic	6.574** (df = 1; 48)
Note:	*p < 0.1; **p < 0.05; ***p < 0.01

The most important thing to get about using dichotomous independent variables is that it does not make sense to interpret the value of b (–8.33) as a slope in the same way we might think of a slope for a numeric variable with multiple categories. There are just two values to the independent variable, 0 (non-South) and 1 (South), so statements like "for every unit increase in `south`, Biden vote drops by 8.33 points" sound a bit odd. Instead, the slope is really saying that the expected level of support for Biden in the South is 8.33 points lower than his support in the rest of the country. Let's plug in the values of the independent variable to illustrate this.

$$\text{When South} = \hat{y} = 50.97 - 8.33(0) = 50.97$$
$$\text{When South} = \hat{y} = 50.97 - 8.33(1) = 42.64$$

These numbers should look familiar to you, as this is exactly what we saw in the `compmeans` results: The mean for Southern states is 42.64, while the mean for other states is 50.97, and the difference between them (-8.33) *is equal to the regression coefficient* for `south`. Also, the t-score and p-value associated with the slope for the South variable in the regression model are exactly the same as in the t-test model.

What this illustrates is that when using a dichotomous independent variable, the "slope" captures the mean difference in the predicted value of the dependent variable between the two categories of the independent variable and the results are equivalent to a difference of means test. In fact, in a bivariate model like this, the slope for a dichotomous variable is really just an addition to or subtraction from the intercept for cases scored 1; for cases scored 0 on the independent variable, the predicted outcome is equal to the intercept.

Interpreting the Vote Share Model

Now, let's return to the multiple regression model to discuss the effects of all three independent variables on Biden's share to the two-party vote in the states in 2020.

```
vote_fit <- lm(data = states20, d2pty20 ~ whitemale + union + south)
stargazer(vote_fit, type = "text",
          dep.var.labels = "Biden % of Two-party Vote",
          covariate.labels = c("Percent White Male", "Percent Union Workforce",
          "Southern State (0,1)"))

=======================================================================
                                    Dependent variable:
                                ---------------------------
                                  Biden % of Two-party Vote
-----------------------------------------------------------------------
Percent White Male                          -0.683***
                                            (0.218)

Percent Union Workforce                      0.690**
                                            (0.293)

Southern State (0,1)                        -7.372**
                                            (3.398)

Constant                                    70.291***
                                            (10.563)

-----------------------------------------------------------------------
Observations                                   50
R2                                           0.417
Adjusted R2                                  0.379
Residual Std. Error                     8.375 (df = 46)
F Statistic                        10.974*** (df = 3; 46)
=======================================================================
Note:                              *p < 0.1; **p < 0.05; ***p < 0.01
```

In this model, all three independent variables are statistically significant ($p < .05$), and the slopes are in the anticipated direction. The model R^2 (.417) and the RMSE (8.38) show that the model fits the data fairly well, although there is still some room for error reduction.

The percentage white male in the state population is negatively related to Biden's vote share. For every one percentage point increase in the white male population, Biden's share of the two-party vote is expected to drop by .68 percentage points, all else held constant.

The percentage union in the state workforce is positively related to Biden's vote share. For every one point increase in percentage union, Biden's vote share is expected to increase by .69 percentage points, all else held constant.

The Southern indicator variable is negatively related to support for Biden, whose vote share is predicted to be 7.37 percentage points lower in Southern states than in all other states, all else held constant. Note that this effect is only a bit weaker than what we found in the simple regression model, where the slope was equal to -8.33.

NEXT STEPS

It's hard to believe we are almost to the end of this textbook. It wasn't so long ago that you were looking at bar charts and frequencies, perhaps even having a bit of trouble getting the R code for those things to work. You've certainly come a long way since then!

You now have a solid basis for using and understanding regression analysis, but there are just a few more things you need to learn about in order to take full advantage of this important analytic tool. The next chapter takes up issues related to making valid comparisons of the relative impact of the independent variables, problems created when the independent variables are very highly correlated with each other or when the relationship between the independent and dependent variables does not follow a linear pattern. Then, Chapter 18 provides a brief overview of a number of important assumptions that underlie the regression model. It's exciting to have gotten this far, and I think you will find the last few topics interesting and important additions to what you already know.

EXERCISES

Concepts and Calculations

1. Answer the following questions regarding the following regression model, which focuses on multiple explanations for county-level differences in internet access, using a sample of 500 counties. The independent variables are the percentage of the county population living below the poverty rate, the percentage of the county population with advanced degrees, and the logged value of population density (logged because density is highly skewed).

```
==========================================================================
                                           Dependent variable:
                                     ---------------------------
                                          % With Internet Access
--------------------------------------------------------------------------
Poverty Rate (%)                                -0.761***
                                                 (0.038)
```

Advanced Degrees (%)	0.561***
	(0.076)
Log10(Population Density)	2.378***
	(0.398)
Constant	79.817***
	(0.922)
Observations	500
R2	0.604
Adjusted R2	0.602
Residual Std. Error	5.686 (df = 496)
F Statistic	252.530*** (df = 3; 496)
Note:	*p < 0.1; **p < 0.05; ***p < 0.01

- Interpret the slope for poverty rate.
- Interpret the slope for advanced degrees.
- Interpret the slope for the log of population density.
- Evaluate the fit of the model in terms of explained variance.
- What is the typical error in the model? Does that seem like a small or large amount of error?

2. Use the information from the model in Question 1, along with the information presented next to generate predicted outcomes for two hypothetical counties, County A and County B. Just to be clear, you should use the values of the independent variable outcomes and the slopes and constant to generate your predictions.

County	Poverty Rate	Adv. Degrees	Log10 (Density)	Prediction?
A	19	4	1.23	
B	11	8	2.06	

- Prediction of County A is . . .
- Prediction of County B is . . .
- What did you learn from predicting these hypothetical outcomes that you could not learn from the model output?

3. The table that follows contains the results of a multiple regression model that uses the Democratic Party feeling thermometer as the dependent variable, and age, sex, race, and number of LGBTQ rights supported are the independent variables.

	Dependent variable:
	Democratic Feeling Thermometer
Age	0.179***
	(0.019)
Race (White = 1, Other = 0)	-19.035***
	(0.718)
Sex (Female = 1, Other = 0)	4.390***
	(0.633)
#LGBTQ Rights Supported (0 to 5)	10.096***
	(0.205)
Constant	13.604***
	(1.350)
Observations	7,422
R2	0.304
Adjusted R2	0.303
Residual Std. Error	27.021 (df = 7417)
F Statistic	809.390*** (df = 4; 7417)
Note:	*p < 0.1; **p < 0.05; ***p < 0.01

- Interpret the slope for age.
- Interpret the slope for race
- Interpret the slope for sex.
- Interpret the slope for support for LGBTQ rights.
- Evaluate the fit of the model in terms of explained variance.
- What is the typical error in the model? Does that seem like a small or large amount of error? Explain.

4. Based on the model results in the previous question, what is the predicted Democratic Party feeling thermometer rating for each of the following:
   - A 35-year-old white man who supports 0 out of 5 LGBTQ rights
   - A 35-year-old non-white woman who supports 4 out of five LGBTQ rights

R Problems

1. I'm trying to extend the Biden vote share regression model used earlier in this chapter to include a couple of additional independent variables, but I can't get it to run. Check out the code I used and let me know what I've done wrong. Produce a nicely organized table of results using `stargazer`. (Hint: there are two errors.)

```
Votes <- lm(data = states20, d2pty20 ~ whitemale + union south + vep_
   turnout20 + early_pct)
```

2. Building on the regression models from the R problems in the last chapter, use the `states20` dataset and run a multiple regression model with infant mortality (`infant_mort`) as the dependent variable and per capita income (`PCincome2020`), the teen birth rate (`teenbirth`), and percentage low birth weight births `lowbirthwt` as the independent variables. Store the results of the regression model in an object called `rhmwrk` and produce a nicely organized table using `stargazer`, one that replaces the R variable names with descriptive labels.
   - Based on the results of this regression model, discuss the determinants of infant mortality in the states. Pay special attention to the direction and statistical significance of the slopes and to all measures of model fit and accuracy.

3. Use the following command to generate predicted values of infant mortality in the states from the regression model produced in Question 1.

```
#generate predicted values
states20$yhat <- predict(rhmwrk)
```

   - Now, produce a scatterplot of predicted and actual levels of infant mortality and replace the scatterplot markers with state abbreviations (see Chapter 15 homework). Make sure this scatterplot includes a regression line showing the relationship between predicted and actual values.
   - Generally, does it look like there is a good fit between the model predictions and the actual levels of infant mortality? Explain.
   - Identify states that stand out as having substantially higher or lower than expected levels of infant mortality. Do you see any pattern among these states?

17 ADVANCED REGRESSION TOPICS

GET READY

This chapter continues the analysis of country-level life expectancy by taking into account other factors that might be related to health conditions in general. We will also address a few important issues related to evaluating multiple regression models. To follow along in R, you should load the `countries2` dataset, as well as the libraries for the `DescTools` and `stargazer` packages.

INCORPORATING ACCESS TO HEALTH CARE

One set of variables that we haven't considered but that should be related to life expectancy are factors that measure access to health care. Generally speaking, countries with greater availability of health care resources (doctors, hospitals, research centers, etc.) should have higher levels of life expectancy. In the table that follows, the regression model is expanded to include two additional variables, percentage of the population living in urban areas and the number of doctors per 10,000 persons. These variables are expected to be positively related to life expectancy, as they are linked to the availability of health care services and facilities. Doctors per 10,000 persons is a direct measure of access to health care, and levels of access to health care are generally lower in rural areas than in urban areas.

These variables are added in the model that follows.

```
#Add % urban and doctors per 10k to the model
fit <- lm(data = countries2, lifexp ~ fert1520 +
          mnschool + log10(pop19_M) + urban + docs10k)
#Use information in 'fit' to create table of results
stargazer(fit, type = "text",
          dep.var.labels = c("Life Expectancy"),
          covariate.labels = c("Fertility Rate",
                               "Mean Years of Education",
                               "Log10 Population", "%Urban",
                               "Doctors per 10,000"))
```

	Dependent variable:
	Life Expectancy
Fertility Rate	-3.237***
	(0.341)
Mean Years of Education	0.413**
	(0.163)
Log10 Population	0.082
	(0.331)
% Urban	0.050***
	(0.015)
Doctors per 10,000	0.049*
	(0.027)
Constant	73.923***
	(2.069)
Observations	179
R2	0.768
Adjusted R2	0.761
Residual Std. Error	3.606 (df = 173)
F Statistic	114.590*** (df = 5; 173)
Note:	*p < 0.1; **p < 0.05; ***p < 0.01

Here's how I would interpret the results:

- The first thing to note is that four of the five variables are statistically significant (counting `docs10k` as significant with a one-tailed test), and the overall model fit is improved in comparison to the fit for the three-variable model from Chapter 16. The only variable that has no discernible impact is population size. The model R^2 is now .768, indicating that the model explains 76.8% of the variation in life expectancy, and the adjusted R^2 is .761, an improvement over the previous three-variable model (adjusted R^2 was .743). Additionally, the RMSE is now 3.606, also signaling less error compared to the previous model (3.768).

Interpretation of the individual variables should be something like the following:

- Fertility rate is negatively and significantly related to life expectancy. For every one unit increase in the value of fertility rate, life expectancy is expected to decline by about 3.24 years, controlling for the influence of the other variables in the model.
- Average years of education is positively related to life expectancy. For every unit increase in education level, life expectancy is expected to increase by about .413 years, controlling for the influence of the other variables in the model.
- Percentage of urban population is positively related to life expectancy. For every unit increase in the percentage of the population living in urban areas, life expectancy is expected to increase by .05 years, controlling for the influence of other variables in the model.
- Doctors per 10,000 population is positively related to life expectancy. For every unit increase in doctors per 10,000 population, life expectancy is expected to increase by about .05 years, controlling for the influence of the other variables in the model.

I am treating the `docs10k` coefficient as statistically significant even though the *p*-value is greater than .05. This is because the reported *p*-value (< .10) is a two-tailed *p*-value, and if we apply a one-tailed test (which makes sense in this case), the *p*-value is cut in half and is less than .05. The two-tailed *p*-value (not shown in the table) is .068, so the one-tailed value is .034. Still, in a case like this, where the level of *significance is borderline*, it is best to assume that the *effect is pretty small*. A weak effect like this is surprising for this variable, especially since there are strong substantive reasons to expect greater access to health care to be strongly related to life expectancy. Can you think of an explanation for this? We explore a couple of potential reasons for this weaker-than-expected relationship in the next two sections.

MULTICOLLINEARITY

One potential explanation for the seemingly tepid role of `docs10k` in the life expectancy model is *multicollinearity*, a problem that can emerge when independent variables are highly correlated with each other. Recall from the discussions in Chapters 14 and 16 that when independent variables are strongly correlated with each, other the simple bivariate relationships are likely overstated and the independent variables lose strength when considered in conjunction with other overlapping explanations. Normally, this is not a major concern. In fact, one of the virtues of regression analysis is that the model sorts out overlapping explanations (this is why we use multiple regression). However, when the degree of overlap is very high, it can make it difficult for substantively important variables to demonstrate statistical significance, which can increase the probability of making a Type II error. Perfect collinearity violates a regression assumption (see next chapter), but generally high levels of collinearity can also cause problems, especially for interpreting significance levels.

Here's how this problem comes about: the t-score for a regression coefficient (b) is calculated as $t = \frac{b}{S_b}$, and the standard error of the coefficient (S_b) for any given variable is directly influenced by how that variable is correlated with other independent variables. The following formula provides a clue to how collinearity can affect the standard error of b_1 in a model with two independent variables:

$$S_{b_1} = \sqrt{\frac{RMSE}{\sum(x_i - \bar{x}_1)^2(1 - R^2_{1 \cdot 2})}}$$

The key to understanding how collinearity affects the standard error, which then affects the t-score, lies in part of the denominator of the formula for the standard error: $(1 - R^2_{1 \cdot 2})$. As the correlation (and R^2) between x_1 and x_2 increases, the denominator of the formula decreases in size, leading to larger standard errors and smaller t-scores. Because of this, high correlations among independent variables can make it difficult for variables to demonstrate statistical significance. This problem can be compounded when there are multiple strongly related independent variables. This problem is usually referred to as high *multicollinearity* or high *collinearity*.

When you have good reasons to expect a variable to be strongly related to the dependent variable and it is not statistically significant in a multiple regression model, you should think about whether collinearity could be a problem. Is this a potential explanation for the borderline statistical significance for `docs10k`? It does make sense that `docs10k` is strongly related to other independent variables, especially since most of the variables reflect an underlying level of economic development. As a first step toward diagnosing this, let's look at the correlation matrix for evidence of collinearity.

```
#Create "logpop" variable for exporting
countries2$lg10pop <- log10(countries2$pop19_M)

#Copy DV and IVs to new data set.
lifexp_corr <- na.omit(countries2[,c("lifexp", "fert1520", "mnschool",
                          "lg10pop", "urban", "docs10k")])
#Use variable in 'lifexp_corr' to produce correlation matrix
cor(lifexp_corr)

             lifexp   fert1520   mnschool    lg10pop      urban   docs10k
lifexp      1.00000  -0.839901   0.768438  -0.036340   0.602412   0.70310
fert1520   -0.83990   1.000000  -0.762732   0.070566  -0.518459  -0.67767
mnschool    0.76844  -0.762732   1.000000  -0.095843   0.578742   0.76661
lg10pop    -0.03634   0.070566  -0.095843   1.000000   0.079091  -0.01679
urban       0.60241  -0.518459   0.578742   0.079091   1.000000   0.55433
docs10k     0.70310  -0.677672   0.766615  -0.016790   0.554334   1.00000
```

Look across the row headed by `docs10k` or down the `docs10k` column to see how strongly it is related the other variables. First, there is a strong bivariate correlation ($r = .70$) between doctors per 10,000 population and the dependent variable, life expectancy, so it is a bit surprising that its effect is so small in the regression model. Further, `docs10k` is also highly correlated with fertility rates (r = -.68), mean years of education (r = .77), and percentage living in urban areas (r = .55). Based on these correlations, it seems plausible that collinearity could be causing a problem for the significance of slope for `docs10k`.

There are two important statistics that help you evaluate the extent of the problem with collinearity: the *tolerance* and *VIF* (variance inflation) statistic. Tolerance is calculated as

$$\text{Tolerance}_{x_1} = 1 - R^2_{x_1,x_2 \cdots x_k}$$

You should recognize this from the denominator of the formula for the standard error of the regression slope; it is the part of the formula that inflates the standard error when the independent variables are highly correlated with each other. Note that as the proportion of variation in x_1 that is accounted for by the other independent variables increases, the tolerance value decreases. The tolerance is the proportion of variance in one independent variable that is not explained by variation in the other independent variables.

The VIF statistic tells us how much the standard error of the slope is inflated due to inter-item correlation. The calculation of the VIF is based on the tolerance statistic:

$$\text{VIF}_{b_1} = \frac{1}{\text{Tolerance}_{b_1}}$$

We can calculate tolerance for `docs10k` stats by regressing it on the other independent variables. In other words, treat `docs10k` as a dependent variable and see how much of its variation is accounted for by the other variables in the model. We can get this information by running the regression model and then using the summary function to extract just the R^2 value.

```
#Use 'docs10k' as the DV to Calculate its Tolerance
fit_tol <- lm(data = countries2, docs10k ~ fert1520 + mnschool
            + lg10pop + urban)

#Use information in 'fit_tol' to create a table of results
summary(fit_tol)$r.square
[1] 0.62321
```

Note that the R^2 = .623, so 62.3% of the variation in `docs10k` is accounted for by the other independent variables. Using this information, we get the following:

$$\text{Tolerance: } 1 - .623 = .377$$
$$\text{VIF (1/tolerance): } 1/.377 = 2.65$$

So, is a VIF of 2.65 a lot? In my experience, no, but it is best to evaluate this number in the context of the VIF statistics for all variables in the model. Instead of calculating all of these ourselves, we can have R get the VIF statistics for us.

```
#Get VIF statistics using information from 'fit'
VIF(fit) #note uppercase VIF

    fert1520       mnschool  log10(pop19_M)        urban        docs10k
      2.5471         3.5018          1.0421       1.6294         2.6540
```

Does collinearity explain the marginally significant slope for `docs10k`? Yes and no. Sure, the slope for `docs10k` would probably have a smaller *p*-value if we excluded the variables that are highly correlated with it, but some loss of impact is *almost always* going to happen when using multiple regression; that's the point of using it! Also, the issue of collinearity is not appreciably worse for `docs10k` than it is for fertility rate, and it is less severe than it is for mean years of education, so it's not like this variable faces a higher hurdle than the others. Based on these results, I would conclude that collinearity is a slight problem for `docs10k`, but that there is probably a better explanation for its level of statistical significance.

While there is no magic cutoff point for determining that collinearity is a problem that needs to be addressed, in my experience, VIF values greater than 5 should get your attention, anything in the 7–10 range signals that there could be a problem worth looking into, and values greater than 10 should be taken seriously. You always have to consider issues like this in the context of the data and model you are using.

Even though collinearity is not a problem here, it is important to consider what to do if there is extreme collinearity in a regression model. Some people suggest dropping a variable, possibly the one with the highest VIF. I don't generally like to do this, but it is something to consider, especially if you have several variables that are measuring different aspects of the same concept and dropping one of them would not hurt your ability to measure an important concept. For instance, suppose this model also included hospital beds per capita and nurses per capita. These variables, along with `docs10k`, all measure different aspects of access to health care and should be strongly related to each other. If there were high VIF values for these variables, we could probably drop one or two of them to decrease collinearity and would still be able to consider access to health care as an explanation of differences in life expectancy.

Another alternative if there are multiple indicators of the same underlying concept in a model is to combine the variables into a single variable. Suppose again that this model included hospital beds per capita, nurses per capita, and `docs10k` as measures of access to health care. If there were a high level of collinearity among these variables, we could probably combine them into a single index of health care access. We did something like this in Chapter 4, where we created a single index of LGBTQ policy preferences based on outcomes from several separate LGBTQ policy questions. Indexing like this minimizes the collinearity problem without dropping any variables and also provides a broader measure of the underlying concept. There are many different ways to combine variables, but expanding on that here is a bit beyond the scope of this book.

CHECKING ON LINEARITY

The relationship between `docs10k` and life expectancy presents an opportunity to explore another important potential problem that can come up in regression analysis. Theoretically, there should be a relatively strong relationship between these two variables, not one whose statistical significance in the multiple regression model depends upon whether you are using a one- or two-tailed test. We have already explored collinearity as an explanation for the weak showing of docs10k, but that does not appear to be the primary explanation. While the bivariate correlation (r = .70) with the dependent variable was previously reported, one thing we did not do earlier was examine a simple scatterplot for this relationship. Let's do that now to see if it offers any clues.

```
#Scatterplot of "lifexp" by "docs10k"
plot(countries2$docs10k, countries2$lifexp, ylab = "Life Expectancy",
      xlab = "Doctors per 10k Population")
#Add linear regression line
abline(lm(countries2$lifexp ~ countries2$docs10k))
```

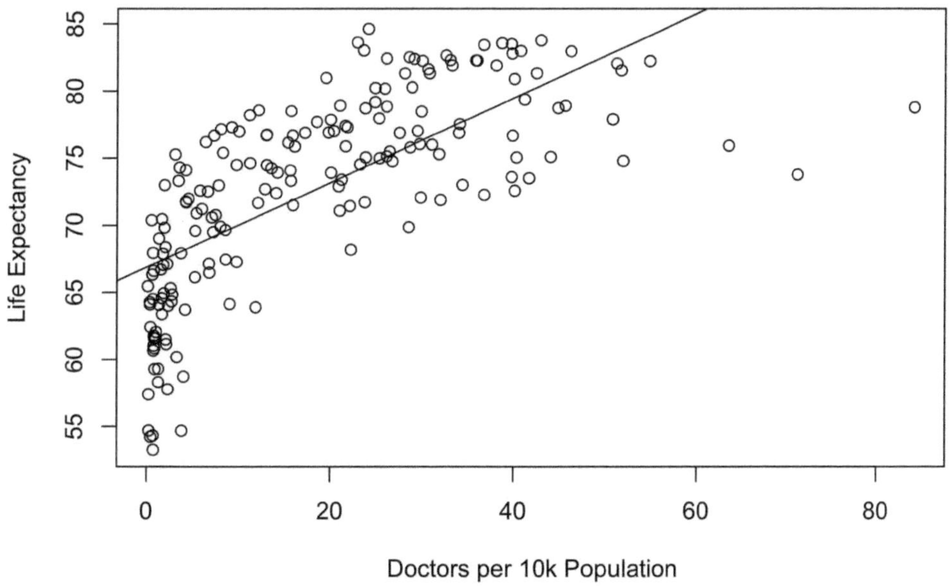

This is interesting. A rather strange-looking pattern, right? Note that the linear prediction line does not seem to fit the data in the same way we have seen in most of the other scatterplots. Focusing on the pattern in the markers, this doesn't look like a typical linear pattern. At low levels of the independent variable, there is a lot of variation in life expectancy, but, on average, it is fairly low. Increases in doctors per 10k from the lowest values to about 15 leads to substantial increases in life expectancy, but then there are diminishing returns from that point on. Looking at this plot, you can imagine that a *curved line* would fit the data points better than the existing straight line. One of the important assumptions of OLS regression is that all relationships are linear, that the expected change in the dependent variable for a unit change in the independent variable is constant across values of the independent variable. This is clearly not the case here.

Sure, you can fit a straight line to the pattern in the data, but if the pattern in the data is not linear, then the line does not fit the data as well as it could, and we are violating an important assumption of OLS regression (see next Chapter).

Just as $y = a + bx$ is the equation for a straight line, there are a number of possibilities for modeling a curved line. Based on the pattern in the scatterplot—one with diminishing returns—I would opt for the following model:

$$\hat{y} = a + b * log_{10}x$$

Here, we transform the independent variable by taking the logged values of its outcomes.

Log transformations are very common, especially when data show a curvilinear pattern or when a variable is heavily skewed. In this case, we are using a log(10) transformation. This means that all of the original values are transformed into their logged values, using a base of 10. As discussed in Chapter 5, a logged value is nothing more than the power to which you have to raise your log base (in this case, 10) in order to get the original raw score. Using logged values has two primary benefits: It minimizes the impact of extreme values, which can be important for highly skewed variables, and it enables us to model relationships as curvilinear rather than linear.

One way to assess whether the logged version of `docs10k` fits the data better than the raw version is to look at the bivariate relationships with the dependent variable using simple regression. The table of regression results shown next offers a side-by-side comparison of the two models.

```
#Life expectancy by "docs10k"
fit_raw <- lm(countries2$lifexp ~ countries2$docs10k,
             na.action = na.exclude)
#Life expectancy by "log10(docs10k)"
fit_log <- lm(countries2$lifexp ~ log10(countries2$docs10k),
             na.action = na.exclude)
#Use Stargazer to compare the results from two models
stargazer(fit_raw, fit_log, type = "text",
          dep.var.labels = c("Life Expectancy"),
          covariate.labels = c("Doctors per 10k",
                               "Log10(Doctors per 10k)"))


==========================================================================
                                          Dependent variable:
                                    ----------------------------
                                           Life Expectancy
                                        (1)              (2)
--------------------------------------------------------------------------
Doctors per 10k                          0.315***
                                          (0.024)
```

Log10(Doctors per 10k)		9.582***
		(0.492)
Constant	66.874***	63.404***
	(0.581)	(0.565)
Observations	186	186
R2	0.487	0.673
Adjusted R2	0.484	0.671
Residual Std. Error(df = 184)	5.294	4.227
F Statistic (df = 1; 184)	174.810***	378.860***
Note:		*p < 0.1; **p < 0.05; ***p < 0.01

All measures of fit, error, and relationship point to the logged (base 10) version of `docs10k` as the superior operationalization. Based on these outcomes, it is hard to make a case against using the log-transformed version of `docs10k`. You can also see this in the scatterplot in Figure 17.1, which includes the prediction lines from both the raw and logged models.

FIGURE 17.1 ■ Alternative Models for the Relationship Between Doctors per 10k and Life Expectancy

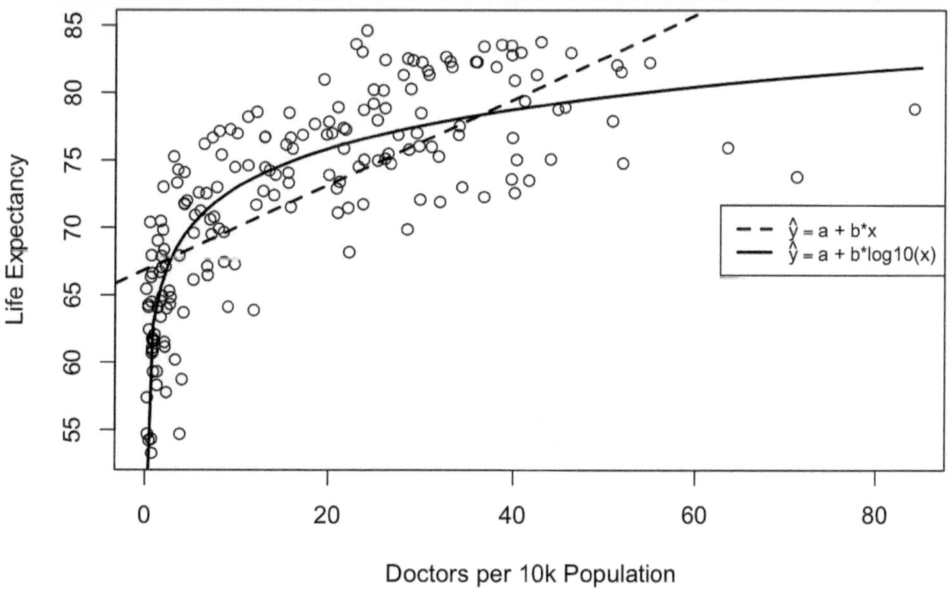

Note that the curved, solid line fits the data points much better than the straight, dashed line, which is what we expect based on the relative performance of the two regression models just examined. The interpretation of the curvilinear relationship is a bit different than for the linear relationship. Rather than concluding that life expectancy increases as doctors per capita increases, we see in the scatterplot that life expectancy increases substantially as we move from nearly 0 to about 15 doctors per 10,000 people, but increases beyond that point have less and less impact on life expectancy.

This is illustrated further in Figure 17.2 where the graph on the left shows the relationship between `docs10k` among countries below the median level (14.8) of doctors per 10,000 population, and the graph on the right shows the relationship among countries with more than the median level of doctors per 10,000 population. This is not quite as artful as Figure 17.1, with the plotted curved line, but it does demonstrate that the greatest gains in life expectancy from increased access to health care come at the low end of the scale, among countries with severely limited access. For the upper half of the distribution, there is almost no relationship between levels of `docs10k` and life expectancy.

FIGURE 17.2 ■ A Segmented View of a Curvilinear Relationship

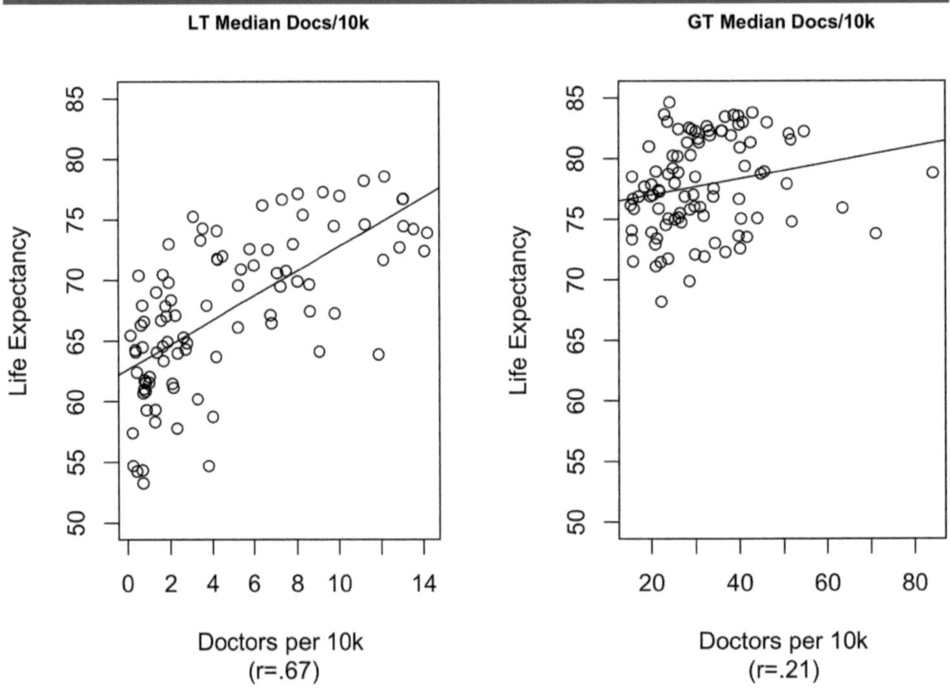

Finally, we can re-estimate the multiple regression model with the appropriate version of `docs10k` to see if this transformation has an impact on its statistical significance.

```
#Re-estimate five-variable model using "log10(docs10k)"
fit <- lm(data = countries2, lifexp ~ fert1520 + mnschool +
         log10(pop19_M) + urban +
         log10(docs10k), na.action = na.exclude)
#Using information in 'fit' to create a table of results
stargazer(fit, type = "text", dep.var.labels = c("Life Expectancy"),
          covariate.labels = c("Fertility Rate",
                               "Mean Years of Education",
                               "Log10(Population)",
                               "% Urban",
                               "log10(Doctors per 10k)"))

========================================================================
                                          Dependent variable:
                                      ---------------------------
                                            Life Expectancy
------------------------------------------------------------------------
Fertility Rate                              -2.808***
                                            (0.399)

Mean Years of Education                      0.356**
                                            (0.163)

Log10(Population)                            0.050
                                            (0.329)

% Urban                                      0.045***
                                            (0.015)

log10(Doctors per 10k)                       2.424**
                                            (0.973)

Constant                                    72.109***
                                            (2.135)

------------------------------------------------------------------------
Observations                                  179
R2                                           0.772
Adjusted R2                                  0.765
Residual Std. Error                     3.577 (df = 173)
F Statistic                         117.000*** (df = 5; 173)
========================================================================
Note:                           *p < 0.1; **p < 0.05; ***p < 0.01
```

Accounting for the curvilinear relationship between `docs10k` and life expectancy made an important difference to the model. First, we now see a statistically significant impact from the logged version of the `docs10k`, whether using a one- or two-tailed test (the *t*-score changes from 1.83 in the original model to 2.49 in the revised model). We need to be a little bit careful here when discussing the coefficient for the logged variable. Although there is a curvilinear relationship between `docs10k` and life expectancy, there is a linear relationship between `log10(docs10k)` and life expectancy. So, we can say that for every unit increase in the *logged value* of `docs10k`, life expectancy increases by 2.4 years, but we need to keep in mind that when we translate this into the impact of the original version of `docs10k`, the relationship is not linear. In addition to doctors per 10,000 population, the fertility rate, mean years of education, and percentage urban population are also statistically significant, and there is less error in predicting life expectancy when using the logged version of `docs10k`: The RMSE dropped from 3.606 to 3.58, and the R^2 increased from .768 to .772.

Stop and Think About Results

These findings underscore the importance of thinking hard about results that don't make sense. There is a strong theoretical reason to expect life expectancy to be strongly related to doctors per capita, as greater access to health care should lead to better health outcomes. However, the initial regression results suggested that doctors per capita was tangentially related to life expectancy. Had we simply accepted this result, we would have come to an erroneous conclusion. Instead, by exploring explanations for the null finding, we were able to uncover a more nuanced and powerful role for the number of doctors per 10k population.

Of course, it is important to point out that one of the sources of "getting it wrong" in the first place was skipping a very important first step: examining the bivariate relationship closely, using not just the correlation coefficient but also the scatterplot. Had I started there, I would have understood the curvilinear nature of the relationship from the get-go.

WHICH VARIABLES HAVE THE GREATEST IMPACT?

As a researcher, one thing you're likely to be interested in is the relative impact of the independent variables; that is, which variables have the greatest and least impact on the dependent variable? If we look at the regression slopes, we might conclude that the order of importance runs from highest-to-lowest in the following order: The fertility rate (b = -2.81) and log of doctors per 10k (b = 2.42) are the most important and of similar magnitude, followed by education level in distant third (b = .36), followed by log of population (b = .05) and percentage urban (b = .045), neither one of which seem to matter very much. But this rank ordering is likely flawed due to differences in the way the variables are measured. A tip-off that something is wrong with this type of comparison is found in the fact that the slopes for urban population and the log of population size are of about the same magnitude, even though the slope for urban population is statistically significant ($p < .01$), while the slope for population size is not and has not been significant in any of the analyses shown thus far. How can a variable whose effect is not distinguishable from zero have an impact equal to one that has been statistically significant in all of the examples shown thus far? Seems unlikely.

Generally, the "raw" regression coefficients are not a good guide to the relative impact of the independent variables because the variables are measured on different scales. Recall that the regression slopes tell us how much y is expected to change for a unit change in x. The problem we encounter when comparing these slopes is that the units of measure for the independent variables are mostly different from each other, making it very difficult to compare the "unit changes" in a fair manner.

To gain a sense of the impact of measurement scale on the size of the regression slope, consider the two plots in Figure 17.3. Both plots show the relationship between the size of the urban population and country-level life expectancy. The plot on the left uses the *percentage* of a country's population living in urban areas, and the plot on the right uses the *proportion* of a country's population living in urban areas. These are the same variables, just using different scales. As you can see, other than the metric on the horizontal axis, the plots are identical. In both cases, there is a moderately strong positive relationship between the size of the urban population and life expectancy: As the urban population increases in size, life expectancy also increases. But look at the linear equations that summarize the results of the bivariate regression models. The slope for the urban population variable in the second model is 100 times the size of the slope in the first model, not because the impact is 100 times greater but because of the difference in the scale of the two variables. This is the problem we encounter when comparing regression coefficients across independent variables measured on different scales.

FIGURE 17.3 ■ The Impact of Scale of the Independent Variable on the Size of Regression Coefficients

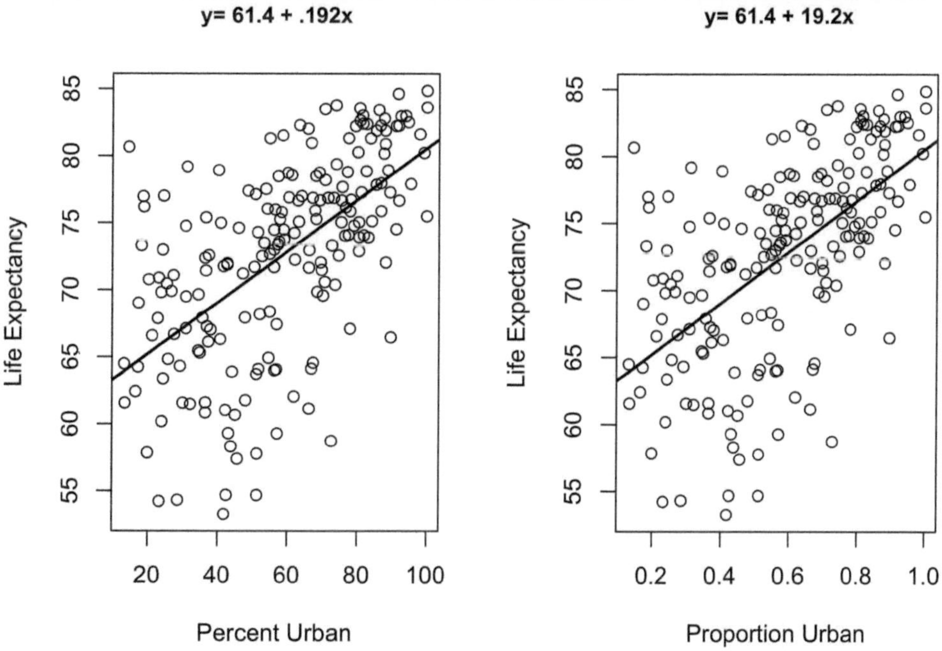

What we need to do is standardize the variables used in the regression model so they are all measured on the same scale. Variables with a common metric (same mean and standard deviation) will produce slopes that are directly comparable. One common way to standardize variables is to transform the raw scores into *z*-scores:

$$Z_i = \frac{x_i - \overline{x}}{S}$$

All variables transformed in this manner will have $\overline{x} = 0$ and $S = 1$. If we do this to all variables (including the dependent variable), we can get the *standardized* regression coefficients (sometimes confusingly referred to as "Beta Weights"). Fortunately, we don't have to make these conversions ourselves since R provides the standardized coefficients. In order to get this information, we can use the `StdCoef` command from the `DescTools` package. The command is then simply `StdCoef(fit)`, where "fit" is the object with the stored information from the linear model.

```
#Produce standardized regression coefficients from 'fit'
StdCoef(fit)

                  Estimate* Std. Error* df
(Intercept)       0.0000000    0.000000 173
fert1520         -0.4809160    0.068382 173
mnschool          0.1501100    0.068652 173
log10(pop19_M)    0.0056696    0.037140 173
urban             0.1390776    0.047084 173
log10(docs10k)    0.2061025    0.082729 173
attr(, "class")
[1] "coefTable" "matrix"
```

The standardized coefficients are in the "Estimate" column. To evaluate relative importance, focus on and compare the absolute values. Based on the standardized coefficients, the ordering of variables from most-to-least in importance is a bit different than when we used the unstandardized coefficients: Fertility rate (−.48) far outstrips all other variables, having more than twice the impact as doctors per 10,000 (.21), which is followed pretty closely by level of education (.15) and urban population (.14), and population size comes in last (.006). These results give a somewhat different take on relative impact than if we relied on the raw coefficients.

We can also make more literal interpretations of these slopes. Since the dependent and independent variables are all transformed into *z*-scores, a "one unit change" in *x* is always a one standard deviation change in *x*. This means that we can interpret the standardized slopes as telling how many standard deviations *y* is expected to change for a one standard deviation change in *x*. For instance, the standardized slope for fertility (−.48) means that for every one standard deviation increase in the fertility rate, we can expect to see a .48 standard deviation decline in life expectancy. Just to be clear, though, the primary utility of these standardized coefficients is to compare the relative impact of the independent variables.

STATISTICS VERSUS SUBSTANCE

The model we've been looking at explains about 77% of the variation in country-level life expectancy using five independent variables. This is a good model fit, but it is possible to account for a lot more variance with even fewer variables. Take, for example, the alternative model that follows, which explains 99% of the variance using just two independent variables, male life expectancy and infant mortality.

```
#Alternative model
fit1 <- lm(countries2$lifexp ~ countries2$malexp + countries2$infant_mort)
#Table with Results
stargazer(fit1, type = "text",
          dep.var.labels = c("Life Expectancy"),
          covariate.labels = c("Male Life Expectancy",
                               "Infant Mortality"))

========================================================================
                                         Dependent variable:
                                     ---------------------------
                                           Life Expectancy
------------------------------------------------------------------------
Male Life Expectancy                          0.810***
                                              (0.018)

Infant Mortality                             -0.081***
                                              (0.007)

Constant                                     17.423***
                                              (1.413)

------------------------------------------------------------------------
Observations                                    184
R2                                             0.989
Adjusted R2                                    0.989
Residual Std. Error                      0.787 (df = 181)
F Statistic                        8,098.000*** (df = 2; 181)
========================================================================
Note:                                *p < 0.1; **p < 0.05; ***p < 0.01
```

We had been feeling pretty good about the model that explains 77% of the variance in life expectancy, and then along comes this much simpler model that explains almost 100% of the variation. Not that it's a contest, but this does sort of raise the question, "Which model is best?"

The new model definitely has the edge in measures of error: The RMSE is just .787, compared to 3.58 in the five-variable model and, as mentioned previously, the $R^2 = .99$.

So, which model is best? Of course, this is a trick question. The second model provides a stronger statistical fit but at significant theoretical and substantive costs. Put another way, the second model explains more variance without providing a useful or interesting substantive explanation of the dependent variable. In effect, the second model is explaining life expectancy with other measures of life expectancy. Not very interesting. Think about it like this. Suppose you are working as an intern at the World Health Organization and your supervisor asks you to give them a presentation on the factors that influence life expectancy around the world. I don't think they would be impressed if you came back to them and said you had two main findings:

1. People live longer in countries where men live longer.
2. People live longer in countries where fewer people die really young.

Even with your impressive R^2 and RMSE, your supervisor would likely tell you to go back to your desk, put on your thinking cap, and come up with a model that provides a stronger substantive explanation, something more along the lines of the original model. With the results of the original model, you can report back that the factors that are most strongly related to life expectancy across countries are fertility rates, access to health care, and education levels—things that policies and aid programs might be able to affect. The share of the population living in urban areas is also related to life expectancy, but this is not something that can be addressed as easily through policies and aid programs.

NEXT STEPS

By now, you should be feeling comfortable with regression analysis and, hopefully, seeing how it can be applied to many different types of data problems. The last step in this journey of discovery is to examine some of the important assumptions that underlie OLS regression. Chapter 18 provides a brief overview of assumptions that need to be satisfied in order to be confident in the results of the life expectancy model that has been the focus of the last three chapters. As you will see, in some ways the model tested in this chapter does a good job of satisfying the assumptions, but in other ways, the model still needs a bit of work. It is important to be familiar with these assumptions, not just so you can think about them in the context of your own work but also so you can use this information to evaluate the work of others.

EXERCISES

Concepts and Calculations

1. The two correlation matrices that follow illustrate the pattern of relationships among variables for two different datasets, A and B.

A					B				
	Y	X_1	X_2	X_3		Y	X_1	X_2	X_3
Y	1.0	0.55	0.32	−0.75	Y	1.0	0.48	0.39	−0.79
X_1	0.55	1.0	0.25	−0.43	X_1	0.48	1.0	0.30	−0.85
X_2	0.32	0.25	1.0	0.02	X_2	0.39	0.30	1.0	−0.58
X_3	−0.75	−0.43	0.02	1.0	X_3	−0.79	−0.85	−0.58	1.0

- In which dataset do you think multicollinearity is most likely to present a problem? Explain your answer.
- For which independent variable in the dataset you chose is collinearity likely to present a problem? Why?
- What test would you advise using to get a better sense of the overall level of collinearity?

2. In which of the following scatterplots does it appear that the relationship between X and Y is not linear? Describe the pattern in that scatterplot. What course of action do you recommend to correct this issue?

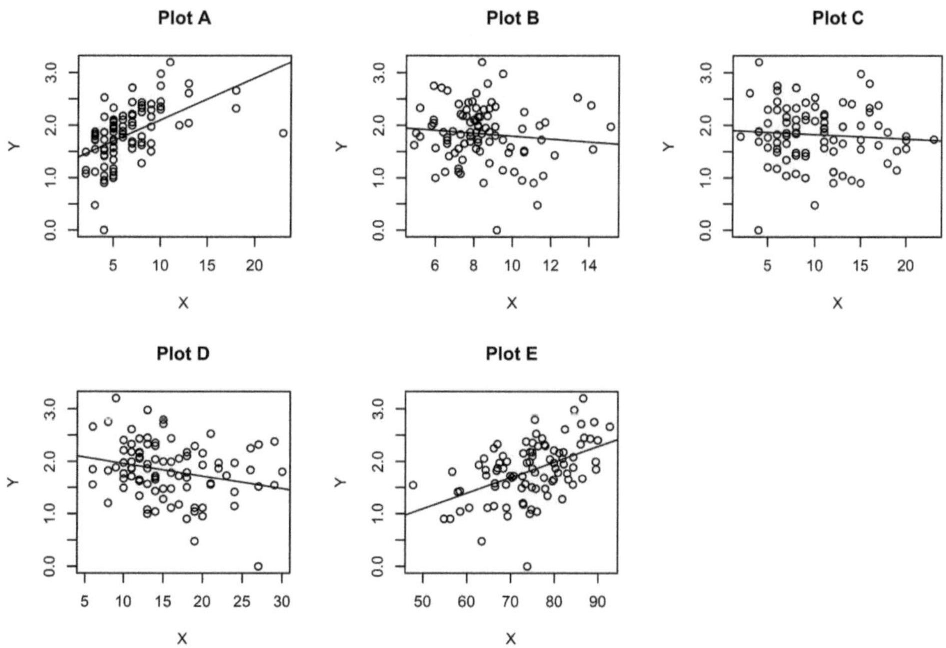

3. The regression output copied in the following table summarizes the impact and statistical significance of four independent variables (X_1, X_2, X_3, and X_4) on a dependent variable (Y).

- Rank-order the independent variables from strongest to weakest in terms of their relative impact on the dependent variable. Explain the basis for your ranking.
- Provide an interpretation of both the *b* value and the standardized coefficient for the variable that you have identified as having the greatest impact on the dependent variable.

	b	*t*-score	Standardized Coefficient
Constant	−13.58	6.05**	
X_1	0.79	2.45**	0.35
X_2	−3.55	1.98**	−0.17
X_3	3.78	2.78**	0.31
X_4	−0.15	3.52**	−0.43

**$p < .05$

R Problems

Building on the regression models from the exercises in the last chapter, use the `states20` dataset and run a multiple regression model with infant mortality (`infant_mort`) as the dependent variable and per capita income (`PCincome2020`), the teen birth rate (`teenbirth`), percentage of low birth weight births (`lowbirthwt`), and Southern region (`south`) as the independent variables.

1. Produce a publication-ready table using `stargazer`. Make sure you replace the R variable names with meaningful labels (see Chapter 15 for instructions on doing this using `stargazer`).
   - Summarize the results of this model. Make sure to address matters related to the slopes of the independent variables and how well the group of variables does in accounting for state-to-state differences in infant mortality.
   - Generate the standardized regression coefficients and discuss the results. Any surprises, given the findings from the unstandardized coefficients? Make it clear that you understand why you need to run this test and what these results are telling you.
   - Run a test for evidence of multicollinearity. Discuss the results. Any surprises? Make it clear that you understand why you need to run this test and what these results are telling you.
2. Use the code listed next to take a sample of 500 counties from `counties20large` and save the sample to a new object named `county500`.

```
set.seed(1234)
#create a sample of 500 rows of data from the county20large data set
county500 <- sample_n(county20large, 500)
```

- Choose one of the following to use as a dependent variable: `vcrim100k`, `gundeath100k`, `uninsured`, and `cvap_turnout20`. Then choose four independent variables that you expect to be related to the dependent variable you selected. Justify your selection and be clear about your expectations. Consult the codebook for the `counties20large` dataset to make sure you understand what these variables are.
- Summarize the results of this model. Make sure to address matters related to the slopes of the independent variables and how well the group of variables does in accounting for state-to-state differences in infant mortality.
- Generate the standardized regression coefficients and discuss the results. Any surprises, given the findings from the unstandardized coefficients? Make it clear that you understand why you need to run this test and what these results are telling you.
- Run a test for evidence of multicollinearity. Discuss the results. Any surprises? Make it clear that you understand why you need to run this test and what these results are telling you.

18 REGRESSION ASSUMPTIONS

GET READY

This chapter explores a set of regression assumptions in the context of the life expectancy regression model developed in Chapter 17. To follow along in R, make sure to load the `countries2` dataset, as well as the libraries for the following packages: `DescTools`, `lmtest`, `sandwich`, and `stargazer`.

REGRESSION ASSUMPTIONS

The level of confidence we can have in regression estimates (slopes and significance levels) depends upon the extent to which certain assumptions are met. These assumptions are outlined briefly in the next section and applied to the life expectancy model, using diagnostic tests when necessary. Several of these assumptions have to do with characteristics of the error term in the population (ϵ_i), which, of course, we cannot observe, so we evaluate them using the sample residuals (e_i) instead. Some of these assumptions have been discussed in some detail in earlier chapters, while others have not.

Linearity

The pattern of the relationship between x and y is **linear**; that is, the rate of change in y for a given change in x is constant across all values of x, and the model for a straight line fits the data best. The example of the curvilinear relationship between doctors per 10,000 and life expectancy shown in Chapter 17 is a perfect illustration of the problem that occurs when a linear model is applied to a non-linear relationship: The straight line is not the best-fitting line, and the model will fall short of producing the least squared error possible for a given set of variables. The best way to get a sense of whether you should test a curvilinear model is to examine the bivariate relationships using scatterplots. This is how we sniffed out the curvilinear relationship for `docs10k` in Chapter 17. The scatterplots for the relationships between life expectancy and all five of the original independent variables are presented in Figure 18.1.

Based on a visual inspection of these scatterplots, `docs10k` is the only variable that appears to violate the linearity assumption. Since this was addressed by using the log (base 10) of its values, no other changes to the model appear to be necessary, at least due to non-linearity.

FIGURE 18.1 ■ Scatterplots Can Be Used to Evaluate Linearity

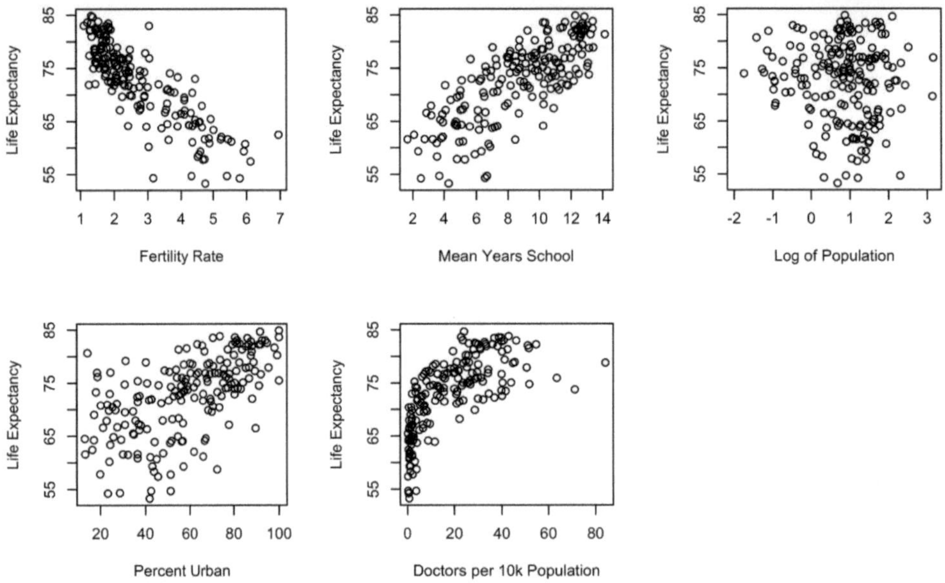

Independent Variables Are Not Correlated With the Error Term

There is no correlation between the error term and the independent variables. This assumption is a bit difficult to understand, at least when stated this way. The real concern here is *omitted variable bias*. The idea is that the error term reflects the unknown and unspecified variables not included in the model, and if those unknown and unspecified variables are strongly related to one of the variables in the model, the slope for that variable is probably biased. Experimental data generally don't have to worry about this assumption, since outcomes on the independent variables are randomly assigned and are unrelated to other potential influences. However, violation of this assumption is a potentially important issue for the type of observational data used in this book. Hence, we need to think seriously about other potential influences on the dependent variable and include them in the model if possible.

Take a look at the other variables in the `countries2` dataset to see if you think there is something that might be strongly related to country-level life expectancy, taking care to avoid variables that are really just other measures of life expectancy. Do you notice any variables whose exclusion might bias the findings for the other variables? One that comes to mind is `food_def`, a measure of the average daily supply of calories as a percentage of the amount of calories needed to eliminate hunger. It makes sense that nutrition plays a role in life expectancy and also that this variable could be related to other variables in the model. If so, then the other coefficients may be biased if `food_def` is not in the model. The scatterplot and correlation (r = .59) both suggest that taking into account this nutritional measure might make an important contribution to the model.

```
#Scatterplot of "lifexp" by "food_def"
plot(countries2$food_def, countries2$lifexp)
#Add regression line to scatterplot
abline(lm(countries2$lifexp ~ countries2$food_def))
```

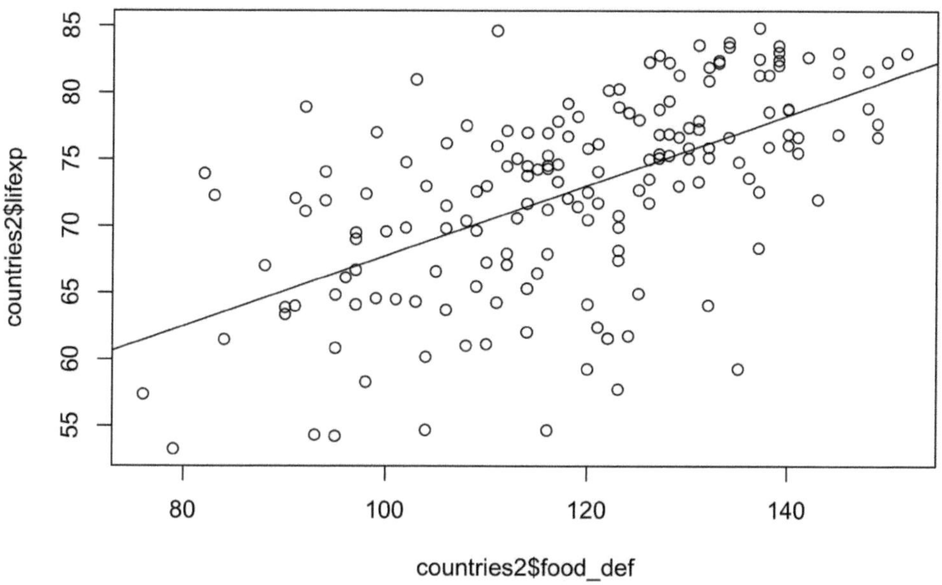


```
#Correlation between "lifexp" and "food_def"
cor.test(countries2$food_def, countries2$lifexp)

     Pearson's product-moment correlation

data: countries2$food_def and countries2$lifexp
t = 9.51, df = 172, p-value < 2e-16
alternative hypothesis: true correlation is not equal to 0
95 percent confidence interval:
 0.48008 0.67662
sample estimates:
   cor
0.58693
```

Let's see how the model is affected when `food_def` is added.

```
#Five-variable model
fit <- lm(countries2$lifexp ~ countries2$fert1520+
          countries2$mnschool + log10(countries2$pop19_M) +
          countries2$urban + log10(countries2$docs10k))
```

```
#Adding" food deficit" to the model
fit2 <- lm(countries2$lifexp ~ countries2$fert1520 +
             countries2$mnschool + log10(countries2$pop19_M) +
             countries2$urban + log10(countries2$docs10k)
           + countries2$food_def)
#Produce a table with both models included
stargazer(fit, fit2, type = "text",
          title = "The Impact of Adding Food Deficit to the Model",
          dep.var.labels = c("Life Expectancy"),
          covariate.labels = c("Fertility Rate",
                               "Mean Years of Education",
                               "Log10 Population", "%Urban",
                               "Log10 Doctors/10k",
                               "Food Deficit"))

The Impact of Adding Food Deficit to the Model
========================================================================
                                        Dependent variable:
                            --------------------------------------------
                                          Life Expectancy
                                      (1)                     (2)
------------------------------------------------------------------------
Fertility Rate                     -2.808***               -2.677***
                                    (0.399)                 (0.409)

Mean Years of Education             0.356**                  0.232
                                    (0.163)                 (0.167)

Log10 Population                     0.050                  -0.281
                                    (0.329)                 (0.341)

% Urban                             0.045***                 0.028*
                                    (0.015)                 (0.016)

Log10 Doctors/10k                   2.424**                 2.394**
                                    (0.973)                 (0.968)

Food Deficit                                                0.093***
                                                            (0.021)
```

```
Constant                        72.109***              63.084***
                                 (2.135)                (2.952)

-----------------------------------------------------------------------
Observations                       179                    167
R2                                0.772                  0.790
Adjusted R2                       0.765                  0.782
Residual Std. Error          3.577(df = 173)        3.434(df = 160)
F Statistic                117.000***(df = 5;173)100.150***(df = 6;160)
=======================================================================
Note:                                  *p < 0.1;**p < 0.05;***p < 0.01
```

Food deficit is significantly related to life expectancy, and adding it to the model had a couple of important consequences: The coefficient for mean years of education is now not statistically significant, the slope for percentage urban was cut almost in half, and the model fit (adjusted R2) increased from .765 to .782. Taking into account this measure of nutritional health showed that the original estimates for mean years of education and percentage urban were biased upward.

The tricky part of satisfying this assumption is to avoid "kitchen sink" models, where you just toss a bunch of variables in to see what works. The best approach is to think hard about which sorts of variables should be taken into account and then see if you have measures of those variables that you can use.

No Perfect Multicollinearity

This assumption is that none of the independent variables is a perfect linear combination of the others. At some level, this is a non-issue when analyzing data, as R and other programs will automatically drop a variable from the model if it is perfectly collinear with one or more other variables. Still, we know from the discussion in Chapter 17 that there is almost always some level of collinearity in a multiple regression model and it can create problems if the level is very high. We know from Chapter 17 that the problem with collinearity is that it inflates the slope standard errors, resulting in suppressed *t*-scores and *p*-values. Because we have added `food_def` to the model since the initial round of collinearity testing in Chapter 17, we should take a look at the VIF statistics for the final model.

```
#Check VIF statistics for new model
VIF(fit2)

  countries2$fert1520      countries2$mnschool  log10(countries2$pop19_M)
               3.6349                   3.8434                     1.0806
    countries2$urban log10(countries2$docs10k)        countries2$food_def
               1.8400                   5.1813                     1.5925
```

The variable with the highest VIF value is `docs10k` (VIF = 5.18), which is still statistically significant in the model, so I would say there is no major problem with collinearity. As mentioned in Chapter 17, there is no universally accepted cutoff point for what constitutes too-high collinearity. One thing to keep in mind is that the impact of collinearity on levels of statistical significance is more acute for small rather than large samples. This is because as the sample size increases, standard errors decrease and *t*-scores increase (Chapter 8). As a result, even though collinearity still leads to decreased *t*-scores in large samples, there is a smaller chance that the reduced size of the *t*-score will affect the hypothesis-testing decision, compared to small samples.

The Mean of the Error Term Equals Zero

If the expected (average) value of the error term does not equal zero, then it is likely that the constant is biased. This is easy enough to test using the sample residuals.

```
#Get the mean prediction error
mean(residuals(fit2), na.rm = T)
[1] 2.8075e-17
```

The mean error in the regression model is .0000000000000000281, pretty darn close to 0.

If you find that the mean of the error term is different from zero, it could be due to specification error, so you should look again at whether you need to transform any of your existing variables or incorporate other variables into the model.

The Error Term Is Normally Distributed

This assumption is important to having confidence in estimates of statistical significance. In practice, when working with a large sample, researchers tend not to worry much about the normality of the error term unless the deviation from normality is quite severe. There are a number of different ways to check on this, including plotting the residuals in a histogram to see if they *appear* to be normally distributed.

```
#Get histogram of stored residuals from fit2
hist(residuals(fit2), prob = T, ylim = c(0, .12))

#Create normal curve using the mean and standard deviation
curve(dnorm(x, mean = mean(residuals(fit2)),
            sd = sd(residuals(fit2))), add = T)
#Add legend
legend("topright", legend = c("Normal Curve"), lty = 1, cex = .7)
```

The contours of this histogram appear to be approaching a normal distribution, but it is hard to tell if the pattern is similar enough to normally distributed that we can be confident that

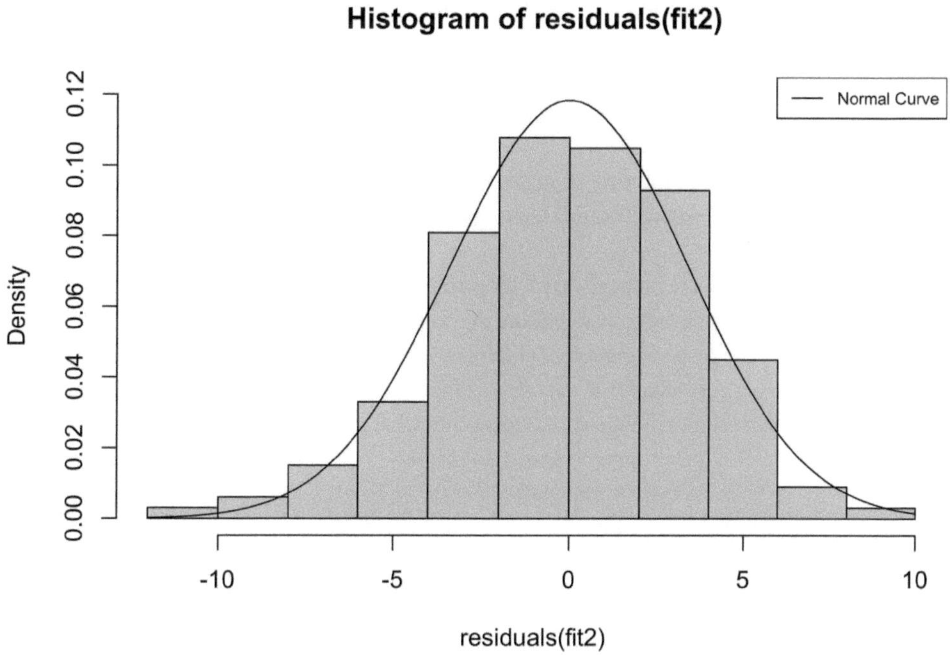

we are not violating the normality assumption. Instead, we can use the Shapiro-Wilk normality test to test the null hypothesis that the residuals from this model are normally distributed. If the *p*-value generated from this test is less than .05, then we reject the null hypothesis and conclude that the residuals are not normally distributed. The results of this test (`shapiro.test`) are presented next.

```
#Test for normality of fit2 residuals
shapiro.test(residuals(fit2))

        Shapiro-Wilk normality test

data: residuals(fit2)
W = 0.99, p-value = 0.29
```

The *p*-value is greater than .05, so we fail to reject the null hypothesis and conclude that the distribution of residuals from this model do not deviate significantly from a normal distribution.

If the distribution of the error term does deviate significantly from normal, it could be the result of outliers or specification error (e.g., omitted variable or non-linear relationship).

Constant Error Variance (Homoscedasticity)

The variance in prediction error (residuals) is the same across all values of the independent variables. In other words, if you predict *y* with *x*, it is expected that the amount of prediction error (spread of data points around the regression line) should be roughly the same as you move from low to high values of x; that the model does not do a better job of predicting y in some ranges of x than in others. This assumption is important because the estimates of statistical significance are based on the amount of error in prediction. If that error varies significantly across observations (*heteroscedasticity*), then you could be over- or underestimating statistical significance. Importantly, violating this assumption does not mean that the regression slopes are biased, just that the reported standard errors of the slopes and, hence, t-scores are likely to be incorrect.

A simple way to evaluate homoscedasticity is to examine a scatterplot of the model residuals across the range of predicted outcomes, looking for evidence of substantial differences in the spread of the residuals over the range of outcomes. The classic pattern for a violation of this assumption is a cone-like pattern in the residuals with a narrow range of error at one end and a wider range at the other. The graph that follows in Figure 18.2 illustrates what this might look like.

FIGURE 18.2 ■ A Non-constant Error Pattern

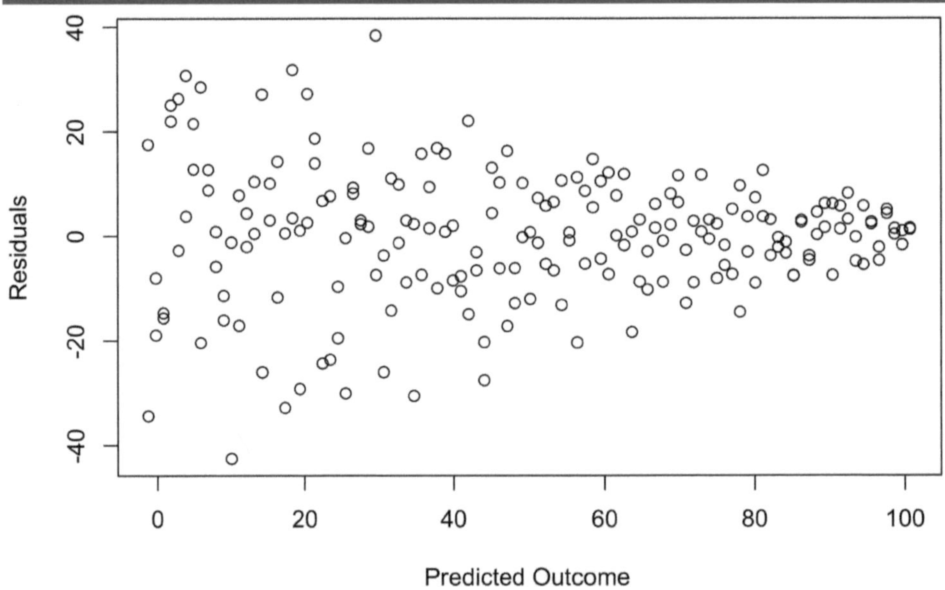

As you can see, the spread in the residuals (errors) is fairly wide for low predicted outcomes and narrows steadily as the predicted outcome increase in value. This is a pretty clear violation of the assumption.

Turning to the current example, using the results from the life expectancy model, the pattern is a bit less clear (as is often the case).

```
#Plot residuals from model as a function of predicted values
plot(predict(fit2), residuals(fit2))
```

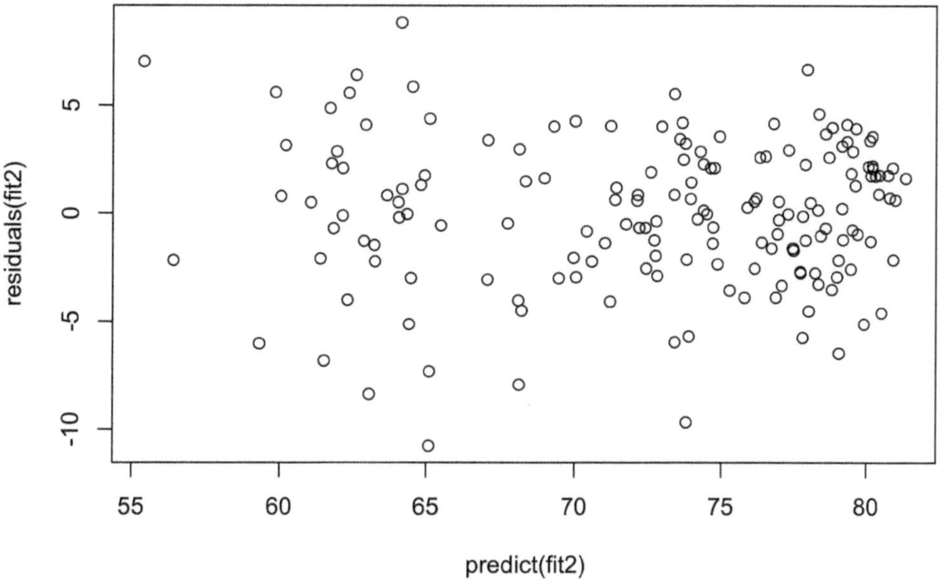

It looks like there is a slight tendency for greater variation in error at the low end of the predicted outcomes than at the high end, but it is hard to tell, based on simply eyeballing the data, if this is a problem. Instead, we can test the null hypothesis that variance in error is constant, using the Breusch-Pagan test (`bptest`) from the `lmtest` package.

```
bptest(fit2)


    studentized Breusch-Pagan test


data:  fit2
BP = 20.4, df = 6, p-value = 0.0023
```

With a *p*-value of .002, we reject the null hypothesis that there is constant variance in the error term. So, what to do about this violation of the homoscedasticity assumption? Most of the solutions, including those discussed next, are a bit beyond the scope of an introductory textbook. Nevertheless, I'll illustrate a set of commands that can be used to address this issue.

There are a number of ways to address heteroscedasticity, mostly by weighting the observations based on the size of their error. This approach, called **weighted least squares**, entails determining if a particular variable (x) is the source of the heteroscedasticity and then weighting the observations by the reciprocal of that variable ($1/x$) or transforming the residuals and weighting the data by the reciprocal of that transformation. Alternatively, you can use the `vcovHC` function, which adjusts the standard error estimates so they account for the non-constant variance in the error. Then, we can incorporate information created by `vcovHC` into the `stargazer` command to produce a table with new standard errors, *t*-scores, and *p*-values. The code that follows shows how to do this.[1]

- First, use the results of the regression model (`fit2`) and method HC1 in `vcovHC` function and store the new covariance matrix in object `hc1`.

```
hc1 <- vcovHC(fit2, type = "HC1")
```

- Then, save the new "robust" standard errors (taken from `hc1`) to a new object.

```
robust.se <- sqrt(diag(hc1))
```

- Finally, integrate the new standard errors into the regression output using `stargazer`. Note that the model (`fit2`) is listed twice, so a comparison can be made between the models with and without the standard error corrections. If you only want to show the corrected model, just list `fit2` once and change `se = list(NULL, robust.se)` to `se = list(robust.se)`.

```
stargazer(fit2, fit2, type = "text", se = list(NULL, robust.se),
          column.labels = c("Original", "Corrected"),
          dep.var.labels = c("Life Expectancy"),
          covariate.labels = c("Fertility Rate",
                               "Mean Years of Education",
                               "Log10 Population", "% Urban",
                               "Log10 Doctors/10k",
                               "Food Deficit"))
```

```
==========================================================================
                                       Dependent variable:
                                   ----------------------------
                                         Life Expectancy
                                   Original              Corrected
                                     (1)                    (2)
--------------------------------------------------------------------------
Fertility Rate                     -2.677***             -2.677***
                                   (0.409)               (0.492)

Mean Years of Education             0.232                 0.232
                                   (0.167)               (0.163)

Log10 Population                   -0.281                -0.281
                                   (0.341)               (0.339)
```

% Urban	0.028*	0.028*
	(0.016)	(0.016)
Log10 Doctors/10k	2.394**	2.394*
	(0.968)	(1.254)
Food Deficit	0.093***	0.093***
	(0.021)	(0.020)
Constant	63.084***	63.084***
	(2.952)	(3.451)
Observations	167	167
R2	0.790	0.790
Adjusted R2	0.782	0.782
Residual Std. Error (df = 160)	3.434	3.434
F Statistic (df = 6;160)	100.150***	100.150***
Note:		*p < 0.1; **p < 0.05; ***p < 0.01

Two important things to note about these new estimates. First, as expected, the standard errors changed at least slightly for virtually all variables. Most of the changes were quite small and had little impact on the *p*-values. The most substantial change occurred for docs10k, whose slope is still statistically significant using a one-tailed test ($t = 1.91$). The other thing to note is that this transformation had no impact on the slope estimates or the model fit statistics, reinforcing the fact that heteroscedasticity does not lead to biased slopes but does render standard errors unreliable.

Independent Errors

The errors produced by different values of the independent variables are unrelated to each other; that is, the errors from one observation do not affect the errors of other observations. One form of correlated errors is referred to as *serial correlation* and is often a problem for time-series data, which we have not worked with much at all in this book. Some academic areas, such as economics and certain subfields of political science, rely heavily on time series data, and scholars in those areas need to examine their models closely in search of serial correlation. When serial correlation is present, it leads to suppressed standard errors, which creates inflated *t*-scores and can lead researchers to falsely conclude that some relationships are statistically significant when they are not (Type I error).

NEXT STEPS

You've made it through the book! Give yourself a pat on the back. You've covered a lot of material and have probably done so during a single semester while taking other courses. Think back to the beginning of the book when you were first learning about a programming language called R and just dipping a toe into data analysis with simple frequency tables. Those were important things to learn, but you've picked up a lot more since then.

There is no "Next Step" from here in this book, but there are some things you can do to build on what you learned. First and foremost, don't let what you've learned sit on the shelf very long. It's not like there is an expiration date on what you've learned, but it will start to fade and be less useful if you don't find ways to use it. The most obvious thing you can do is follow up with another course. One type of course that would be particularly useful is one that focuses primarily on regression analysis and its extensions. While the last four chapters in this book focused on regression analysis, they really just scratched the surface. Taking a stand-alone course on regression analysis will reinforce what you've learned and provide you with a stronger base and a deeper understanding of this important method. Be careful, though, to make sure you find a course that is pitched at the right level for you. For instance, if you felt a bit stretched by some of the technical aspects of this book, you should not jump right into a regression course offered in the math or statistics department. Instead, a regression-based course in the social sciences might be more appropriate for you. But if you are comfortable with a more math-stats orientation, then by all means find a regression course in the math or statistics department at your school.

There are other potentially fruitful follow-up courses you might consider. If you were particularly taken with the parts of this book that focused on survey analysis using the 2020 ANES data, you should look for a course on survey research. These courses are usually found in political science, sociology, or mass communications departments. At the same time, if you found the couple of very brief discussions of experimental research interesting, you could follow up with a course on analyzing experimental data. The natural sciences are a "natural"[2] place to find these courses, but if your tastes run more toward the social sciences, you should check out the psychology department on your campus for these courses.

Since you've already invested some valuable time learning how to use R, you should also find ways to build on that part of the course. If you take another data analysis course, try to find one that uses R so you can continue to improve your programming skills. This brings me to one of the most important things you should look for in other data analysis courses: Make sure the course requires students to do "hands-on" data analysis problems utilizing real-world data. It would be great if the course also required the use of R, but one of the most important things is that you are required to use some type of data analysis program (e.g., R, SPSS, Stata, Minitab, Excel, etc.) to do independent data analysis. Absent this, you might *learn about* data analysis, but you will not *learn to do* data analysis.

EXERCISES

Concepts and Calculations

1. Do any regression assumptions appear to be violated in this scatterplot? If so, which one? Explain your answer. If you think an assumption was violated, discuss what statistical test you would use to confirm your suspicion.

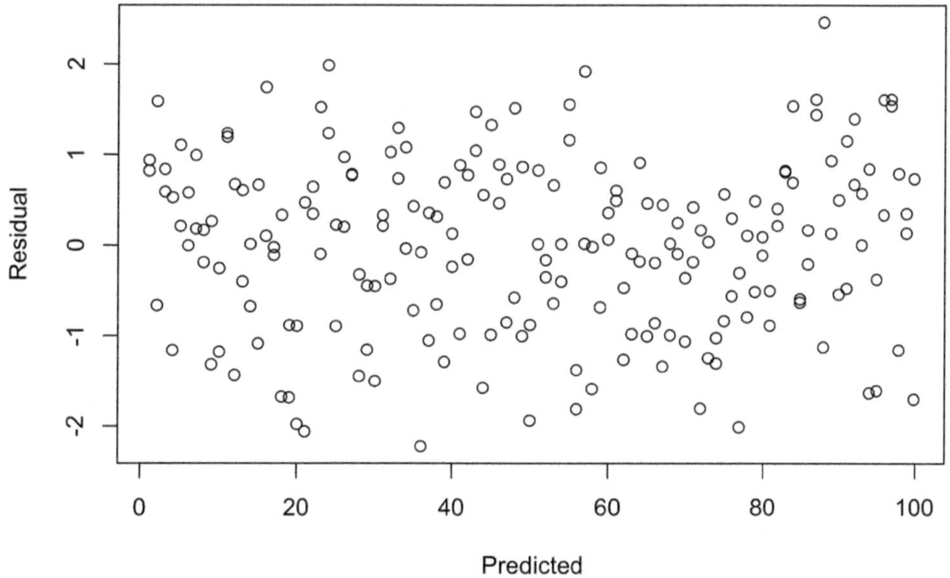

2. Do any regression assumptions appear to be violated in this histogram? If so, which one? Explain your answer. If you think an assumption was violated, discuss what statistical test you would use to confirm your suspicion.

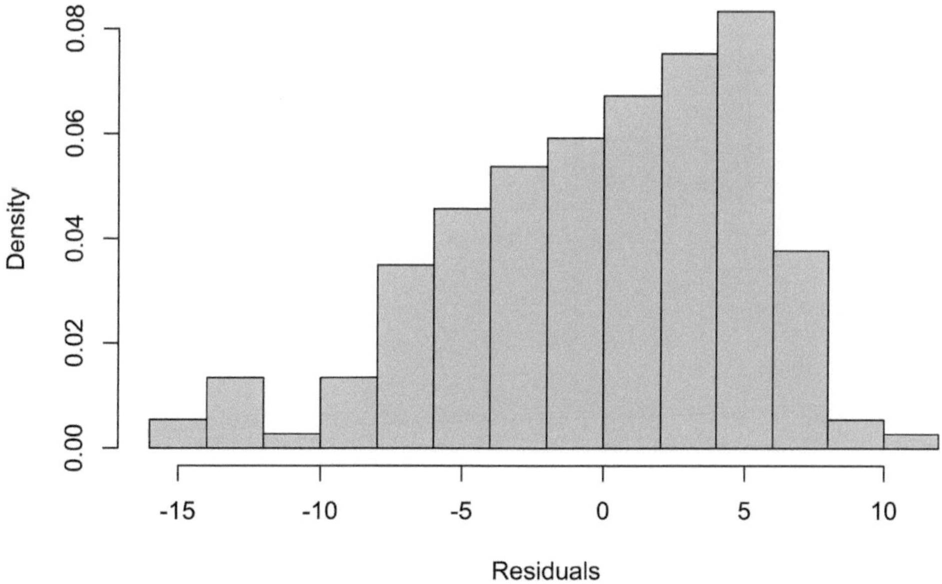

3. Do any regression assumptions appear to be violated in this histogram? If so, which one? Explain your answer. If you think an assumption was violated, discuss what statistical test you would use to confirm your suspicion.

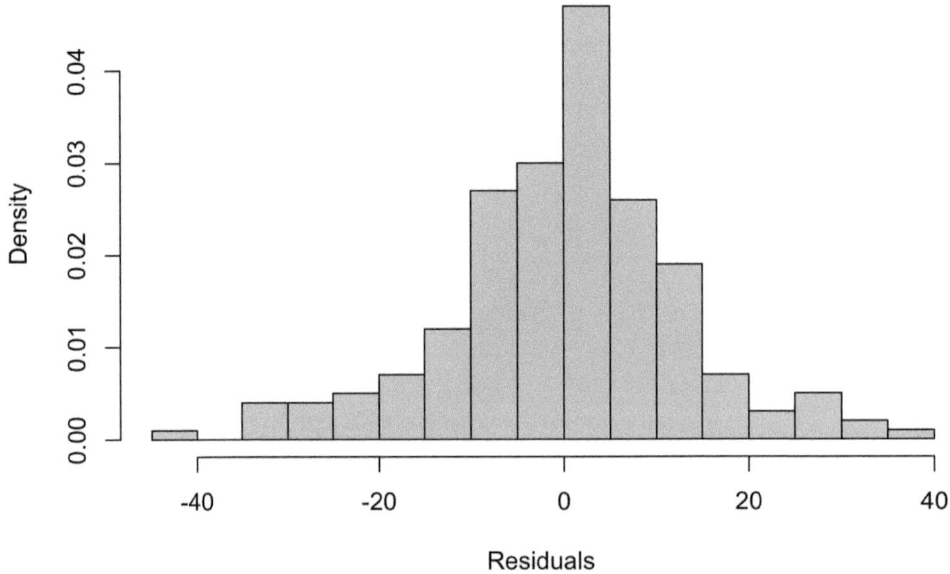

4. Do any regression assumptions appear to be violated in this scatterplot? If so, which one? Explain your answer. If you think an assumption was violated, discuss what statistical test you would use to confirm your suspicion.

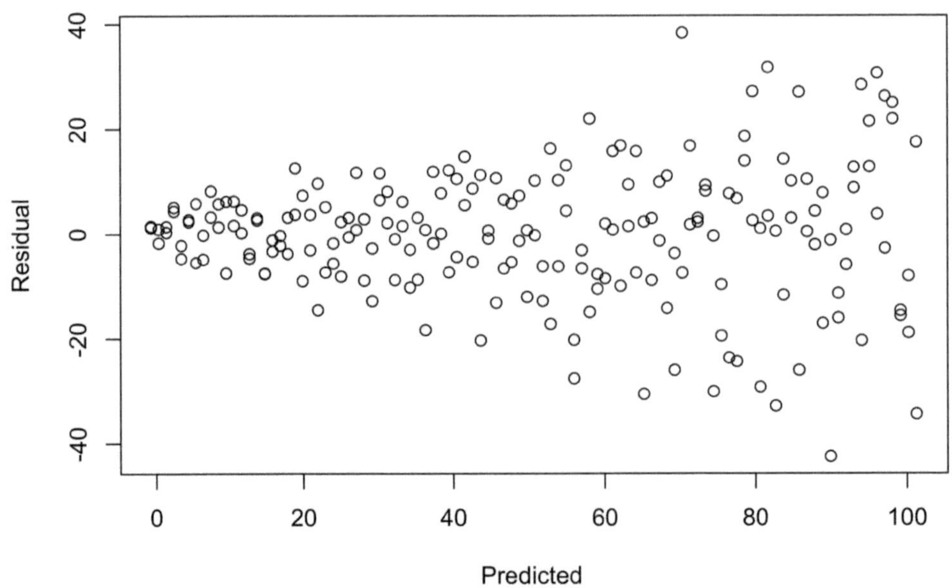

5. Do any regression assumptions appear to be violated in this histogram? If so, which one? Explain your answer. If you think an assumption was violated, discuss what statistical test you would use to confirm your suspicion.

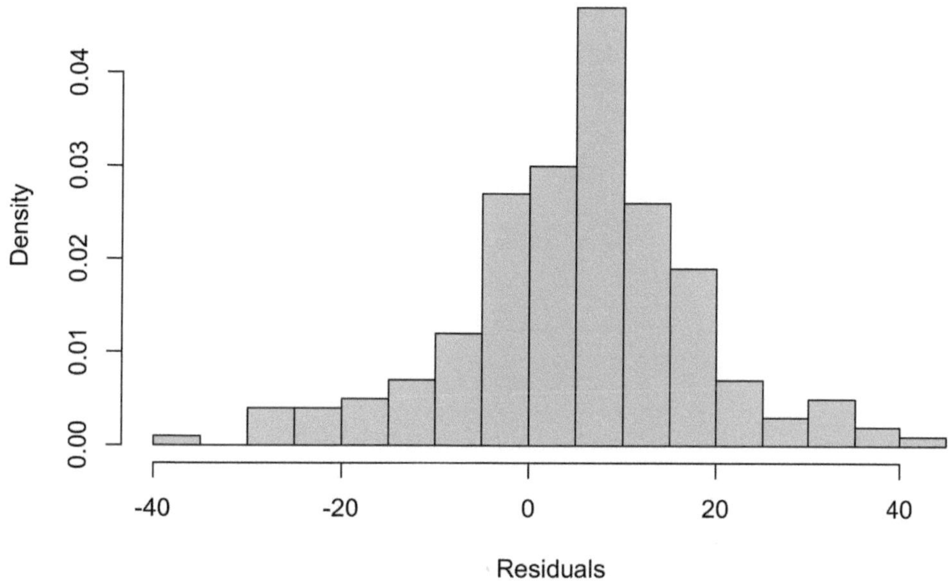

R Problems

These problems focus on the same regression model used for the R exercises in Chapter 17. With the `states20` dataset, run a multiple regression model with infant mortality (`infant_mort`) as the dependent variable and per capita income (`PCincome2020`), the teen birth rate (`teenbirth`), percentage of low-birth-weight births (`lowbirthwt`), Southern region (`south`), and percentage of adults with diabetes (`diabetes`) as the independent variables.

1. Use a scatterplot matrix to explore the possibility that one of the independent variables violates the linearity assumption. If it looks like there might be a violation, what should you do about it?

2. Test the assumption that the mean of the error term equals zero.

3. Use both a histogram and the `shapiro.test` function to test the assumption that the error term is normally distributed.

4. Plot the residuals from the infant mortality model against the predicted values and evaluate the extent to which the assumption of constant variation in the error term is violated. Use the `bptest` function as a more formal test of this assumption. Interpret both the scatterplot and the results of the `bptest`.

5. The infant mortality model includes five independent variables. While parsimony is a virtue, it also creates the potential for omitted variable bias. Look at the other variables in the `states20` dataset and identify one of them that you think might be an important influence on infant mortality and whose exclusion from the model might bias the estimated effects of the other variables. Justify your choice. Add the new variable to the model, use `stargazer` to display the new model alongside the original model, and discuss the differences between the models and whether the original estimates were biased.

APPENDIX A: CODEBOOKS

anes20

Variable	Variable Descriptions
version	Version of ANES 2020 Time Series Release
V200001	2020 Case ID
V200010a	Full sample pre-election weight
V200010b	Full sample post-election weight
V200010c	Full sample variance unit
V200010d	Full sample variance stratum
V201006	PRE: How interested in following campaigns
V201114	PRE: Are things in the country on right track
V201119	PRE: How happy R feels about how things are going in the country
V201120	PRE: How worried R feels about how things are going in the country
V201121	PRE: How proud R feels about how things are going in the country
V201122	PRE: How irritated R feels about how things are going in the country
V201123	PRE: How nervous R feels about how things are going in the country
V201129x	PRE: SUMMARY: Approve or disapprove President handling job
V201151	PRE: Feeling Thermometer: Joe Biden, Democratic Presidential candidate
V201152	PRE: Feeling Thermometer: Donald Trump, Republican Presidential candidate
V201153	PRE: Feeling Thermometer: Kamala Harris, Democratic Vice-Presidential candidate
V201154	PRE: Feeling Thermometer: Mike Pence, Republican Vice-Presidential candidate
V201155	PRE: Feeling Thermometer: Barack Obama
V201156	PRE: Feeling Thermometer: Democratic Party
V201157	PRE: Feeling Thermometer: Republican Party
V201200	PRE: 7pt scale liberal-conservative self-placement

(Continued)

(Continued)

Variable	Variable Descriptions
V201201	PRE: If R had to choose liberal or conservative self-placemt
V201202	PRE: 7pt scale liberal-conservative: Democratic Presidential candidate
V201203	PRE: 7pt scale liberal-conservative: Republican Presidential candidate
V201217	PRE: Who does R think will be elected President
V201219	PRE: Which Presidential candidate will carry state
V201231x	PRE: SUMMARY: Party ID
V201232	PRE: Party identity importance
V201233	PRE: How often trust government in Washington to do what is right [revised]
V201237	PRE: How often can people be trusted
V201249	PRE: 7pt scale defense spending: self-placement
V201252	PRE: 7pt scale gov-private medical insurance scale: self-placement
V201262	PRE: 7pt scale environment-business tradeoff: self-placement
V201308x	PRE: SUMMARY: Federal Budget Spending: Tightening border security
V201314x	PRE: SUMMARY: Federal Budget Spending: welfare programs
V201320x	PRE: SUMMARY: Federal Budget Spending: aid to the poor
V201327x	PRE: SUMMARY: National economy better or worse in last year
V201330x	PRE: SUMMARY: Economy better or worse in next 12 months
V201333x	PRE: SUMMARY: Unemployment better or worse in last year
V201336	PRE: STD Abortion: self-placement
V201337	PRE: Importance of abortion issue to R
V201340	PRE: Abortion rights Supreme Court
V201346	PRE: During last year, US position in world weaker or stronger
V201351	PRE: Votes counted accurately
V201352	PRE: Trust election officials
V201353	PRE: How often people denied right to vote
V201354	PRE: Favor or oppose vote by mail
V201359x	PRE: SUMMARY: Favor/oppose requiring ID when voting
V201360	PRE: Favor or oppose allowing felons to vote
V201377	PRE: How much trust in news media

Variable	Variable Descriptions
V201382x	PRE: SUMMARY: Corruption increased or decreased since Trump
V201384	PRE: Favor or oppose House impeachment decision
V201387	PRE: Favor or oppose Senate acquittal decision
V201390	PRE: Federal government response to COVID-19
V201393	PRE: Limits placed on public activity due to COVID-19 too strict or not
V201397	PRE: Income gap today more or less than 20 years ago
V201401	PRE: Government action about rising temperatures
V201405x	PRE: SUMMARY: Require employers to offer paid leave to parents of new children
V201406	PRE: Services to same sex couples
V201409	PRE: Transgender policy
V201412	PRE: Does R favor/oppose laws protect gays/lesbians against job discrimination
V201415	PRE: Should gay and lesbian couples be allowed to adopt
V201416	PRE: R position on gay marriage
V201418	PRE: Favor or oppose ending birthright citizenship
V201421	PRE: Should children brought illegally be sent back or allowed to stay
V201424	PRE: Favor or oppose building a wall on border with Mexico
V201433	PRE: Is religion important part of R life [revised]
V201435	PRE: What is present religion of R
V201436	PRE: If R has no particular religion does R mean atheist or agnostic
V201452	PRE: Ever attend church or religious services
V201453	PRE: Attend religious services how often
V201457x	PRE: SUMMARY: Full religion summary
V201458x	PRE: SUMMARY: Major group religion summary
V201462	PRE: Religious identification
V201502	PRE: R how much better or worse off financially than 1 year ago
V201507x	PRE: SUMMARY: Respondent age
V201508	PRE: Marital status
V201510	PRE: Highest level of Education

(Continued)

(*Continued*)

Variable	Variable Descriptions
V201511x	PRE: SUMMARY: Respondent 5 Category level of education
V201516	PRE: Armed forces active duty
V201544	PRE: Anyone in HH belong to labor union
V201545	PRE: Who in HH belongs to union
V201546	PRE: R: Are you Spanish, Hispanic, or Latino
V201549x	PRE: SUMMARY: R self-identified race/ethnicity
V201553	PRE: Native status of parents
V201554	PRE: Rs: born US, Puerto Rico, or some other country
V201567	PRE: How many children in HH age 0-17
V201569	PRE: Does R use Internet at home
V201570	PRE: Does R use Internet at any other location
V201594	PRE: How worried is R about current financial situation
V201600	PRE: What is your (R) sex? [revised]
V201601	PRE: Sexual orientation of R [revised]
V201602	PRE: Justified to use violence
V201606	PRE: Money invested in Stock Market
V201617x	PRE: SUMMARY: Total (family) income
V201620	PRE: Does R have health insurance
V201622	PRE: R concerned about paying for health care
V201624	PRE: Anyone in household tested pos for COVID-19
V201627	PRE: How often self censor
V201628	PRE: How many Guns owned
V202013	POST: R attend online political meetings, rallies, speeches, fundraisers
V202014	POST: R go to any political meetings, rallies, speeches, dinners
V202015	POST: R wear campaign button or post sign or bumper sticker
V202016	POST: R do any (other) work for party or candidate
V202017	POST: R contribute money to individual candidate running for public office
V202019	POST: R contribute money to political party during this election year
V202021	POST: R contribute to any other group that supported or opposed candidates

Variable	Variable Descriptions
V202022	POST: R ever discuss politics with family or friends
V202023	POST: How many days in past week discussed politics with family or friends
V202051	POST: R registered to vote (post-election)
V202072	POST: Did R vote for President
V202073	POST: For whom did R vote for President
V202143	POST: Feeling thermometer: Democratic Presidential candidate: Joe Biden
V202144	POST: Feeling thermometer: Republican Presidential candidate: Donald Trump
V202158	POST: Feeling thermometer: Dr. Anthony Fauci
V202159	POST: Feeling thermometer: Christian fundamentalists
V202160	POST: Feeling thermometer: feminists
V202161	POST: Feeling thermometer: liberals
V202162	POST: Feeling thermometer: labor unions
V202163	POST: Feeling thermometer: big business
V202164	POST: Feeling thermometer: conservatives
V202165	POST: Feeling thermometer: U.S. Supreme Court
V202166	POST: Feeling thermometer: gay men and lesbians
V202167	POST: Feeling thermometer: congress
V202168	POST: Feeling thermometer: Muslims
V202169	POST: Feeling thermometer: Christians
V202170	POST: Feeling thermometer: Jews
V202171	POST: Feeling thermometer: police
V202172	POST: Feeling thermometer: transgender people
V202173	POST: Feeling thermometer: scientists
V202174	POST: Feeling thermometer: Black Lives Matter
V202175	POST: Feeling thermometer: journalists
V202178	POST: Feeling thermometer: National Rifle Association (NRA)
V202183	POST: Feeling thermometer: #MeToo movement
V202185	POST: Feeling thermometer: Planned Parenthood
V202232	POST: What should immigration levels be

(*Continued*)

(Continued)

Variable	Variable Descriptions
V202234	POST: Favor or oppose allowing refugees to come to US
V202237	POST: Effect of illegal immigration on crime rate
V202240	POST: Favor or oppose providing path to citizenship
V202243	POST: Favor or oppose returning unauthorized immigrants to native country
V202246	POST: Favor or oppose separating children of detained immigrants
V202253	POST: Less government better OR more that government should be doing
V202255x	POST: SUMMARY: Less or more government
V202261	POST: We'd be better off if worried less about equality
V202300	POST: Agree/disagree: blacks should work their way up without special favors
V202301	POST: Agree/disagree: past slavery & discrimination make it difficult for blacks
V202302	POST: Agree/disagree: blacks have gotten less than they deserve
V202303	POST: Agree/disagree: if blacks tried harder they'd be as well off as whites
V202325	POST: Favor or oppose tax on millionaires
V202337	POST: Should federal government make it more difficult or easier to buy a gun
V202342	POST: Favor or oppose banning 'assault-style' rifles
V202352	POST: How would R describe social class [EGSS]
V202353	POST: Is R lower middle class, middle class, upper middle class? [EGSS]
V202371	POST: Does increasing diversity made US better or worse place to live
V202375	POST: Favor or oppose federal program giving all citizens $12K/year (STRENGTH)
V202384	POST: Attention to sexual harrassment has gone too far or not far enough
V202388	POST: Favor/oppose allowing transgender people to serve in military
V202416	POST: CSES5-Q05a: Out-group attitudes: minorities should adapt
V202418	POST: CSES5-Q05c: Out-group attitudes: immigrants good for America's economy
V202419	POST: CSES5-Q05c: Out-group attitudes: America's culture harmed by immigrants
V202475	POST: Does R consider themself a feminist or anti-feminist
V202477	POST: Feeling thermometer: Asian-Americans
V202478	POST: Feeling thermometer: Asians

Variable	Variable Descriptions
V202479	POST: Feeling thermometer: Hispanics
V202480	POST: Feeling thermometer: blacks
V202481	POST: Feeling thermometer: illegal immigrants
V202482	POST: Feeling thermometer: whites
V202491	POST: Do police treat blacks or whites better
V202549	POST: Did Russia try to interfere in 2016 presidential election or not
V202553	POST: Does most scientific evidence show vaccines cause autism or not
V203000	SAMPLE: Sample location FIPS state
V203001	SAMPLE: Sample location state postal abbreviation
V203003	SAMPLE: Census region

county20large

Variable	Variable Description
statefips	State numeric ID
countyfips	County numeric ID
fips	State/County numeric ID
stateab	State abbreviation
CountyName	County name
low_brthwt	% low-weight births
adult_smoke	% adult smokers
adult_obesity	% obese adults
teenbirth1k	teen (15–19yrs) births/10,000 births
docs_10k	doctors per 10,000 population
flu_vacc_elder	% Medicare patients with flu vaccine
kid_poverty	% minors in poverty
vcrim100k	Violent crimes per 100,000
airpollution	Average daily density of fine particulate matter in micrograms per cubic meter

(Continued)

(Continued)

Variable	Variable Description
lifeexp	Life expectancy at birth
kid_mort100k	Death per 100,000 minors
Infmort1k	Infant deaths per 1000 live births
diabetes	% adults w/diabetes
food_insec	% who did not have access to a reliable source of food during the past year
drug_death100k	Drug overdose deaths per 100,000
mv_death100k	Motor vehicle deaths per 100,000
med_HH_inc	Median household income
kidlunch	% of students eligible for free/discounted lunch
wht_nwht_seg	Higher values indicate greater residential segregation between non-white and white county residents
homicide_100k	Homicides per 100,000
suicide100k	Suicides per 100,000
gundeath100k	Gun deaths per 100,000
traffic_volume	Average traffic volume per meter of major roadways in the county
internet	% households with subscription broadband internet connection
rural	% living in rural area
County type	County classification
pop_2019	Population, 2019
case_sept821	COVID-19 cases through 9/8/21
cases100k_sept821	COVID-19 cases per 100,000 through 9/8/21
deaths_sept821	COVID-19 deaths through 9/8/21
deaths100k_sept21	COVID-19 deaths per 100,000 through 9/8/21
uninsured	% under age 65 without health insurance
povtyAug21	Poverty rate August 2021
SVI_sept21	Vulnerability human suffering and financial loss in the event of a disaster (CDC)
CVD_vulerability	COVID-19 vulnerability index
totalvote	Total votes, 2016
hrcvote	Clinton votes, 2016
djtvote	Trump votes, 2016

Variable	Variable Description
d2pty16	Democratic % of two-party vote, 2016
thirdpct16	Third-party % of total vote, 2016
county_state	County name/state name
hs	% HS or equiv. degree
4yrcoll	% 4-year college degree
postgrad	% with post-graduate degree
whitepct	% white (non-Hispanic)
blackpct	% Black (non-Hispanic)
nativepct	% Native American (non-Hispanic)
apipct	% Asian Pacific Islander (non-Hispanic)
otherpct	% Other race/ethnicity (non-Hispanic)
latinopct	% Latino/Hispanic
under6_pvty	% kids under 6 in poverty
senior_pvty	% 65+ years-old in poverty
poverty_15	% in poverty 2015
gini_coefficient	Income inequality
kid_poverty15	% minors in poverty, 2015
manage_prof	% workers in managerial/professional sector
service_occupations	% workers in service sector
sales_office	% workers in sales/office sector
farming_	% workers in agricultural sector
construction	% workers in construction sector
production_transp	% workers in manufacturing/transport sector
pop2010	Population 2010
pop2000	Population 2000
nomove_1yr	One-year residential stability
foreign_born	% foreign born
foreign_lang	% speaking non-English at home
veterans	# military veterans

(Continued)

(*Continued*)

Variable	Variable Description
home_own	% owner-occupied housing
pc_income15	Per-capita income 2015
density	Population per square mile
cvap16	Citizen voting-age population
cvap_turnout16	% turnout 2016, CVAP
State	State name
pop_2019	Population, 2019
Cdeaths_jan521	COVID-19 deaths through 1/5/21
cases_jan521	COVID-19 cases through 1/5/21
votes_gop20	Republican votes 2020
votes_dem20	Democratic votes 2020
total_votes20	Total votes, 2020
other_vote20	Third-party votes, 2020
djtpct20	Trump % 2020
jrbpct20	Biden % 2020
d2pty20	Democratic % of two-party vote, 2020
cvap18	Citizen voting-age population, 2018
cvap_turnout20	% turnout 2020, CVAP

countries2

Variable	Variable Description
wbcountry	Country name
ccode	Country code
hdi_rank	HDI rank
hdi	Human Development Index (HDI)
lifexp	Life expectancy at birth
mnschool	Mean years of schooling
gini1019	Gini coefficient: Income inequality
femexp	Female life expectancy at birth

Variable	Variable Description
malexp	Male life expectancy at birth
fem_mnschool	Female mean years of schooling
male_mnschool	Male mean years of schooling
gender_inequality	Gender Inequality Index: Inequality in reproductive health, empowerment and the labor market
matmort	Maternal mortality ratio: Number of deaths due to pregnancy-related causes per 100,000 live births
teen_fert	Adolescent birth rate: Number of births to women ages 15–19 per 1,000
fem_leg	% female share of seats in parliament
fem_seced	% of female population with at least some secondary education
male_seced	Male population with at least some secondary education
fem_labor	Female labor force participation rate
male_labor	Male labor force participation rate
chg_pop	Population average annual growth
urban	% urban population
fert1520	Total fertility rate 2015–2020
inf_no_dtp	% infants lacking immunization against DTP
inf_no_measel	% infants lacking immunization against measles
infant_mort	Infant mortality rate: Likelihood of dying between birth and exactly age 1, expressed as number of deaths per 1,000 live births
kid_mort	Under-five mortality rate: Likelihood of dying between birth and exactly age 5, expressed as number of deaths per 1,000 live births
TB100k	Tuberculosis incidence: New and relapse tuberculosis cases per 100,000 people
health_exp	Current health expenditure (% of GDP)
sec_ed	Population with at least some secondary education
educ_exp	Government expenditure on education (% of GDP)
jail100k	Prison population per 100,000 people
homicide100k	Homicides per 100,000 people
fem_suicie	Female suicide rate per 100,000 females
male_suicide	Male suicide rate per 100,000 males

(Continued)

(*Continued*)

Variable	Variable Description
food_def	Average supply of calories for food consumption, expressed as a percentage of the average dietary energy requirement
net_mig	Net migration rate: Ratio in-migrants and out-migrants per 1,000 people
internet	Internet users: % of people with access to the worldwide network
mobile_phone	Mobile phone subscriptions: Number of subscriptions for the mobile phone service, expressed per 100 people
docs10k	Physicians: Number of medical doctors, both generalists and specialists, expressed per 10,000 people
hosp10k	Hospital beds: Number of hospital beds available, expressed per 10,000 people
rural_electric	% rural population with access to electricity
fem_loc_gvt	% seats held by women in local government
fem_finance	% women with account at financial institution or with mobile money-service provider
redlist	Red List Index: Measure of the aggregate extinction risk across groups of species
skilled1019	Skilled labor force: Percentage of the labor force ages 15 and older with intermediate or advanced education
mil_exp	Military expenditures: All current and capital expenditures on the armed forces (% of GDP)
pop19_M	Population (millions) 2019
gdp_billions	Gross domestic product (billions of U.S. $)
gdp_pc	Gross domestic product per capita

states20

Variable Name	Variable Description
state	State name
stateab	State abbreviation
abortion_laws	# abortion restrictions in state law (pre-*Dobbs*)
age1825	% 18 to 24 years old
age65plus	% older than 65
Approve	Biden approval, January–June 2021
Approve_T	Trump approval, January–May 2020

Variable Name	Variable Description
attend_low	% who rarely or never attend religious services
aug_cases	#COVID-19 cases through 8/30/21
aug_CVD_deathrate	COVID-19 deaths/cases through 8/30/21
aug_deaths	Cum. #COVID-19 deaths 8/30/21
beer	Gallons of beer consumed per capita
blknh	% non-Hispanic Black
bornagain18	% born-again Christian, 2018
Cases10k	COVID-19 cases per 10,000, through October 2020
closing_date20	# days prior to election for end of voter registration
coll_4yr	% with 4-year college degree
cong10k	# religious congregations per 10,000 pop
covid_hosp	COVID-19 hospitalizations per 100K pop 9/2/2021
d2pty16	% of two-party vote for Clinton
d2pty20	% of two-party vote for Biden
DemLeg2020	% Democratic state legislators, 2020
density	Population per square mile
diabetes	% with diabetes in population
Disapprove	Biden disapproval, January–June 2021
Disapprove_T	Trump disapproval, January–May 2020
docs100k	#physician per 100,000 population
early_pct20	Early votes as % of total votes 2020
environment	Environmental liberalism (0–1) high = liberal
fb	% foreign born
FemLeg2020	% female sate legislators, 2020
gaymarr	% support for legal gay marriage
gini16	Gini index 2016
gtba	% with advanced degrees
gundeath100k	Gun deaths per 100,000 population
immig	Immigration liberalism: high = more liberal

(Continued)

(*Continued*)

Variable Name	Variable Description
infant_mort	# infant deaths per 1000 live births
jun_cases	#COVID-19 cases through 6/01/21
jun_CVD_deathrate	COVID-19 deaths/cases through 6/01/21
jun_death	#COVID-19 deaths through 6/01/21
latino	% Latino population
legprof	Legislative professionalism (high = more professional
lg10_density	log10 population density
lgbtq_rgts	# legal protection for LGBTQ persons (Human Rights Campaign)
lifeexp	Life expectancy
lowbirthwt	% low birth-weight births
margin20	Margin of victory (Winner % minus loser %), 2020 presidential race
metro	% living in metro area
mngt	% with management jobs
netage	Age 65+ minus age 18–25
netdem	% Democratic minus % Republican identifiers
netlib	% liberal minus % conservative
nonwhite	% non-white
NonWhtLeg2020	% non-white in state legislature
nurses1k	# nurses per thousand population
obesity	% obese in population
other_occ	% other occupations
othnh	% non-Hispanic, white, Black, or API
PCincome2020	Per capita income
pct_gun_owners	% of adults who own at least one gun
PerPupilExp	Per pupil $, K–12 schools
pol_cult	Political culture classification
policy_innov	Policy innovation score: High values = innovative states
policy_lib	Policy liberalism: high scores –> liberal policies
poverty1819	% living in poverty, 2018–2019

Variable Name	Variable Description
prison100k	Incarcerated people per 100,000 pop
prochoice18	% pro-choice, 2018
prof	% with professional jobs
race_att	Racial liberalism (0–1) high = liberal
relig_imp	% who say religion is important to them
single	% single
south	South = 1, other = 0
suicide100k	Suicides per 100k population
Taxburden	Total state taxes as % of total state income
taxprog	Tax progressivity (rich pay greater share than poor) –> high = progressive
teenbirth	% of births to teen mothers
union	% union in workforce
vacc_rate	% with full COVID-19 vaccine (8/30/21)
vcrime100k	Violent crimes per 100,000 population
vep_turnout16	% eligible voter turnout, 2016
vep_turnout20	% eligible voter turnout, 2020
whitemale	% white male
whtmale_nocol	% white male, no college
whtnh	% non-Hispanic white population

APPENDIX B: QUARTO TUTORIAL

Quarto is a wonderful tool available in the RStudio framework for integrating text, R code, and statistical and graphic output into neat and readable documents. It can be integrated into classrooms, both as a tool for presenting material to students and as a resource for student to use for submitting R homework assignments. Quarto documents can be built using either source code or via the visual editor. This brief tutorial uses the visual editor.

The first thing you need to do is open a Quarto document in RStudio. Go to the + tab in the upper-left corner of the script window and select Quarto document.

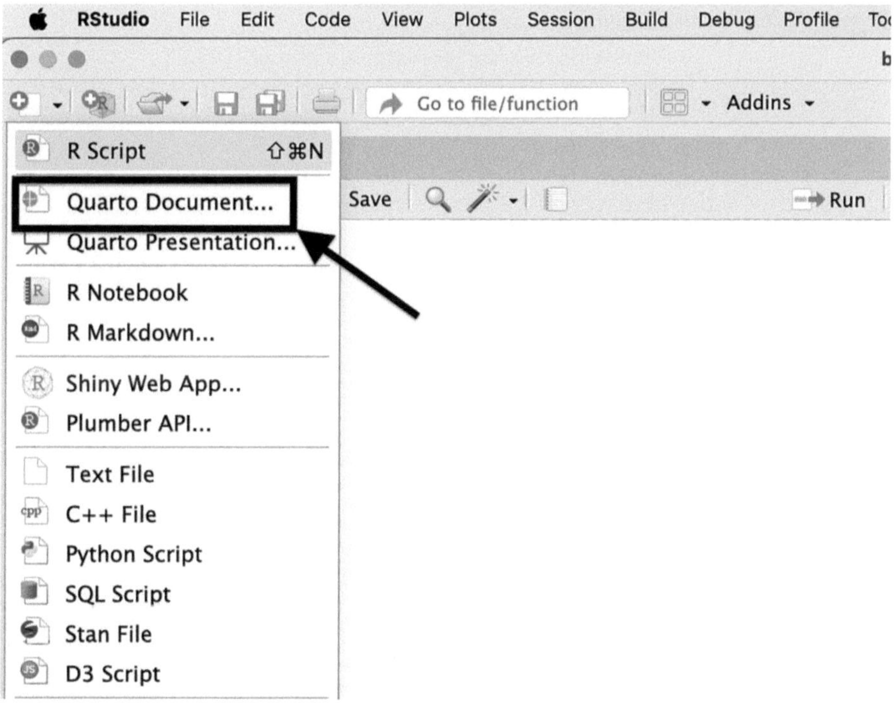

When you do this, box will appear giving you a few options. You should provide a title for your document, probably identifying which homework assignment, providing your name, selecting the PDF output (unless you want a Word or html document), and click on `Create Empty Document`.

QUARTO ENVIRONMENT

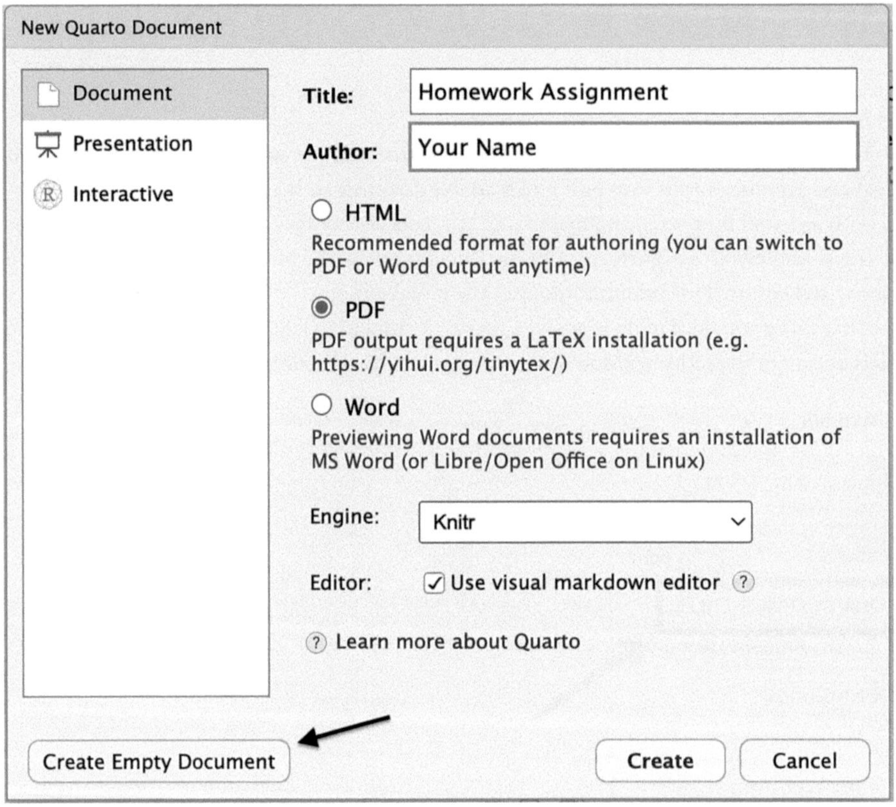

Once you do this, the Quarto editor will appear. It includes both a command (top) and an editing (second) toolbar and should look like the following figure, except without all the annotations.

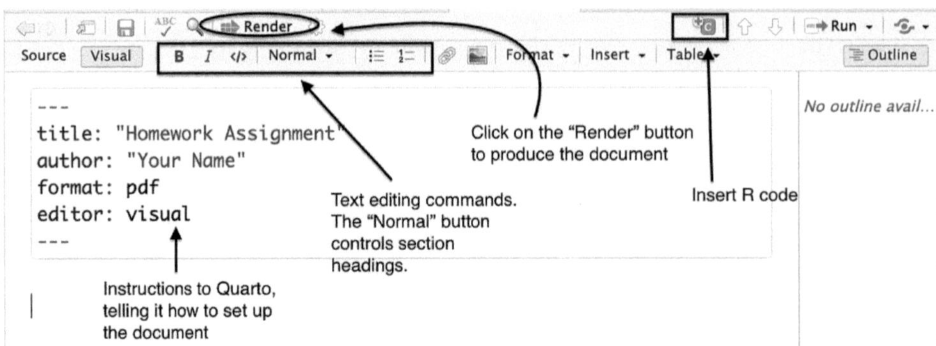

I recommend toggling on the `Outline` button to remove the outline from the right side and give you a wider screen for editing your document.

When writing text, the default is to use the "Normal" font, but you can click on the Headings tab to create differently sized section headings.

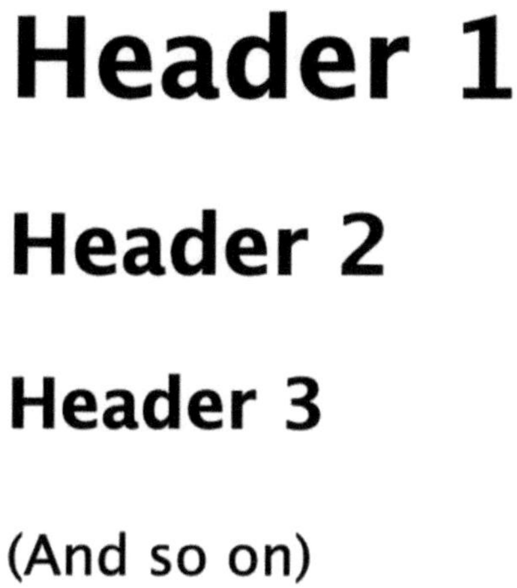

RUNNING R CODE

Click on the +C button in the command toolbar to open a code chunk window in your document, and then enter your code in the window. When you have everything you need, click the arrowhead button on the right to preview the results. This is where you find out if everything looks the way you think it should, or if there are mistakes.

You can then add text below the code window to discuss and interpret the results.

Here is an example, using a bit of code from Chapter 1. You can see that I used the code window to tell R to load the `Approve21` dataset and get the names of all of the variables. When I clicked the upper-right arrowhead, a preview of the results was added.

Sometimes, you may be interested in including the results of the R commands in your documents without including the code. To do this, you need to add `echo = FALSE` as a code execution option.

```
{r echo=FALSE}
load("Approve21.rda")
names(Approve21)
```

One thing I recommend is that you suppress the output when attaching libraries or installing packages in your Quarto document. This is because sometimes the messages displayed when libraries are loaded can take up a fair amount of space in your document and can be somewhat inelegant, compared to the rest of the document. This is easy to fix by adding `message = FALSE` as an option.

```
{r message=FALSE}
library(Hmisc)
```

PRODUCING THE DOCUMENT

Once you have all of the R code you need along with whatever text you need to explain and interpret the results, you can produce the PDF document by clicking on the `Render` tab in the command toolbar. If there are no errors, the document will appear on your screen and a copy will be saved to your working directory.

I have found that working with Quarto is an iterative process. I produce a document, check to see how it looks, go back and edit it in the Quarto editor, produce another copy, edit it, and so on.

Quarto has many more features than just those discussed here. I encourage you to explore and find ways in which Quarto can help you produce the most functional and attractive documents possible. A good place to start is with the resources provided at quarto.org.

APPENDIX C: HIDDEN R CODE

The code chunks copied in this section can be used to replicate the figures from *Introduction to Political and Social Data Analysis (With R)*. In many cases, you need to load one of the datasets used in the textbook, so make sure you have copied them all to your working directory.

```
load("anes20.rda")
load("cces20.rda")
load("countries2.rda")
load("county20.rda")
load("county20large.rda")
load("gdpvotes.rda")
load("votes1620.rda")
```

You should also make sure you have installed all of the packages used in the textbook, plus the `MASS` and `plotrix` packages.

Figure 1.2

```
par(mfrow = c(1, 2))
set.seed(135)
samples = 25
r = .85

library('MASS')
data = mvrnorm(n = samples, mu = c(0, 0),
               Sigma = matrix(c(1, r, r, 1), nrow = 2),
               empirical = TRUE)
X = data[, 1] # standard normal (mu = 0, sd = 1)
Y = data[, 2] # standard normal (mu = 0, sd = 1)

samples = 25
r = .35

data2 = mvrnorm(n = samples, mu = c(0, 0),
                Sigma = matrix(c(1, r, r, 1), nrow = 2),
                empirical = TRUE)
```

```
X1 = data2[, 1] # standard normal (mu = 0, sd = 1)
Y1 = data2[, 2] # standard normal (mu = 0, sd = 1)

plot(X1, Y1, xlab = "Independent Variable A",
     ylab = "Dependent Variable", main = "Weak Relationship",
     sub = "(Correlation = .35)")
abline(lm(Y1 ~ X1))

plot(X, Y, xlab = "Independent Variable B",
     ylab = "Dependent Variable", main = "Strong Relationship",
     sub = "(Correlation = .85)")
abline(lm(Y ~ X))

par(mfrow = c(1, 1))
```

Figure 1.3

```
plot(gdpvotes$gdpchng, gdpvotes$vote, cex = .5,
     xlab = "Percentage Change in Real GDP Per Capita, Q1 to Q3",
     ylab = "Incumbent Percent of Two party Vote",
     ylim = c(45, 65))
abline(lm( gdpvotes$vote ~ gdpvotes$gdpchng))
text(gdpvotes$gdpchng, gdpvotes$vote, gdpvotes$year, pos = 3, cex = .5)
```

Figure 1.4

```
par(mfrow = c(1, 2))
plot(gdpvotes$gdpchng[gdpvotes$openseat == 0],
     gdpvotes$vote[gdpvotes$openseat == 0], cex = .5,
     ylim = c(40,65), main = "Incumbent Running",
     xlab = "Change in GDP Per Capita, Q1 to Q3",
     ylab = "Incumbent Percent of Two party Vote",
     sub = "(Correlation = .68)")
abline(lm(gdpvotes$vote[gdpvotes$openseat == 0] ~
            gdpvotes$gdpchng[gdpvotes$openseat == 0]))
text(gdpvotes$gdpchng[gdpvotes$openseat == 0],
     gdpvotes$vote[gdpvotes$openseat == 0],
     gdpvotes$year[gdpvotes$openseat == 0], pos = 3, cex = .5)

plot(gdpvotes$gdpchng[gdpvotes$openseat == 1],
     gdpvotes$vote[gdpvotes$openseat == 1], cex = .5,
     ylim = c(40,65), main = "Open Seat",
     xlab = "Change in GDP Per Capita, Q1 to Q3",
```

```
     ylab = "Incumbent Percent of Two party Vote",
     sub = "(Correlation = .16)")
abline(lm(gdpvotes$vote[gdpvotes$openseat == 1] ~
            gdpvotes$gdpchng[gdpvotes$openseat == 1]))
text(gdpvotes$gdpchng[gdpvotes$openseat == 1],
     gdpvotes$vote[gdpvotes$openseat == 1],
     gdpvotes$year[gdpvotes$openseat == 1],
     pos = 3, cex = .5)
par(mfrow = c(1, 1))
```

Figure 5.3

```
par(mfrow = c(2, 2))
#skewed distributions
set.seed(5)
y <- rnorm(10000)^2
ydens = density(y)
plot(ydens, xlim = c(-.5, 6), cex.main = .8,
     main = "Positive (Right) Skew: Mean > Median",
     lwd = 2,
     xlab = "X")
abline(v = mean(y))
abline(v = median(y), lty = 2)
legend("topright", legend = c("Mean", "Median"),
       lty = 1:2, cex = .7)

set.seed(5)
y <- (-1*rnorm(10000)^2)
y <- 6 + y
ydens = density(y)
plot(ydens, cex.main = .8,
     main = "Negative (Left) Skew: Mean < Median",
     xlab = "X", xlim = c(0, 6.5), lwd = 2)
abline(v = mean(y))
abline(v = median(y), lty = 2)
legend("topleft", legend = c("Mean", "Median"),
       lty = 1:2, cex = .7)

#No Skew
set.seed(12)
x <- seq(0, 6, length = 1000)
y <- dnorm(x, mean = 3, sd = .75)
plot(x, y, type = "l", lwd = 2,
```

```
     main = "No Skew (Mean = Median)",
     cex.main = .8, xlab = "X",
     ylab = "Density")
abline(v = 3)
legend("topright", legend = c("Mean, Median"),
       lty = c(1), cex = .7)
par(mfrow = c(1, 1))
```

Figure 6.1

```
par(mfrow = c(2, 2))
x <- seq(4, 16, length = 500)
y <- dnorm(x, mean = 10, sd = .5)
plot(x, y, type = "l", ylab = "Density",
     lwd = 1, main = "Mean = 10, Median = 10",
      cex.axis = .7, cex.main = 1.3)
x <- seq(4, 16, length = 500)
y <- dnorm(x, mean = 10, sd = 1.1)
plot(x, y, type = "l", ylab = "Density",
     lwd = 1, main = "Mean = 10, Median = 10",
      cex.axis = .7, cex.main = 1.3)

x <- seq(4, 16, length = 500)
y <- dnorm(x, mean = 10, sd = 2.5)
plot(x, y, type = "l", ylab = "Density",
     lwd = 1, main = "Mean = 10, Median = 10",
      cex.axis = .7, cex.main = 1.3)

par(mfrow = c(1, 1))
```

Figure 6.4

```
#First, create cces20$age, using cces20$birthyr
cces20$age <- 2020-cces20$birthyr
hist(cces20$age, xlab = "Age of respondent",
     main = "",
     prob = T,
     xlim = c(0,100), ylim = c(0, .03))
abline(v = c(66), lty = 2, lwd = 2)
curve(dnorm(x, mean = mean(cces20$age,
                           na.rm = T),
           sd = sd(cces20$age, na.rm = T)),
      lwd = 2, add = T)
```

```
legend("topright", legend = c("Age = 66", "Normal Curve"),
       lty = c(2, 1), lwd = c(2, 2), cex = .9)
```

Figure 7.2

```
#Set seed to ensure identical results
set.seed(2020)
#Create a new object, "Coin", with two outcomes, Heads and Tails
coin < - c("Heads", "Tails")
#Take a random sample of outcomes from ten tosses
coin10 = sample(coin, 10, rep = T)
#Show the breakdown of coins tosses
tosses < - c(10, 20, 60, 100, 150, 300, 500, 1000,
             2000, 400, 750, 1500, 1250, 1750)
Heads < - c(30, 33.3, 40, 58, 50, 51, 52, 50.9, 50.5,
            50, 52, 49.7, 48.2, 52.6)
coin_tosses <- data.frame(tosses, Heads)

plot(coin_tosses$tosses, coin_tosses$Heads,
     xlab = "Number of Coin Tosses",
     ylab = "Percent Heads", ylim = c(25, 65))
abline(h = 50)
```

Figure 7.3

```
set.seed(2020)
#Create a new object, "die", outcomes 1, 2, 3, 4, 5, 6
die <- c(1, 2, 3, 4, 5, 6)
die100 = sample(die, 100, replace = T)
die2000 = sample(die, 2000, replace = T)
par(mfrow = c(1, 2))
barplot(prop.table(table(die100)),
     main = "100 Rolls", ylab = "Proportion",
     xlab = "Ouctome",
     ylim = c(0, .25))
abline(h = .167, lwd = 2)
barplot(prop.table(table(die2000)),
     main = "2000 Rolls", ylab = "Proportion",
     xlab = "Outcome",
     ylim = c(0, .25))
abline(h = .167, lwd = 2)

par(mfrow = c(1, 1))
```

Figure 8.1

```
#normal dist
x <- seq(30, 38, by = .025)
# mean as 34.04 and standard deviation (sd/sqrt(300)).
y <- dnorm(x, mean = 34.04,
           sd = sd(county20$d2pty20)/sqrt(300))

plot(x,y, type = "l", ylab = "Density",
     lwd = 2, xlab = "Distribution of Sample Means")
axis(1, at = c(30:38))
abline(v = 34.04)
legend("topright",
       legend = c("Mu = 34.04"), lty = 1, cex = .8)
```

Figure 8.2

```
set.seed(251)
#create an object (sample_means50) with space to
#store fifty sample means
sample_means50 <- rep(NA, 50)
# Run through the data 50 times, getting 50-county samples
#Store the fifty samples in "sample_means50
for(i in 1:50){
  samp <- sample(county20$d2pty20, 50)
  sample_means50[i] <- mean((samp), na.rm = T)
}

##Create Plots
d50 <- density(sample_means50)
plot(d50, ylim = c(0, .19),xlab = "Sample Means",
     lwd = 2, xlim = c(26, 42), xaxt = "n", main = "")
curve(dnorm(x, mean = mean(sample_means50),
             sd = sd(sample_means50)),
      lwd = 2, lty = 2, add = T)
axis(1, at = c(26:42))
legend("topright", legend = c("Sampling Dist", "Normal Dist"),
       lty = 1:2, cex = .8)
```

Figure 8.3

```
#Gather sample means from 500 samples of 50 counties
set.seed(251)
```

```
sample_means500 <- rep(NA, 50)
for(i in 1:500){
  samp <- sample(county20$d2pty20, 50)
  sample_means500[i] <- mean((samp), na.rm = T)
}


d500 <- density(sample_means500)
plot(d500, xlab = "Sample Means", lwd = 2,
     main = "", ylim = c(0, .18),
     xlim = c(26, 42), xaxt = "n")
axis(1, at = c(26:42))
curve(dnorm(x, mean = mean(sample_means500),
            sd = sd(sample_means500)),
      lwd = 2, lty = 2, add = T)
legend("topright", legend = c("Sampling Dist", "Normal Dist"),
       lty = 1:2, cex = .8)
```

Figure 8.4

```
#Generate CI data for fifty samples
samp_mean <- rep(NA, 50)
samp_sd <- rep(NA, 50)
n <- 100

# obtain a sample of size n = 100 from the population
set.seed(250)
for(i in 1:50){
  samp <- sample(county20$d2pty20, n)
  samp_mean[i] <- mean(samp) # save sample mean of each sample
  samp_sd[i] <- sd(samp) # save sample sd of each sample
}

lower_vector <- samp_mean - samp_sd / sqrt(n)
upper_vector <- samp_mean + samp_sd / sqrt(n)
y <- samp_mean

cidata <- data.frame(y, lower_vector, upper_vector)

#Generate plots
library(plotrix)
plotCI(1:50, y, ui = upper_vector, li = lower_vector,
       ylab = "Sample Estimates", xlab = "Sample")
```

```
#Identify "misses"
points(x = 2, y = 34.04, pch = 16)
points(x = 4, y = 34.04, pch = 16)
points(x = 6, y = 34.04, pch = 16)
points(x = 11, y = 34.04, pch = 16)
points(x = 14, y = 34.04, pch = 16)
points(x = 17, y = 34.04, pch = 16)
points(x = 20, y = 34.04, pch = 16)
points(x = 21, y = 34.04, pch = 16)
points(x = 36, y = 34.04, pch = 16)
points(x = 39, y = 34.04, pch = 16)
points(x = 43, y = 34.04, pch = 16)
points(x = 48, y = 34.04, pch = 16)
abline(h = 34.04)
```

Figure 8.5

```
obs <- seq(100, 3000, 100)
width <- 2*(1.96 * (15.89/sqrt(obs)))
plot(width ~ obs,
     ylab = "Width of Confidence Interval",
     xlab = "Sample Size", type = "l")
```

Figure 8.6

```
# Create samples
set.seed(251)
demwin <- as.numeric(county20$d2pty20 > 50)
#Create an object with space to store 50 sample means
sample_prop500 <- rep(NA, 50)
#run through the date 500 times, 50-county samples of 'demwin'
#each time. Store the mean of each sample in 'sample_prop500'
for(i in 1:500){
  samp <- sample(demwin, 50)
  sample_prop500[i] <- mean((samp), na.rm = T)
}

#Plot data
dwin500 <- density(sample_prop500)
plot(dwin500, lwd = 2, main = "",
     xlab = "Sample Proportions")
curve(dnorm(x, mean = mean(sample_prop500),
```

```
            sd = sd(sample_prop500)), lwd = 2, lty = 2, add = T)
legend("topright",
       legend = c("Sampling Dist", "Normal Dist"),
       lty = 1:2, cex = .8)
```

Figure 9.1

```
x <- seq(-4, 4, by = .05)
yz <- dnorm(x)
plot(x, yz, type = "l",
     ylab = "Density", xlab = "z-score",
     ylim = c(0, .43))
segments(x0 = -1.96, y0 = -.01, x1 = -1.96, y1 = .06, lty = 2)
segments(x0 = 0, y0 = -.01, x1 = 0, y1 = .396, lwd = 3)

text(0, .415, expression(mu))
text(-1.3, 0, expression(z[bar(x)] == -1.96), cex = .8)

legend(-4.5, .35, legend = c("What is the probability",
                             "of getting a sample",
                             "outcome of this",
                             "value..."),
       bty = "n", cex = .85)
legend("topright", legend = c("...if the population value",
                              "is equal to this?"),
       bty = "n", cex = .85)

arrows(x0 = -3,
       x1 = -1.96,
       y0 = .21,
       y1 = .06, lwd = 1.5)
arrows(x0 = 2.2,
       x1 = 0,
       y0 = .36,
       y1 = .25, lwd = 1.5)
```

Figure 9.2

```
x <- seq(53, 65, length = 1000)
y <- dnorm(x, mean = 59.2, sd = 1.538)
plot(x, y, type = "l", lwd = 1, main = "",
     xlab = "Average Annual Sick Leave Hours",
     ylab = "Density", cex = .8, yaxs = "i")
```

```
segments(x0 = 59.2, y0 = -.03, x1 = 59.2, y1 = .2594)
segments(x0 = 56.19, y0 = -.07, x1 = 56.19, y1 = .05, lty = 2, lwd = 3)
segments(x0 = 54.8, y0 = -.05, x1 = 54.8, y1 = .02, lwd = 3)
legend("topright",
       legend = c("Mu", "C.V.", "Sample Mean"),
       lty = c(1, 2, 1), lwd = c(1, 2, 3), cex = .8)
```

Figure 9.3

```
x <- seq(-3.5, 3.5, length = 1000)
y <- dnorm(x, mean = 0, sd = 1)
plot(x, y, type = "l", lwd = 1, main = "",
     xlab = "Critical Values for p = .05, (one-tailed and two-tailed)", ylab =
"Density",yaxs = "i")
legend("topright",
       legend = c("One-tailed, z = -1.65", "Two-tailed, z = + /-1.96"),
       lty = 1:2, cex = 0.8)

segments(x0 = -1.65, y0 = -.02, x1 = -1.65, y1 = .098, )
segments(x0 = -1.96, y0 = -.02, x1 = -1.96, y1 = .06, lty = 2)
segments(x0 = 1.96, y0 = -.02, x1 = 1.96, y1 = .06, lty = 2)
```

Figure 9.4

```
x <- seq(-5, 5, by = .05)
yz <- dnorm(x)
y3 <- dt(x, df = 3)
plot(x, y3, type = "l", lty = 2, lwd = 2 , ylim = c(0, .40),
      ylab = "Density",
     xlab = "t(z)-score")
lines(x, yz, lty = 1, lwd = 2)
legend("topright",
       legend = c("t-distribution", "Normal Distribution"),
       lty = c(2, 1), lwd = c(2, 2), cex = .75)
```

Figure 9.5

```
x <- seq(-3, 3,by = .05)
yz <- dnorm(x)
y3 <- dt(x, df = 3)
y5 <- dt(x, df = 5)
```

```
y7 <- dt(x, df = 7)
plot(x, y7, type = "l", lty = 2, ylim = c(0, .40),
     ylab = "Density", xlab = "t-score")
lines(x, y5, lty = 3)
lines(x, y3, lty = 4)
lines(x, yz, lty = 1)
legend("topright",
       legend = c("Normal", "df = 7", "df = 5", "df = 3"),
       lty = c(1,2,3,4), cex = .8)
```

Figure 10.1

```
x <- seq(-4, 4,by = .05)
yt <- dt(x, df = 500)
plot(x, yt, type = "l",
     ylab = "Density", xlab = "t-score",
     ylim = c(0,.43))
segments(x0 = -1.96, y0 = -.01, x1 = -1.96, y1 = .06, lty = 2)
segments(x0 = 0, y0 = -.01, x1 = 0, y1 = .396, lwd = 3)

text(0,.41, expression(mu["1"] - mu["2"]))
text(-1.1,0, expression(t[(bar(x)["1"]-bar(x)["2"])] == -1.96), cex =
   .8)

legend(-4.5, .35,legend = c("What is the probability",
                            "of getting a sample",
                            "difference of this",
                            "size..."),
       bty = "n", cex = .85)
legend("topright",legend = c("...if the difference in the",
                             "population is equal to this?"),
       bty = "n", cex = .85)

arrows(x0 = -3.2,
       x1 = -1.96,
       y0 = .21,
       y1 = .06, lwd = 1.5)
arrows(x0 = 2,
       x1 = 0,
       y0 = .36,
       y1 = .25, lwd = 1.5)
```

Figure 11.1

```
df1 <- 3
df2 <- 179
curve(df(x, df1 = df1, df2 = df2),
      from = 0, to = 7, xlab = "F-Ratio",
      ylab = "Density", main = "",
      ylim = c(0,.75), cex.main = .8, yaxs = "i")
#abline(v = 2.66)
segments(x0 = 2.66, y0 = .00, x1 = 2.66, y1 = .067, lwd = 2 )

legend("topright", legend = c("C.V. = 2.66"), lwd = 2, cex = .9)
```

Figure 12.1

```
library(descr)
#Create religious importance variable
anes20$relig_imp <- anes20$V201433
#Recode Religious Importance to three categories
levels(anes20$relig_imp) <- c("High", "High", "Moderate", "Low", "Low")
#Change order to Low-Moderate-High
anes20$relig_imp <- ordered(anes20$relig_imp,
                            levels = c("Low", "Moderate", "High"))

#Use "plot = T" to get mosaic plot, add graphing details
tab_plots <- crosstab(anes20$relig_imp, anes20$V203003, prop.c = T,
        plot = T,
        chisq = T,
        xlab = "Region of Residence",
        ylab = "Religious Importance",
        main = "Mosaic Plot: Religious Importance by region")
```

Figure 12.2

```
curve(dchisq(x, df = 6), from = 0,
      to = 25, xlab = "Chi-square",
      ylab = "Density", yaxs = "i",
      xaxs = "i", cex.main = .8)
segments(x0 = 12.592, y0 = 0, x1 = 12.592, y1 = .017, lwd = 2)
#abline(v = 12.59)
legend("topright", legend = c("C.V. = 12.592"),
       lty = 1, lwd = 2, cex = .8)
```

Figure 14.1

```
library(MASS)
set.seed(35)
par(mfrow = c(2,3))
samples = 200
r = -.5

data = mvrnorm(n = samples, mu = c(0, 0),
               Sigma = matrix(c(1, r, r, 1), nrow = 2),
               empirical = TRUE)
Xneg.5 = data[, 1] # standard normal (mu = 0, sd = 1)
Yneg.5 = data[, 2] # standard normal (mu = 0, sd = 1)

samples = 200
r = -.25

data = mvrnorm(n = samples, mu = c(0, 0),
               Sigma = matrix(c(1, r, r, 1), nrow = 2),
               empirical = TRUE)
Xneg.25 = data[, 1] # standard normal (mu = 0, sd = 1)
Yneg.25 = data[, 2] # standard normal (mu = 0, sd = 1)

samples = 200
r = -.75

data = mvrnorm(n = samples, mu = c(0, 0),
               Sigma = matrix(c(1, r, r, 1), nrow = 2),
               empirical = TRUE)
Xneg.75 = data[, 1] # standard normal (mu = 0, sd = 1)
Yneg.75 = data[, 2] # standard normal (mu = 0, sd = 1)

samples = 200
r = .5

data = mvrnorm(n = samples, mu = c(0, 0),
               Sigma = matrix(c(1, r, r, 1), nrow = 2),
               empirical = TRUE)
Xpos.5 = data[, 1] # standard normal (mu = 0, sd = 1)
Ypos.5 = data[, 2] # standard normal (mu = 0, sd = 1)


samples = 200
```

```
r = .25

data = mvrnorm(n = samples, mu = c(0, 0),
               Sigma = matrix(c(1, r, r, 1), nrow = 2),
               empirical = TRUE)
Xpos.25 = data[, 1] # standard normal (mu = 0, sd = 1)
Ypos.25 = data[, 2] # standard normal (mu = 0, sd = 1)

samples = 200
r = .75

data = mvrnorm(n = samples, mu = c(0, 0),
               Sigma = matrix(c(1, r, r, 1), nrow = 2),
               empirical = TRUE)
Xpos.75 = data[, 1] # standard normal (mu = 0, sd = 1)
Ypos.75 = data[, 2] # standard normal (mu = 0, sd = 1)

plot(Xneg.25, Yneg.25, main = "r = -.25",
     xlab = "X", ylab = "Y", cex = .8, cex.main = 1.3, pch = 20)
plot(Xneg.5, Yneg.5, main = "r = -.50",
     xlab = "X", ylab = "Y", cex = .8, cex.main = 1.3, pch = 20)
plot(Xneg.75, Yneg.75, main = "r = -.75",
     xlab = "X", ylab = "Y", cex = .8, cex.main = 1.3, pch = 20)

plot(Xpos.25, Ypos.25, main = "r = .25",
     xlab = "X", ylab = "Y", cex = .8, cex.main = 1.3, pch = 20)
plot(Xpos.5, Ypos.5, main = "r = .50",
     xlab = "X", ylab = "Y", cex = .8, cex.main = 1.3, pch = 20)
plot(Xpos.75, Ypos.75, main = "r = .75",
     xlab = "X", ylab = "Y", cex = .8, cex.main = 1.3, pch = 20)

par(mfrow = c(1, 1))
```

Figure 15.1

```
plot(votes1620$vote16, votes1620$vote20,
     xlab = "Dem two-party % 2016",
     ylab = "Dem two-party % 2020", ylim = c(35,60))
abline(lm(votes1620$vote20 ~ votes1620$vote16))
arrows(x0 = 39.3,
       x1 = 39.3,
       y0 = 34.1,
       y1 = 39.5, lwd = 1.5)
```

```
arrows(x0 = 39.3,
       x1 = 34.7,
       y0 = 39.5,
       y1 = 39.5, lwd = 1.5)
legend(38, 50,
       legend = expression(hat(y) == "-4.289 + 1.115x"), bty = "n", cex
          = .8)
```

Figure 15.2

```
plot(votes1620$vote16, votes1620$vote20,
     xlab = "Dem two-party % 2016",
     ylab = "Dem two-party % 2020", ylim = c(35,60))
abline(lm(votes1620$vote20 ~ votes1620$vote16))
segments(x0 = 35.6, y0 = 35.43, x1 = 35.6, y1 = 37.1 )
segments(x0 = 49.4, y0 = 50.81, x1 = 49.4, y1 = 48.3 )
segments(x0 = 51.5, y0 = 53.15, x1 = 51.5, y1 = 54.7 )
segments(x0 = 50.2, y0 = 51.70, x1 = 50.2, y1 = 53.8 )
segments(x0 = 58.3, y0 = 60.74, x1 = 58.3, y1 = 60.6 )
segments(x0 = 39.3, y0 = 39.55, x1 = 39.3, y1 = 37.6 )
segments(x0 = 49.6, y0 = 51.03, x1 = 49.6, y1 = 50.3 )
```

Figure 15.3

```
plot(votes1620$vote16, votes1620$vote20,
     xlab = "Dem two-party % 2016",
     ylab = "Dem two-party % 2020", ylim = c(35, 60))
abline(h = mean(votes1620$vote20))
segments(x0 = 35.6, y0 = 48.914, x1 = 35.6, y1 = 37.1 )
segments(x0 = 49.4, y0 = 48.914, x1 = 49.4, y1 = 48.3 )
segments(x0 = 51.5, y0 = 48.914, x1 = 51.5, y1 = 54.7 )
segments(x0 = 50.2, y0 = 48.914, x1 = 50.2, y1 = 53.8 )
segments(x0 = 58.3, y0 = 48.914, x1 = 58.3, y1 = 60.6 )
segments(x0 = 39.3, y0 = 48.914, x1 = 39.3, y1 = 37.6 )
segments(x0 = 49.6, y0 = 48.914, x1 = 49.6, y1 = 50.3 )
legend(41,52, legend = "Mean = 48.914", bty = "n", cex = .85)
```

Figure 17.1

```
#Life expectancy by "docs10k"
fit_raw <- lm(countries2$lifexp ~ countries2$docs10k,
              na.action = na.exclude)
```

```
#Life expectancy by "log10(docs10k)"
fit_log <- lm(countries2$lifexp ~ log10(countries2$docs10k),
              na.action = na.exclude)
countries2$yhat <- predict(fit_raw)
countries2$yhatlog <- predict(fit_log)

plot(countries2$docs10k,countries2$lifexp,
     xlab = "Doctors per 10k Population",
     ylab = "Life Expectancy")
curve(63.4 + 9.5824*log10(x), .01, 85,
      lwd = 2, add = T)
abline(fit_raw, lty = 2, lwd = 2)
legend("right",
       legend = c(expression(hat(y) == "a + b*x"),
  expression(hat(y) == "a + b*log10(x)")),
       lty = c(2,1), lwd = c(2,2), cex = .8)
```

Figure 17.2

```
lowdoc <- subset(countries2,docs10k <= 14.781)
hidoc <- subset(countries2,docs10k > 14.781)
par(mfrow = c(1, 2))
plot(lowdoc$docs10k, lowdoc$lifexp,
     ylim = c(50, 85),
     xlab = "Doctors per 10k",
     ylab = "Life Expectancy", sub = " (r = .67)",
     main = "LT Median Docs/10k",
     cex.main = .8)
abline(lm(lowdoc$lifexp ~ lowdoc$docs10k))
plot(hidoc$docs10k, hidoc$lifexp,
     ylim = c(50, 85), xlab = "Doctors per 10k",
     ylab = "Life Expectancy", sub = "(r = .21)",
     main = "GT Median Docs/10k",
     cex.main = .8)
abline(lm(hidoc$lifexp ~ hidoc$docs10k))
par(mfrow = c(1, 1))
```

Figure 17.3

```
countries2$urban_prop <- countries2$urban/100
par(mfrow = c(1, 2))
plot(countries2$urban, countries2$lifexp,
```

```
     xlab = "Percent Urban", ylab = "Life Expectancy",
     main = "y = 61.4 + .192x", cex.main = .9)
abline(lm(countries2$lifexp ~ countries2$urban), lwd = 2)
plot(countries2$urban_prop, countries2$lifexp,
     xlab = "Proportion Urban",
     ylab = "Life Expectancy",
     main = "y = 61.4 + 19.2x", cex.main = .9)
abline(lm(countries2$lifexp ~ countries2$urban_prop), lwd = 2)

par(mfrow = c(1, 1))
```

Figure 18.1

```
par(mfrow = c(2, 3))
plot(countries2$fert1520, countries2$lifexp,
     ylab = "Life Expectancy",
     xlab = "Fertility Rate")
plot(countries2$mnschool, countries2$lifexp,
     ylab = "Life Expectancy",
     xlab = "Mean Years School")
plot(log10(countries2$pop19_M), countries2$lifexp,
     ylab = "Life Expectancy",
     xlab = "Log of Population")
plot(countries2$urban, countries2$lifexp,
     ylab = "Life Expectancy",
     xlab = "Percent Urban")
plot(countries2$docs10k, countries2$lifexp,
     ylab = "Life Expectancy",
     xlab = "Doctors per 10k Population")
par(mfrow = c(1, 1))
```

Figure 18.2

```
set.seed(135)
n      = rep(1:100, 2)
a      = 0
b      = 1
sigma2 = n^1.3
eps    = rnorm(n,mean = 0, sd = sqrt(sigma2))
y      = a + b*n + eps
mod    = lm(y ~ n)
```

```
yhat <- (-1*predict(mod)) + 100
#Plot residuals from model as a function of predicted values
plot(yhat, residuals(mod),
     xlab = "Predicted Outcome",
     ylab = "Residuals")
```

ENDNOTES

CHAPTER 1

1. Fiorina, M. (1981). *Retrospective voting in American national elections*. Yale University Press.
2. Other examples include SAT scores, credit scores, and IQ scores.

CHAPTER 2

1. You could write this using = instead of <- and you would get the same result. I favor using <- rather than = because I think it is more intuitive ("Assign the thing on the right to the thing on the left").

CHAPTER 3

1. American National Election Studies. (2021). *ANES 2020 time series study full release*. July 19, 2021 version. www.electionstudies.org
2. Note that the data for this variable do not reflect the sweeping changes to abortion laws in the states that took place in 2021 and 2022.
3. There are other graphing methods that provide this type of information, but they require a bit more knowledge of measures of central tendency and variation, so they will be presented in later chapters.
4. One option not shown here is to reduce the size of the labels using `cex.names`. Unfortunately, you have to cut the size in half to get them to fit, rendering them hard to read.

CHAPTER 4

1. One alternative if this does happen is to go back and download the data from the original source and redo all of the transformations.
2. The ".3" extension is to remind me that this is the three-category version of the variable.
3. That's right, this is another example of a numeric variable for which a bar chart works fairly well.

CHAPTER 5

1. The original version of this variable is used here instead of the transformed version (`anes20$poor_spnd`), just in case you weren't assigned Chapter 4 or didn't save the transformations you made there.
2. Vishal (2016, March 30). *Cross validated: Explaining mean, median, mode in layman's terms*. Stack exchange. https://stats.stackexchange.com/questions/200282/explaining-mean-median-mode-in-laymans-terms
3. For instance, a U-shaped distribution or any number of multi-modal distributions could have no skewness.
4. $Skewness = \frac{(x_i - \bar{x})^3}{n * S^3}$, where n = number of cases, and S = standard deviation.

CHAPTER 6

1. Outliers are defined $Q1 - 1.5 * IQR$ and $Q3 + 1.5 * IQR$.
2. The standard normal distribution has a mean of 0 and a standard deviation of 1. Other normal distributions share the characteristics listed but have different means and standard deviations.
3. Charpentier, A. (2013, October 15). *Generating your own normal distribution table*. Freakonometrics. https://freakonome

trics.hypotheses.org/9404

CHAPTER 7

1. Kachigan, S. K. (1986). *Statistical analysis: An interdisciplinary introduction to univariate & multivariate methods*. Radius Press.
2. By "simulated," I mean that I created an object named `coin`, with two values, `Heads` and `Tails` (`coin <- c("Heads," "Tails")`) and told R to choose randomly from these two values and store the results in a new object, `coin10`, where the "10" indicates the number of tosses (`coin10 = sample(coin, 10, rep = T)`). The frequencies of the outcomes for object `coin10` can be used to show the number of `Heads` and `Tails` that resulted from the ten tosses. I then used the same process to generate results from larger samples of tosses.
3. You will see in just a bit, in the section on conditional probabilities, that the probabilities of candidate outcomes are not the same across levels of education.

CHAPTER 8

1. Presidential votes cast in Alaska are not tallied at the county level (officially, Alaska does not use counties), so this dataset includes the forty electoral districts that Alaska uses, instead of counties.
2. You might look at this distribution and wonder how the average Biden share of the two-party vote could be roughly 34% when he captured 52% of the national two-party vote. The answer, of course, is that Biden won where more people live while Trump tended to win in counties with many fewer voters. In fact, just over 94 million votes were cast in counties won by Biden, compared to just over 64 million votes in counties won by Trump.
3. Of course there are reasons other than minimizing sampling error to favor large samples, such as providing larger sub-samples for relatively small groups within the population or in anticipation of missing information on some of the things being measured.
4. Recall from Chapter 6 that you can calculate the standard deviation of a proportion based on the value of value of p: $S = \sqrt{p*(1-p)}$. This means that we can calculate the standard error of the proportion as: $S = \frac{\sqrt{p*(1-p)}}{\sqrt{n}}$.

CHAPTER 9

1. Kachigan, S. K. (1986). *Statistical analysis: An interdisciplinary introduction to univariate & multivariate methods*. Radius Press.
2. The code for generating this table comes from Ben Bolker *via stackoverflow*. https://stackoverflow.com/questions/31637388/
3. Remember that 3e-16 is scientific notation and means that you should move the decimal point 16 places to the left of 3. This means that p = .0000000000000003.

CHAPTER 10

1. Here, it is important to acknowledge multiple other forms of gender identity and gender expression, including but not limited to transgender, gender-fluid, nonbinary, and intersex. The survey question used in the 2020 ANES, as well as in most research on "gender gap" issues, utilizes a narrow sense of biological sex, relying on response categories of "Male" and "Female."
2. Kachigan, S. K. (1986). *Statistical analysis: An interdisciplinary introduction to univariate & multivariate methods*. Radius Press.
3. You also can calculate with information provided in the R `t.test`

output: $D = \frac{2 * t}{\sqrt{df}}$. This formula uses the *t*-score as an estimate of impact but then deflates it by taking into account sample size via *df*.

CHAPTER 11

1. The negative *t*-score and Cohen's *D* value are correct here, since they are based on subtraction of the mean from the second group (High GDP) from the mean of the first group (Low GDP).
2. This part of the first line, `[countries2$gdp_pc! = "NA"]`, ensures that the mean of the dependent variable is taken just from countries that have valid observations on the independent variable.
3. The term "error" may be confusing to you. In data analysis, people often use "error" and "variation" interchangeably, since both terms refer to deviation from some predicted outcome, such as the mean.
4. Remember, the format for `cut2` is `data$new variable <- cut2 (data$oldvariable, g = # of groups.`

CHAPTER 12

1. Knorre, A. (2017, May 26). *How to build a chi-square distribution table in R.* Stack Overflow. https://stackoverflow.com/questions/44198167/how-to-build-a-chi-square-distribution-table-in-r.

CHAPTER 13

1. Another popular measure of association for crosstabs based on chi-square is phi, which takes into account sample size but not table size.

   $$phi = \sqrt{\frac{\chi^2}{N}}$$

   I'm not adding phi to the discussion here because it is most appropriate for 2x2 tables, in which case it is equivalent to Cramer's *V*.
2. If you are itching for another formula, this one shows how tau-*b* integrates tied pairs:

   $$\text{tau-}b = \frac{N_{similar} - N_{different}}{\sqrt{\left(N_s + N_d + N_{ty}\right)\left(\left(N_S + N_d + N_{tx}\right)\right)}},$$

   where and are the number of tied pairs on *y* and *x*, respectively.

CHAPTER 14

1. You may have noticed that I did not include the code for these basic descriptive statistics. What R functions would you use to get these statistics? Go ahead, give it a try!
2. Recall from Chapter 5 that when variables such as population are severely skewed, we sometimes use the logarithm of the variable to minimize the impact of outliers. In the case of country-level population size, the skewness statistic is 8.3.
3. Heiss, A. (2021). *Exploring R2 and regression variance with Euler/Venn diagrams*. https://doi.org/10.59350/t57vy-p5115.

CHAPTER 15

1. The term "predict" is used a lot in the next few chapters. This term is used rather loosely here and does not mean we are forecasting future outcomes. Instead, it is used here and throughout the book to refer to guessing or estimating outcomes based on information provided by different types of statistics.
2. You might get this warning message when you use stargazer: "length of NULL cannot be changed." This does not affect your table so you can ignore it.
3. Why do I call this a fairly strong relationship? After all $r^2 = .422$ is pretty small compared to those shown earlier in the chapter. One useful way to think about the r^2 statistic is to convert it to a correlation coefficient ($\sqrt{r^2}$), which in this case, gives us $r = .65$, a fairly strong correlation.

CHAPTER 16

1. na.action = na.exclude is included in the command because I want to add some of the hidden information (the predicted outcomes) from `fit` to the `countries2` dataset that we will use later. Without adding this to the regression command, R would skip all observations with missing data when generating predictions, and the rows of data for the predictions would not match the appropriate rows in the original dataset. This is a bit of a technical point. Ignore it if it makes no sense to you.

CHAPTER 18

1. This explanation and solution are a bit more "black box" in nature than I usually prefer, but the topic and potential solutions are complicated enough that I think this is the right way to go about it in an introductory textbook.
2. Sorry, I can't resist the chance for an obvious pun.

INDEX

If you are based in the European Union and have safety concerns related to physical Sage products, in compliance with GPSR, please contact our Authorized Representative:

EC REP

International Associates Auditing & Certification Limited
The Black Church, St Mary's Place,
Dublin 7, D07 P4AX Ireland
EUAR@ie.ia-net.com